Two Manuals Combined into One Book

Medical Evacuation

And

FM 7-85 Ranger Unit Operations

Department of Defense

Pentagon
Publishing

Field Manual
No. 4-02.2

Headquarters
Department of the Army
Washington, DC, 8 May 2007

MEDICAL EVACUATION

Contents

Distribution Restriction: Approved for public release; distribution is unlimited.

***This publication supersedes Chapters 1–7 and Appendixes A, B, D–F, K, L, and N of FM 8-10-6 dated 14 April 2000, and FM 8-10-26 dated 16 February 1999.**

Preface

This field manual (FM) provides doctrine, as well as techniques and procedures for conducting medical evacuation and medical regulating operations. Medical evacuation encompasses both the evacuation of Soldiers from the point of injury (POI) or wounding to a medical treatment facility (MTF) staffed and equipped to provide essential care in theater and further evacuation from the theater to provide definitive, rehabilitative, and convalescent care in the continental United States (CONUS) and the movement of patients between MTFs or to staging facilities. Medical evacuation entails the provision of en route medical care; supports the joint health service support (JHSS) system; and links the continuum of care. In addition, it discusses the difference between medical evacuation and casualty evacuation (CASEVAC), as well as coordination requirements for and the use of nonmedical transportation assets to accomplish the CASEVAC mission. This publication is intended for use by medical commanders and their staffs, command surgeons, and nonmedical commanders involved in medical evacuation operations.

Users of this publication are encouraged to submit comments and recommendations to improve this publication. Comments should include the page, paragraph, and line(s) of the text where the change is recommended. The proponent for this publication is the United States (US) Army Medical Department (AMEDD) Center and School (USAMEDDC&S). Comments and recommendations should be forwarded, in letter format, directly to **Commander, USAMEDDC&S, ATTN: MCCS-FCD-L, 1400 East Grayson Street, Fort Sam Houston, Texas 78234-5052** or by using the e-mail address: medicaldoctrine@amedd.army.mil.

This publication applies to the Active Army, the Army National Guard (ARNG)/Army National Guard of the United States (ARNGUS), and the U.S. Army Reserve (USAR), unless otherwise stated.

Unless this publication states otherwise, masculine nouns and pronouns do not refer exclusively to men.

Use of trade or brand names in this publication is for illustrative purposes only and does not imply endorsement by the Department of Defense (DOD).

The staffing and organization structure presented in this publication reflects those established in base tables of organization and equipment (TOE) and are current as of the publication print date. However, such staffing is subject to change to comply with manpower requirements criteria outlined in Army Regulation (AR) 71-32. Those requirements criteria are also subject to change if the modification table of organization and equipment (MTOE) is significantly altered.

This publication implements the following North Atlantic Treaty Organization (NATO) International Standardization Agreements (STANAGs) and American, British, Canadian, and Australian (ABCA) Quadripartite Standardization Agreements (QSTAGs):

TITLE	STANAG	QSTAG
Stretchers, Bearing Brackets, and Attachment Supports	2040	
Medical Employment of Air Transport in the Forward Area	2087	
Medical and Dental Supply Procedures	2128	
Minimum Labeling Requirements for Medical Materiel		436
Documentation Relative to Medical Evacuation, Treatment, and Cause of Death of Patients	2132	470
Road Movements and Movement Control	2454	

TITLE	STANAG	QSTAG
Orders for the Camouflage of the Red Cross and Red Crescent on Land in Tactical Operations	2931	
Aeromedical Evacuation	3204	

When amendment, revision, or cancellation of this publication is proposed which will affect or violate the international agreements concerned, the preparing agency will take appropriate reconciliatory action through international standardization channels. These agreements are available on request from the Standardization Documents Order Desk, 700 Robins Avenue, Building 4, Section D, Philadelphia, Pennsylvania 19111-5094.

The AMEDD is in a transitional phase with terminology. This publication uses the most current terminology; however, other FM 4-02-series and FM 8-series may use the older terminology. Changes in terminology are a result of adopting the terminology currently used in the joint and/or NATO and ABCA Armies publication arenas. Also, the following terms are synonymous and the current terms are listed first:

- Medical logistics (MEDLOG); health service logistics (HSL); and combat health logistics (CHL).
- Roles of care, echelons of care, and level of care.
- Combat and operational stress control (COSC) and combat stress control (CSC).
- Behavioral health (BH) and mental health (MH).
- Chemical, biological, radiological and nuclear (CBRN) and nuclear, biological, and chemical (NBC).
- Stability operations; stability, security, transition, and reconstruction (SSTR) operations; and stability operations and support operations.

Introduction

The Army Health System (AHS) is a complex system of interrelated and interdependent systems which provides a continuum of medical treatment from the POI or wounding through successive roles of health care to definitive, rehabilitative, and convalescent care in the CONUS, as required. *Medical evacuation* is the system which provides the vital linkage between the roles of care necessary to sustain the patient during transport. This is accomplished by providing en route medical care and emergency medical intervention, if required, and to enhance the individual's prognosis and to reduce long-term disability.

Medical evacuation occurs at the tactical, operational, and strategic levels and requires the synchronization and integration of service component medical evacuation resources and procedures with the DOD worldwide evacuation system operated by the United States Transportation Command (USTRANSCOM).

Army medical evacuation is a multifaceted mission accomplished by a combination of dedicated ground and air evacuation platforms synchronized to provide direct support (DS), general support (GS), and area support within the joint operations area (JOA). At the tactical level, organic or DS medical evacuation resources locate, acquire, treat, and evacuate Soldiers from the POI or wounding to an appropriate MTF where they are stabilized, prioritized, and, if required, prepared for further evacuation to an MTF capable of providing required essential care within the JOA.

Although the most recognized mission of Army medical evacuation assets is the evacuation and provision of en route medical care to battlefield wounded, the essential and vital functions of medical evacuation resources encompass many additional missions and tasks that support the JHSS system. Medical evacuation resources are used to transfer patients between MTFs within the JOA and from MTFs to United States Air Force (USAF) mobile aeromedical staging facilities (MASFs) or aeromedical staging facilities (ASFs); emergency movement of Class VIII, blood and blood products, medical personnel and equipment; and serve as messengers in medical channels.

Medical regulating provides the interface with the DOD worldwide medical evacuation system by determining the patient's destination (the MTF best suited to provide the required care) and scheduling the means to transport the patient with the required en route medical care. Formal medical regulating begins at Role 3, however technological advances in information management (IM)/information technology (IT) are permitting this capability to be used at Role 1 and Role 2 MTFs in some situations.

Chapter 1

Overview of Army Health System Operations and Medical Evacuation

The AHS is the Army component of the military health system (MHS). Its capabilities are focused on delivering health care across the continuum of military operations—from the POI or wounding, through the JOA, to the CONUS support base. The two missions of the AHS are to provide health service support (HSS) (casualty care, medical evacuation, and medical logistics) and force health protection (FHP) (casualty prevention). The AHS is focused on promoting wellness, preventing casualties due to disease and nonbattle injuries (DNBIs), and providing timely and effective casualty care and management. Medical evacuation is the key factor in ensuring the continuity of care provided to our Soldiers by providing en route medical care during evacuation, facilitating the transfer of patients between MTFs to receive the appropriate specialty care, and ensuring that scarce medical resources (personnel, equipment, and supplies [to include blood]) can be rapidly transported to areas of critical need on the battlefield.

SECTION I — ARMY HEALTH SYSTEM

1-1. The provision of AHS is governed by well established and time-tested principles and rules which ensure the care provided to our Soldiers is timely and effective. For an in-depth discussion of these principles, rules, and roles of medical care, refer to FM 4-02.

1-2. The AHS is comprised of 10 medical functions. They are—

- Medical command, control, communications, computers, and intelligence.
- Medical treatment.
- Medical evacuation.
- Hospitalization.
- Dental services.
- Preventive medicine (PVNTMED) services.
- Combat and operational stress control (COSC).
- Veterinary services.
- Medical logistics.
- Medical laboratory support.

1-3. Army health system resources are arrayed across the battlefield in successive levels of support. These successive levels have increased medical care capabilities at each higher level. Medical evacuation and the provision of en route medical care ensures an uninterrupted continuum of care is maintained while Soldiers are moved through the roles of medical care to the MTF best suited to treat the patient's specific injuries.

PRINCIPLES OF ARMY HEALTH SYSTEM

1-4. The principles of AHS provide a framework in which medical planners can ensure that comprehensive plans are developed to support the tactical commander's operation plan (OPLAN). The

MAINTAIN THE HEALTH OF THE COMMAND

1-16. Ensure that medical evacuation assets are in close proximity to the MTF. The closer evacuation assets are to an MTF, the more rapidly care may be rendered.

SAVE LIVES

1-17. Ensure medical evacuation assets are the primary means of evacuating patients. Medical evacuation assets provide en route medical care that provides a continuum of care that is instrumental in preserving life and reducing long-term disability. The use of CASEVAC should only be used in extreme emergencies or when the medical evacuation system is overwhelmed. Refer to paragraphs 1-27 through 1-33 for a discussion on the differences between medical evacuation and CASEVAC operations.

PROVIDE STATE-OF-THE ART MEDICAL CARE

1-18. Ensure sufficient medical evacuation assets are available to facilitate the movement of patients between MTFs (intratheater evacuation) and the emergency movement of Class VIII, blood and blood products, and medical personnel and equipment. This capability permits the medical planner to maximize the use of scarce medical resources and exercise economy of force without sacrificing state-of-the-art medical care.

EARLY RETURN TO DUTY OF THE SOLDIER

1-19. Ensure seamless integration with the intertheater medical evacuation system. This allows rapid evacuation and continuous medical treatment throughout the continuum of care resulting rapid return to duty (RTD) rates.

SECTION II — MEDICAL EVACUATION

PURPOSE

1-20. An efficient and effective medical evacuation system—

- Minimizes mortality by rapidly and efficiently moving the sick, injured, and wounded to an MTF.
- Serves as a force multiplier as it clears the battlefield enabling the tactical commander to continue his mission with all available combat assets.
- Builds the morale of Soldiers by demonstrating that care is quickly available if they are wounded.
- Provides en route medical care that is essential in improving the prognosis and reducing disability of the wounded, injured, or ill Soldiers.
- Provides medical economy of force.
- Provides connectivity of the AHS as appropriate to the MHS.

Tactical	Operational	Strategic
Corps Division BCT Emergency Class VIII Resupply Emergency Movement of Medical Personnel/Equipment Stability Operations	Coalition Joint Class VIII Resupply Joint Blood Program DOD Civilians/Contractors Medical Personnel/Equipment Military Working Dogs Homeland Security Operations Shore-to-Ship Stability Operations	Host Nation Nongovernment Organizations Private Volunteer Organizations DOD support to Stability Operations International Disaster Relief

Figure 1-1. Army medical evacuation as it supports full spectrum operations

1-21. The AHS is established in roles of increasing capability from the POI to definitive care. On the battlefield, casualties are evacuated rearward from one role of care to the next, the sequencing of this movement is dependent upon METT-TC factors. In a contiguous battlefield, well established lines of communication (LOC), large numbers of casualties, and wide array of MTFs can result in very deliberate evacuation from one sequential role to the next higher. However in many situations, such as noncontiguous battlefield, the array of medical resources across the battlefield, the availability of medical evacuation resources, and the number of patients being evacuated may facilitate procedures that permit bypassing roles of care in order to ensure the timely treatment and care of casualties. This evacuation plan will be established by the appropriate level of command in coordination with the command surgeon to ensure the best treatment is provided to all casualties.

ATTRIBUTES

DEDICATED RESOURCES

1-22. The Army medical evacuation system is comprised of dedicated air and ground evacuation platforms. These platforms have been designed, staffed, and equipped to provide en route medical care to patients being evacuated and are used exclusively to support the medical mission. The focus of the medical evacuation mission coupled with the dedicated platforms permit a rapid response to calls for support. The dedicated nature of this mission dictates that Army medical evacuation unit's posture themselves in a "ready alert" status, ready to rapidly respond to evacuation missions and not diverted to perform any other task. Medical evacuation resources are protected under the provisions of the Geneva Conventions from intentional attack by the enemy. (For a discussion of the Geneva Conventions in relation to air and ground evacuation operations, refer to Appendix A.)

EN ROUTE MEDICAL CARE

1-23. En route care is provided on all Army medical evacuation platforms. This care is essential for minimizing mortality, enhancing survival rates, and reducing disability of wounded, injured, or ill Soldiers. Refer to paragraphs 1-27 through 1-33 for a discussion on the differences between medical evacuation and CASEVAC operations.

CONNECTIVITY

1-24. The Army provides connectivity to the MHS. The synchronized employment of medical evacuation resources provides and maintains the seamless continuum of care from the POI through successive roles of essential care within the theater. In addition to evacuating patients and providing en route medical care, medical evacuation resources provide for the emergency movement of scarce medical resources such as critical Class VIII, blood, medical personnel, and medical equipment. Further, medical evacuation resources are used to transfer patients from one MTF to another within the theater, to facilitate specialty care as well as transferring patients from an MTF to a MASF to facilitate intertheater evacuation. Figure 1-1 depicts the connectivity provided by medical evacuation platforms between the roles of care within theater. In addition to providing connectivity within the AHS, Army medical evacuation resources provide medical evacuation support and interface with MTFs of the other services deployed in the theater. In joint and multinational operations, the geographic combatant commander (GCC) may direct Army assets to provide this support and connectivity to joint, allied, and coalition forces within the JOA.

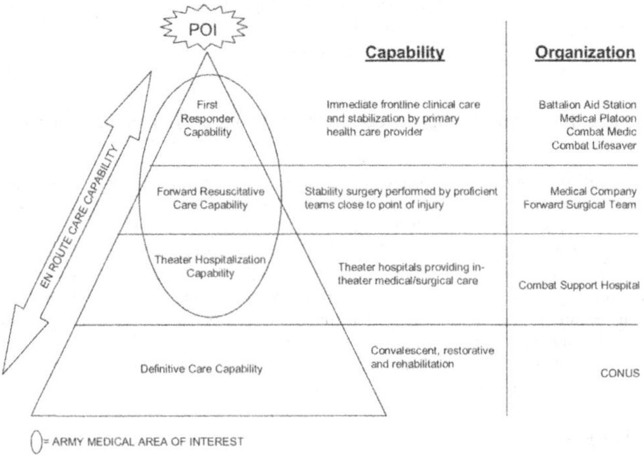

Figure 1-2. Army air and ground evacuation platforms provide connectivity to assure a seamless continuum of medical care

MEDICAL ECONOMY OF FORCE

1-25. The flexibility and versatility which medical evacuation resources afford the AHS to respond to urgent requirements throughout the battlefield is essential in ensuring the provision of seamless medical care. Medical economy of force is achieved for low density, high demand medical specialties (such as a neurosurgeon), medical supplies, and medical equipment (such as computer tomography scans or magnetic resonance imaging equipment) by having the capability to move the patient to the required care over long distances and by permitting the emergency of cross-leveling on medical supplies reducing the need for large Class VIII stockpiles. By providing medical economy of force, the deployed medical footprint is reduced without negatively impacting on the care provided to the Soldier.

FORCE MULTIPLIER

1-26. A highly effective, efficient, and responsive AHS increases Soldiers' confidence that if wounded in battle, they will receive timely and appropriate medical care to enhance their prognosis, speed their recovery, and reduce long-term disabilities. Medical evacuation resources locate and acquire wounded Soldiers and facilitate their entry into the AHS. These resources clear the battlefield of casualties which enables the tactical commander the ability to capitalize on and exploit battlefield opportunities with all available assets in the execution of his warfighting mission.

SECTION III — MEDICAL EVACUATION VERSUS CASUALTY EVACUATION

MEDICAL EVACUATION

1-27. Medical evacuation is performed by dedicated, standardized medical evacuation platforms, with medical professionals who provide the timely, efficient movement and en route care of the wounded, injured, or ill persons from the battlefield and/or other locations to MTFs. Medical evacuation is an AMEDD function that supports and is an integral part of the AHS. The provision of en route care on medically equipped vehicles or aircraft greatly enhances the patient's potential for recovery and may reduce long-term disability by maintaining the patient's medical condition in a more stable manner.

1-28. Medical evacuation ground/air ambulance platforms are defined as: Platforms especially for the medical evacuation mission with allocated medical equipment specifically designed for the purpose of enroute care and by trained medical personnel.

1-29. The gaining MTF in coordination with the losing MTF, is responsible for arranging for the evacuation of patients from the lower role of care. For example, Role 2 medical units are responsible for evacuating patients from Role 1 MTFs.

1-30. Medical evacuation begins when a request is initiated and continues until the patient is released from the AHS by medical authorities.

CASUALTY EVACUATION

1-31. Casualty evacuation is a term used to refer to the movement of casualties aboard nonmedical vehicles or aircraft.

WARNING

Casualties transported in this manner may not receive proper en route medical care or be transported to the appropriate MTF to address the patient's medical condition. If the casualty's medical condition deteriorates during transport, or the casualty is not transported to the appropriate MTF, an adverse impact on his prognosis and long-term disability or death may result.

1-32. If dedicated medical evacuation platforms (ground and air) are available, casualties should be evacuated on these conveyances to ensure they receive proper en route medical care.

1-33. Since CASEVAC operations can reduce combat power and degrade the efficiency of the AHS, units should only use CASEVAC to move Soldiers with less severe injuries when medical evacuation assets are overwhelmed. Medical planners should ensure CASEVAC operations are addressed in the OPLAN/OPORD as a separate operation, as these operations require preplanning, coordination, synchronization, and rehearsals. The CASEVAC plan should ensure casualties with severe or life-threatening injuries are prioritized for evacuation and are evacuated on dedicated medical evacuation platforms.

1-34. When possible, nonmedical vehicles/aircraft transporting casualties should be augmented with a trauma specialist or combat lifesaver (CLS). (On nonmedical aircraft, sufficient space may not be available to permit a caregiver to accompany the casualties.) The type of en route monitoring and medical care/first aid provided is limited by the following factors:

Number and Type of Patients

1-48. Another factor affecting the policy is the number and types of patients anticipated and the rate of patient RTD. Admission rates vary widely in different geographical areas of the world and in different types of military operations.

Evacuation Means

1-49. The means (quantity and type of transportation) available for evacuation of patients from the theater to CONUS is an essential factor impacting on the evacuation policy.

Availability of Replacements

1-50. The capability of CONUS to furnish replacements to the theater is another consideration. For each patient who is evacuated from the theater to CONUS, a fully trained and equipped replacement must be provided. During a small-scale conflict overseas, the CONUS replacement capability is much greater than when compared to a large-scale conflict such as World War II.

Availability of In-Theater Resources

1-51. Limitations of all AHS resources (such as insufficient number and types of medical units in the division/corps to support the brigade combat teams [BCTs] and an insufficient amount of medical and nonmedical logistics) will have an impact on the theater evacuation policy. The more limitations (or shortages), the shorter the theater evacuation policy.

IMPACT OF THE EVACUATION POLICY ON ARMY HEALTH SYSTEM REQUIREMENTS

Shorter Evacuation Policy

1-52. A shorter theater evacuation policy—

- Results in fewer hospital beds required in the theater and a greater number of beds required elsewhere.
- Creates a greater demand for intertheater USAF evacuation resources. (A shortened intratheater evacuation policy would likewise increase the number of airframes required in the theater.)
- Increases the requirements for replacements to meet the rapid personnel turnover which could be expected, especially in combat units. (The impact this would have on both intra- and intertheater transportation and other requirements must also be considered.)

Longer Evacuation Policy

1-53. A longer theater evacuation policy—

- Results in a greater accumulation of patients and a demand for a larger AHS infrastructure in the theater. It decreases bed requirements elsewhere.
- Increases the requirements for medical logistics (medical supplies, equipment, and equipment maintenance) and nonmedical logistics support.
- Increases the requirements for hospitals, engineer support, and all aspects of base development for deployed AHS force. (It demands the establishment of a larger number of hospitals within the theater and may require medical specialty augmentation.)
- Provides for a greater proportion of patients to RTD within the theater and, thus, reduces the loss of experienced manpower.
- A longer intratheater evacuation policy may decrease the demand on the intratheater evacuation assets and system.

1-54. The concept of essential care in theater does not support longer evacuation policies as the deployed hospitals are not designed to provide definitive, rehabilitative, and convalescent care/services. If the theater evacuation policy is extended in theaters operating under the essential care in theater concept, augmentation of medical specialty resources will be required.

PATIENT STABILIZATION

1-55. The evacuation policy has no impact on the patient stabilization period for movement. This period is known as the *evacuation delay*. It is the period of time planned for between the time of patient reporting and the time of medical evacuation of the patient to the next role of care. Evacuation delays normally range from 24 to 72 hours and are designated by the Army service component command (ASCC) surgeon.

Chapter 2

Medical Evacuation Resources

This chapter discusses the mission, functions, and capabilities of medical evacuation units and elements, as specified in the unit's TOE. It also discusses the command and control (C2) headquarters to which they are assigned. The discussion of each organization also includes the basis of allocation (BOA), limitations, and dependencies.

SECTION I — MANEUVER BATTALION MEDICAL PLATOON AMBULANCES

MEDICAL PLATOON AMBULANCE SQUADS

2-1. The ground ambulance team is the basic medical evacuation element used within the BCTs and at the corps and theater AOs. These ground ambulance teams provide medical evacuation from the POI to supporting MTFs while ensuring the continuity of care en route. Ambulance squads consisting of two ambulance teams organic to the medical platoons of movement and maneuver and fires units and to the evacuation platoon of the brigade support medical companies (BSMCs).

2-2. The primary mission of the maneuver medical platoon ambulance squads is to provide ground ambulance evacuation support from supported infantry/armor companies or from POI back to a casualty collection point (CCP) or to the Role 1 MTF/battalion aid station (BAS). Maneuver medical platoon ambulance teams are assigned to heavy BCT (HBCT), infantry BCT, Stryker BCT (SBCT), reconnaissance squadrons, and reconnaissance and target acquisition (RSTA) squadrons. They also provide area support to other elements (which do not have organic medical evacuation resources) operating in their AO. Ground ambulance support provided is consistent with evacuation precedence, tactical standing operating procedures (TSOPs), and other operational considerations.

2-3. For definitive information on maneuver battalion medical platoon operations, see FM 4-02.4, FM 4-02.6, and FM 4-02.21.

MANEUVER UNIT MEDICAL PLATOON

ORGANIZATION

2-4. The medical platoon consists of a platoon headquarters section, a treatment squad, combat medic section, and an ambulance squad. The number of ambulance squads in an ambulance platoon varies and is based on the type of parent organization. The infantry, airborne, and air assault battalions' ambulance platoons have two ambulance squads equipped with high mobility multipurpose wheeled vehicle (HMMWV) ambulances. The mechanized infantry and armor combat maneuver battalions' ambulance platoons have four ambulance squads equipped with M-113 track ambulances. The SBCT infantry battalion's and RSTA squadron's ambulance platoons have two ambulance squads equipped with Stryker wheeled armored ambulances (referred to as the Stryker medical evacuation vehicle [MEV]).

Medical Platoon Headquarters

2-5. The medical platoon leader is a physician and also serves as the battalion surgeon. He is assisted by the medical operations officer (field medical assistant) in the operational, administrative, and logistical support aspects of the platoon.

Ambulance Section

2-6. The ambulance section of the medical platoon is organized into ambulance squads and is supervised by the platoon sergeant. Each squad contains a noncommissioned officer (NCO) squad leader, three combat medics, and two ambulances.

Ambulance Squad Leader

2-7. The ambulance squad leader ensures that ambulance team members perform operator maintenance on assigned vehicles. An ambulance team consists of one vehicle and two medical personnel (an aide/evacuation NCO and a combat medic).

Ambulance Team

2-8. Ambulance team members are also responsible for performing operator maintenance on assigned ambulances. Specific duties of the ambulance team are to—

- Administer EMT as required.
- Initiate the Department of Defense (DD) Form 1380, US Field Medical Card (FMC), as required.
- Prepare patients for ground and air medical evacuation.
- Provide medical evacuation of wounded or injured Soldiers from the POI to supporting MTF or ambulance exchange point (AXP).
- Provide patient care en route.
- Provide emergency movement of medical personnel and emergency delivery of blood, medical supplies, and medical equipment.
- Assist in the care of COSR casualties.
- Maintain operational readiness of the ambulance.
- Resupply company and platoon trauma specialists, when required.
- Operate the vehicle and maintain contact with supported and supporting elements.
- Find and collect the wounded.
- Serve as messengers within medical channels.

PRIMARY FUNCTIONS OF AMBULANCE TEAMS

2-9. The primary function of the ambulance team is to collect and treat the sick, injured, and wounded Soldiers on the battlefield and to provide medical evacuation support from the POI, CCP, or AXP to the supporting MTF.

SECTION II — AMBULANCE/EVACUATION PLATOON—FORWARD SUPPORT, BRIGADE SUPPORT, OR AREA SUPPORT MEDICAL COMPANY

MEDICAL COMPANIES

2-10. Medical companies provide Role 2 MTF and medical evacuation. Medical companies assigned to or in DS of a BCT are referred to as either a forward support medical company (FSMC) or a BSMC. Area support medical companies (ASMCs) are assigned to multifunctional medical battalions and operate Role 2 MTF and provide medical evacuation support for division, and theater units. In addition, all medical companies provide Role 1 area medical support to units without organic medical personnel and may augment/reinforce maneuver medical platoons/sections with treatment and medical evacuation support.

Ambulance/Evacuation Platoon

2-25. The ambulance/evacuation platoon provides C2 for ambulance platoon operations. The ambulance platoon headquarters element maintains communications to direct ground ambulance evacuation of patients. It provides ground ambulance evacuation support for supported maneuver battalions and for supported units operating in the sustainment areas. The ambulance headquarters element performs route reconnaissance and develops and issues graphic overlays to all its ambulance teams. It also coordinates and establishes AXPs for both air and ground ambulances, as required.

2-26. Ambulance squads provide ground ambulance evacuation of patients from supported BASs/unit aid stations back to the Role 2 MTF that is located in the brigade support area (BSA). An ambulance squad consists of two ambulance teams (two ambulances, wheel or tracked vehicles). Ambulance squad personnel—

- Perform EMT, evacuate patients, and provide for their continued care en route.
- Operate and maintain assigned communication and navigational equipment.
- Perform preventive maintenance checks and services (PMCS) on ambulances and associated equipment.
- Maintain supply levels for the ambulance medical equipment sets (MES).
- Ensure that appropriate property exchange of medical items (such as litters and blankets) is made at sending and receiving MTF.
- Prepare patients for ground and air medical evacuation.
- Provide medical evacuation of wounded or injured Soldiers from the POI to supporting MTF, CCP or AXP.
- Initiate the DD Form 1380, FMC, as required.
- Provide emergency movement of medical personnel and emergency delivery of blood, medical supplies, and medical equipment.
- Assist in the care of COSR casualties.
- Maintain operational readiness of the ambulance.
- Resupply company and platoon trauma specialists, when required.
- Operate the vehicle and maintain contact with supported and supporting elements.
- Find and collect the wounded.
- Serve as messengers within medical channels.

Preventive Medicine Section

2-27. Preventive medicine sections are organic to the divisions and the BSMC of the BCT. The PVNTMED section is also found in the multifunctional medical battalion (MMB) headquarters. The PVNTMED section assigned to medical companies has a primary responsibility for supervising the unit's PVNTMED program as described in AR 40-5. The section—

- Ensures measures are implemented to protect personnel against food, water, and arthropodborne diseases, as well as environmental injuries.
- Provides advice and consultation in the area of health threat assessment, FHP, environmental sanitation, epidemiology, sanitary engineering, and pest management.
- Assists the higher headquarters in determining requirements for medical intelligence assessments, particularly with respect to CBRN and disease prevalence.
- Coordinates with supporting veterinary teams for conducting and implementing food safety and quality assurance surveillance and assisting in foodborne and zoonotic disease surveillance and control.

Behavioral Health Section

2-28. Combat and operational stress is controlled through vigorous prevention, consultation, and restoration programs by the BSMC BH section. These programs are designed to maximize the RTD rate of

COSR Soldiers by providing rest/restoration within or near their unit areas. The BSMC BH section has a primary responsibility for assisting commanders in controlling combat stress and serves as a consultant to the commander, staff, and others involved with providing prevention and intervention services to unit Soldiers and their families. For additional information on the operations and functions of BSMCs, and ASMCs, refer to FMs 4-02.6 and 4-02.24.

SECTION III — MEDICAL COMPANY (GROUND AMBULANCE)

OPERATIONAL INFORMATION

2-29. The AHS is a continuum of increasing roles of care extending from the POI through the CONUS base. All sick, injured, and wounded patients must be evacuated from the battlefield in the shortest possible time to the MTFs that can provide the required treatment. Ground ambulances serve as one of the primary means of evacuating patients from the battlefield.

MISSION

2-30. The mission of the medical company, ground ambulance, TOE 08453A00, is to provide ground evacuation within the theater.

ASSIGNMENT

2-31. The medical company, ground ambulance is normally assigned or attached to MMB (TOE 08485G00) or a medical brigade (MEDBDE) for C2.

EMPLOYMENT

2-32. The medical company, ground ambulance, provides DS to BCTs and is employed in the division, corps, and EAC to provide area support. It is tactically located where it can best control its assets and execute its patient evacuation mission.

BASIS OF ALLOCATION

2-33. The BOA within the combat zone is one per division supported and one per two divisions supported within the theater.

CAPABILITIES

2-34. At Level I, the unit is capable of providing:

- A single-lift capability for evacuation of 96 litter patients or 192 ambulatory patients.
- Medical evacuation from BCT/division medical companies and ASMCs to supporting hospitals.
- Reinforcement of BCT/division medical company evacuation assets.
- Reinforcement of covering force and unassigned area operations.
- Movement of patients between hospitals and aeromedical staging squadrons/MASFs, railheads, or seaports in the division, corps, and EAC.
- Area evacuation support beyond the capabilities of the ASMC.
- Emergency movement of medical personnel and supplies.
- Preparation of patients for ground and air medical evacuation.
- Medical evacuation of wounded or injured Soldiers from the POI to supporting MTF, CCP, or AXP.
- Field feeding and vehicle refueling support when collocated with the MMB.

DEPENDENCY

2-35. This unit is dependent upon the appropriate elements of the division, corps, EAC, or ASCC for—

- Religious, financial management, legal, personnel, mortuary affairs, and administrative services.
- Viable communications systems for C2 and adequate road networks.
- Laundry, shower, and clothing repair.
- Generator, communication equipment, and communication security (COMSEC) equipment maintenance.
- Health service support to include hospitalization.

ORGANIZATION AND FUNCTIONS

2-36. This company is organized into a company headquarters section, two ambulance platoon headquarters, and two evacuation sections (Figure 2-1).

2-37. The company headquarters provides C2, communications, administration, field feeding, and logistical support (to include maintenance) for the subordinate ambulance platoons.

2-38. The ambulance platoon headquarters provides C2 for the subordinate ambulance squads.

2-39. The evacuation section operates ambulances and provides en route medical care for patients. Each evacuation section consists of 12 ambulances, each with a two-man crew. The members of the squad maintain the level of expendable Class VIII supplies in the ambulance MES by reconstituting supplies from medical companies or hospitals when they pick up or drop off patients. They are also responsible for performing operator maintenance on assigned vehicles.

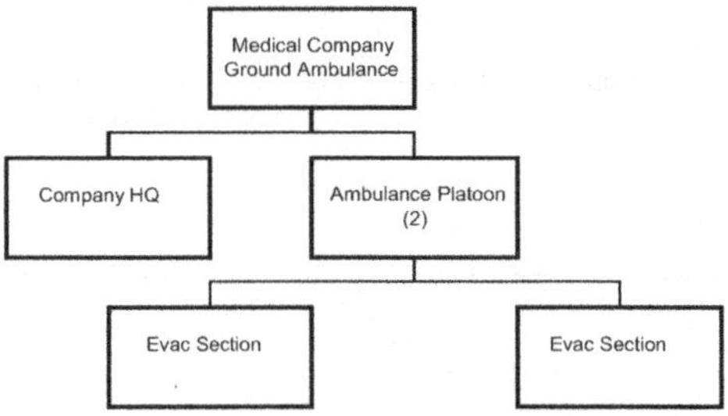

Figure 2-1. Medical company, ground ambulance

SECTION IV — MULTIFUNCTIONAL MEDICAL BATTALION

MODULAR FORCE

2-40. A modular force is a key aspect in the transformation of the Army. The modular force model is based on brigade-sized elements that are more responsive than division-sized elements and can perform joint and expeditionary-type missions. Using modular units, Army planners can tailor force structure, reduce strategic lift requirements, and create flexible forces with specialized capabilities based on everchanging mission requirements. To facilitate the synchronization and provision of AHS services and support, the AMEDD C2 organizations have been redesigned into a modular design. The MMB will replace the functional battalions (area support, evacuation, and medical logistics).

2-41. The new modular design of medical units gives the medical command (MEDCOM) the ability to—

and attached medical organizations (companies, detachments, and teams) task-organized for support of deployed forces.

ASSIGNMENT

2-51. This EEE is assigned to an MMB.

EMPLOYMENT

2-52. Normally employed in a MEDCOM/MEDBDE AO. It can also be deployed to provide C2 to expeditionary forces in early entry operations and facilitates the reception, staging, onward movement, and integration (RSOI) of medical forces.

BASIS OF ALLOCATION

2-53. The BOA is one per MMB.

CAPABILITIES

2-54. At Level I, this unit provides—

- Command and control, staff planning, supervision of operations, medical and general logistics support as required, and administration of the assigned and attached units conducting AHS operations in its supported AO.
- Task organization of medical assets.
- Advice to senior commanders in the AO on the medical aspects of their operations.
- Coordination of medical regulating and patient movement within the AO.
- Monitoring, planning, and coordinating ground and air evacuation within the battalion AO. Coordinating air evacuation support requirements with the supporting aviation unit, and synchronizing the air evacuation plan into the overall medical evacuation plan.
- Consultation and technical advice on PVNTMED (medical entomology, medical and OEH surveillance, and sanitary engineering), COSC and BH, medical records administration, veterinary services, nursing practices and procedures, dental services, and automated medical information systems to supported units.
- Guidance for facility site selection and area preparation, as required.
- Monitoring and supervising medical logistics operations, to include Class VIII supply/resupply.
- Consolidated property book maintenance.
- Unit-level maintenance for wheeled vehicles and power generation equipment and wheeled vehicle recovery operations support to assigned or attached units.
- Organizational communications equipment maintenance support for the battalion.
- Food service support for staff and assigned/attached medical units.

DEPENDENCY

2-55. This unit is dependent on the—

- Corps for legal, administrative, financial management, human resources, religious support, transportation services, area damage control, CBRN decontamination assistance, mortuary affairs, and laundry and bath services.
- Quartermaster supply company or equivalent for Class I rations.
- Engineer company or equivalent for site selection, waste disposal, and minor construction.
- Movement control battalion or equivalent for supplemental transportation requirements.

ORGANIZATION AND FUNCTIONS

2-56. The battalion command section provides C2 of assigned and attached medical companies, detachments, and teams. (See Figure 2-3.)

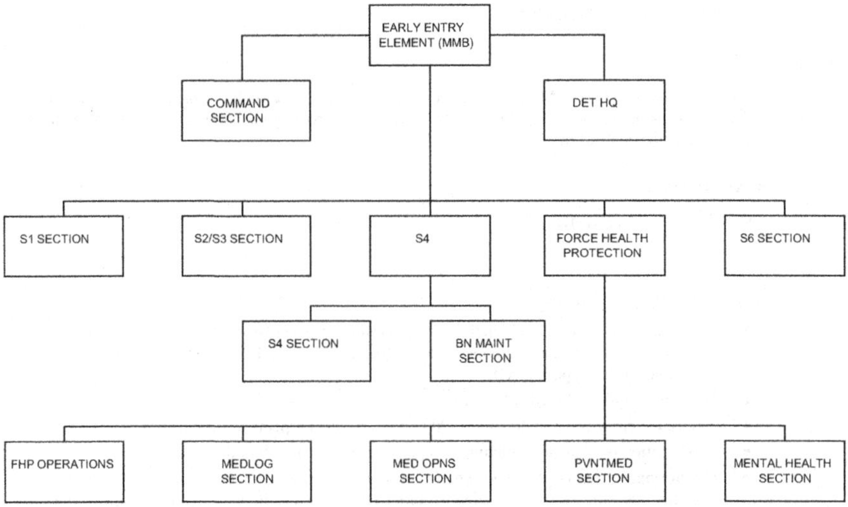

Figure 2-3. Early entry element, multifunctional medical battalion

2-57. The Adjutant (US Army) (S1) operates according to METT-TC and is the principle coordinating staff officer responsible for the delivery of human resources support. The S1 is responsible for the execution of all of the human resources (HR) core competencies to include: personnel readiness management (PRM); personnel accountability and strength reporting (PASR); personnel information management (PIM); reception, replacement, RTD, rest and recuperation (R&R), redeployment (R5) operations; casualty operations; essential personnel services (EPS); and postal operations. Morale, welfare, and recreation (MWR) operations, HR planning; and staff operations. The S1 also coordinates with elements of supporting agencies for finance, legal, religious, and administrative services.

2-58. The Intelligence Officer (US Army)/Operations and Training Officer (US Army) (S2/S3) section is responsible for security, plans and operations, deployment, relocation, and redeployment of the battalion and its assigned and attached units. The battalion's primary net control station is in this paragraph.

2-59. The Supply Officer (US Army) (S4) section is responsible for coordination, control, and management of logistics for assigned and attached units. The battalion maintenance section is under the staff supervision of the battalion S4. This section provides unit level maintenance for wheeled vehicles assigned to the HHD and assigned or attached units without unit level maintenance capability.

2-60. The FHP operations section is responsible for the planning, coordination, and execution of the AHS mission within the battalion's AO. This section supervises the operations of the medical logistics, operations, PVNTMED, and BH sections as well as coordinates medical regulation and patient movement within the AO. The medical operations section is responsible for—

- Management of area medical support (Roles 1 and 2).
- Planning, coordination, and execution of the Class VIII mission within the battalion's AO. This includes blood and medical maintenance management.

- Situational awareness for locations of ground and air ambulance assets.
- Planning and coordination for evacuation of patients using ground and air ambulance assets.
- Planning and coordination between air evacuation elements and the maneuver brigade S3. Information provided to the maneuver brigade S3 should include, but not be limited to the location of—
  - Medical treatment facilities, CCPs, and AXPs.
  - Forward air refueling equipment.
  - Location of supported units and liaison requirements. Air evacuation corridors, ground evacuation routes, and recommendations on usage.
- The PVNTMED section is responsible for the planning, coordination, and execution of the PVNTMED mission within the battalion's AO. This includes management of PVNTMED and veterinary assets.
- The BH section is responsible for the planning, coordination, and execution of the COSC mission within the battalion's AO. This section also collects and records social and psychological data.

2-61. The S6 section is responsible for all aspects of information management, automation, and communications-electronic (CE) support to assigned and attached units. The battalion's alternate net control station is in this paragraph.

2-62. The detachment headquarters provides for billeting, field feeding, discipline, security, training, and administration for personnel assigned to the EEE.

2-63. The battalion maintenance section is under the staff supervision of the battalion S4. This section provides unit level maintenance for wheeled vehicles assigned to the HHD and assigned or attached units without unit level maintenance capability.

CAMPAIGN SUPPORT ELEMENT, MULTIFUNCTIONAL MEDICAL BATTALION

MISSION

2-64. The mission of the campaign support element, MMB, TOE 08486GB00, is to complete the staffing of the headquarters of the MMB to enhance the delivery of campaign quality health care to deployed forces. To provide a scalable, flexible, and modular C2, administrative assistance, logistical support, and technical supervision capability for assigned and attached medical organizations (companies, detachments, and teams) task-organized for support of deployed forces.

ASSIGNMENT

2-65. This unit is assigned to an MMB.

EMPLOYMENT

2-66. Normally employed with the MEDBDE or a MEDCOM. If the EEE is deployed first, the campaign element falls in on this deployed module to provide a more robust C2 element and additional operational and planning capabilities, increased medical logistical staff, and a more diverse and robust clinical staff.

BASIS OF ALLOCATION

2-67. The BOA is one per MMB.

CAPABILITIES

2-68. At Role 1, this unit provides—

- Augmentation of C2 in personnel, logistics, HSS, FHP, medical operations, PVNTMED, BH, automation, maintenance, religious support, and food service to the EEE.
- Unit maintenance on all organic equipment except COMSEC equipment.

DEPENDENCY

2-69. This unit is dependent on the HHD, MMB for legal, administrative, financial management, human resources, food service, transportation services, area damage control, CBRN decontamination assistance, mortuary affairs, and laundry and bath services.

ORGANIZATION AND FUNCTIONS

2-70. The command section provides C2 for assigned and attached medical companies, detachments, and teams. (See Figure 2-4.)

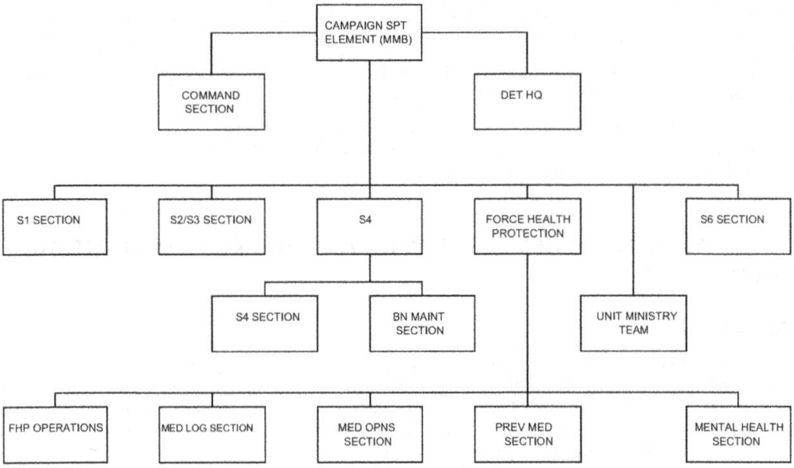

Figure 2-4. Campaign support element, multifunctional medical battalion

2-71. The S1 operates according to METT-TC and is the principle coordinating staff officer responsible for the delivery of HR support. The S1 is responsible for the execution of all of the HR core competencies to include: PRM, PASR, PIM, R5 operations, and casualty operations, EPS, postal operations, MWR operations, and HR planning and staff operations. The S1 also coordinates with elements of supporting agencies for finance, legal, religious, and administrative services.

2-72. The S2/S3 section coordinates with the EEE S2/S3 section of the MMB in the plans and operations support to subordinate organizational elements.

2-73. The S4 section is responsible for coordination, control, and management of logistics for assigned and attached units.

2-74. The FHP operations section plans and coordinates the HSS/FHP mission within the battalion's AO when collocated with the EEE. This section also provides advice on the operation of medical logistics, operations, PVNTMED, BH, medical regulating, and patient movement to supported organizations.

2-75. The medical logistics section plans and organizes Class VIII supply support (including blood and medical equipment management) in coordination with the medical logistics section of the EEE.

2-76. The medical operations section is responsible for—

- Management of area medical support (Roles 1 and 2).
- Situational awareness for locations of ground and air ambulance assets.
- Planning and coordination for evacuation of patients using ground and air ambulance assets.
- Planning and coordination between air evacuation elements and the maneuver brigade S3. Information provided to the maneuver brigade S3 should include, but not be limited to the location of—
  - Medical treatment facilities, CCPs, and AXPs.
  - Forward air refueling equipment.
  - Supported units and liaison requirements.
  - Evacuation corridors and evacuation routes and recommendations on usage.

2-77. The PVNTMED section plans and coordinates PVNTMED/veterinary support to subordinate organizations.

2-78. The BH section plans and coordinates BH support to subordinate organizations. This section also collects and records social and psychological data when collocated with the EEE.

2-79. The S6 section provides advice and support to subordinate organizations for CE support and information management.

2-80. The unit ministry team provides and coordinates religious support for all MMB staff, assigned/attached medical units and casualties in MMB subordinate unit MTFs.

2-81. The detachment headquarters provides coordination with the EEE in support of the field feeding and administration mission of the unit.

2-82. The battalion maintenance section provides unit level maintenance management support to assigned and attached units in coordination with the battalion S4 and the battalion maintenance section of the EEE.

SECTION V — GENERAL SUPPORT AVIATION BATTALION, COMBAT AVIATION BRIGADE

GENERAL SUPPORT AVIATION BATTALION

MISSION

2-83. The mission of the general support aviation battalion (GSAB) (Figure 2-5) is to provide the combat aviation brigade with aerial sustainment and maneuver support.

Basis Of Allocation

2-93. The BOA is one per GSAB and two per theater of operations (TOEs 01125G100, 01225G100, and 01305G100).

Capabilities

2-94. At Level I, this unit provides—

- Twelve helicopter ambulances to evacuate critically wounded or other patients consistent with evacuation priorities and operational considerations, from points as far forward as possible, to division, corps, and EAC hospitals. Single patient lift capability for the company is 72 litter patients or 84 ambulatory patients, or some combination thereof.
- Four forward support medical evacuation teams (three helicopters each) that can be deployed as a team or an independent aircraft.
- Air crash rescue support.
- Expeditious delivery of whole blood, biological, and medical supplies to meet critical requirements.
- Rapid movement of medical personnel and accompanying equipment/supplies to meet the requirements for mass casualty (MASCAL), reinforcement/reconstitution, or emergency situations.
- Movement of patients between hospitals, ASFs, hospital ships, casualty receiving and treatment ships, seaports and railheads in the theater of operations.
- Military working dog evacuation and aeromedical evacuation support to a combat search and rescue task force.

Dependency

2-95. This unit is dependent upon—

- Appropriate elements of the division or corps for religious, legal, financial management, personnel and administrative services, and CE and COMSEC maintenance.
- The headquarters and headquarters company of the GSAB for religious, legal, HSS, FHP, personnel and administrative service, and unit CBRN support.
- The forward support company of the GSAB for Class III, automotive and generator maintenance, and field feeding.
- The aviation support company of the GSAB for unit level field maintenance of organic aircraft, including unit level supply support for aircraft Class IX.
- United States Air Force weather team in the headquarters and headquarters company of the division aviation brigade for air weather service support.

Organization and Functions

2-96. The company headquarters provides C2, logistical support, and administration for the company.

2-97. The four forward support MEDEVAC teams (FSMTs) provide a task-organized means for medical evacuation at the DS and GS level. They also, provide emergency movement of medical personnel and emergency delivery of whole blood, biological, and medical supplies and equipment.

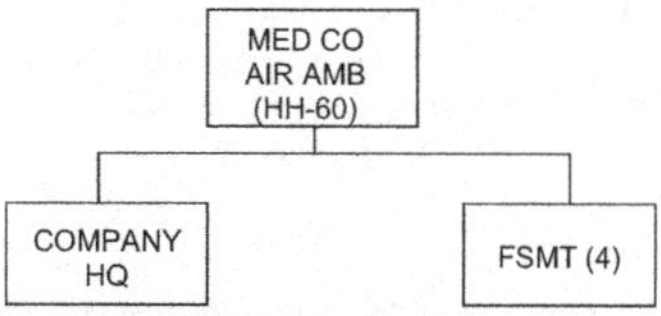

Figure 2-6. Medical company, air ambulance (HH-60)

AVIATION MAINTENANCE COMPANY, AVIATION UNIT MAINTENANCE, GENERAL SUPPORT AVIATION BATTALION, COMBAT AVIATION BRIGADE

MISSION

2-98. The mission of the aviation maintenance company is to perform aviation unit maintenance for the GSAB.

ASSIGNMENT

2-99. This unit is organic to a GSAB (TOE 01305G100).

EMPLOYMENT

2-100. The aviation maintenance company is employed as part of combined forces and provides unit level field maintenance support to the GSAB. The aviation maintenance company is located and its operations are normally conducted immediately behind the main battle area (MBA) in the EAB. The aviation maintenance company operations in forward areas are usually limited to daily inspections, minor adjustments, emergency and contact maintenance, and aircraft recovery. Contact teams and slice elements could be utilized as necessary.

BASIS OF ALLOCATION

2-101. The BOA is one per GSAB (TOE 01305G100).

CAPABILITIES

2-102. The aviation maintenance company provides unit level field maintenance support for 8 UH-60, 12 HH-60, and 12 CH-47 aircraft. Individuals of this organization can assist in the coordinated defense of the unit's area or installation. This unit performs unit level field maintenance on only the supported aircraft and some associated maintenance equipment.

DEPENDENCY

2-103. This unit depends on—

- Appropriate elements of the division or corps for religious, legal, HSS, FHP, financial management, personnel and administrative services, and logistical support (including supplemental transportation).
- Headquarters and headquarters company of the GSAB (TOE 01306G100) for religious, legal, HSS, FHP, financial management, personnel, and administrative services, unit-level logistical support, and communications maintenance.

- Forward support company (TOE 63317G200) for Class III, automotive and generator maintenance, and field feeding service support.
- Aviation maintenance battalion (TOE 01918G200), for unit level field maintenance support, backup unit level field maintenance support, DS maintenance for assigned AGPU, and support for aviation and common Class IX repair parts.

ORGANIZATION AND FUNCTIONS

2-104. This unit inspects aircraft systems, subsystems, components, and the maintenance performed by crew chiefs and repairers to determine airworthiness and compliance with applicable technical publications; performs operational checks and troubleshoots aircraft systems to isolate malfunctions; participates in maintenance test flights and maintains historical maintenance files.

2-105. This unit provides company maintenance operations. It maintains and regulates uniform flow of aircraft and associated components. Further, it maintains the company's aircraft logbooks, provides unit level Class IX support, and performs maintenance on aviation life support equipment (ALSE).

2-106. This unit provides unit level maintenance on UH/HH-60/CH-47 helicopters and associated equipment; services and lubricates helicopter systems and subsystems; and diagnoses and troubleshoots operational malfunctions.

MEDICAL COMMAND

2-107. The MEDCOM serves as the senior medical command within the theater, and works directly for the theater commander. With augmentation, MEDCOM is joint capable. In the absence of a theater commander the MEDCOM works for the regional combatant commander during a major combat operation (MCO). It possesses a centralized capability to effectively and efficiently task-organize medical elements based on theater specific medical requirements. The MEDCOM is depicted in Figure 2-7.

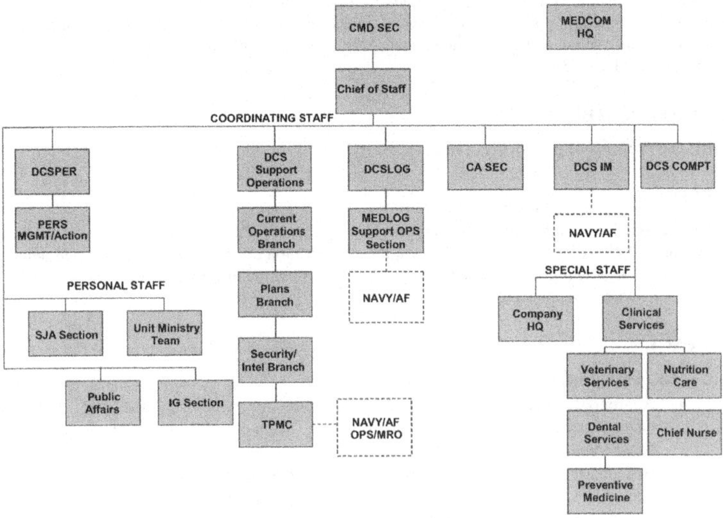

Figure 2-7. Medical command

2-108. The MEDCOM serves as the medical force provider (Figure 2-8) for the corps and EAC and provides the requisite medical C2 necessary to provide campaign-quality health care to deployed forces. The MEDCOM also focuses on theater medical operations plans and medical contingency plans.

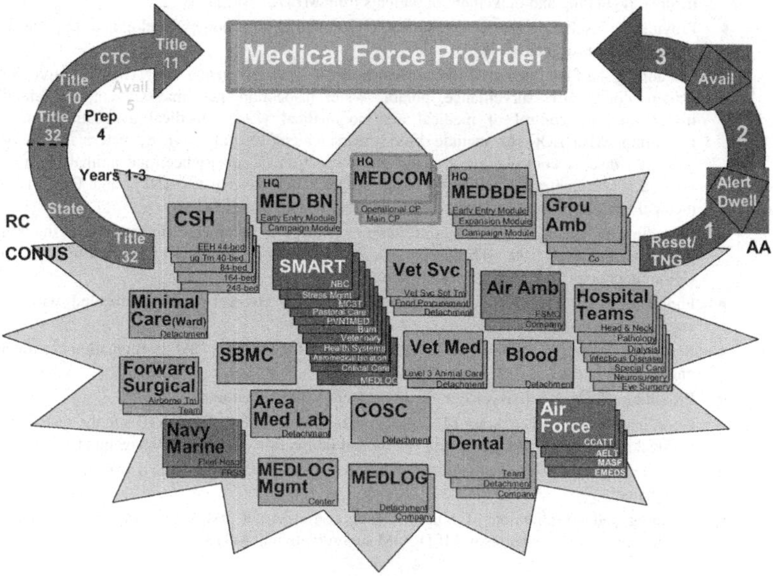

Figure 2-8. Medical force pool

2-109. The ability of the MEDCOM staff to develop accurate patient estimates based on real-time medical data ensures the right mix of medical specialties and resources are in the right place and at the right time to treat casualties on the battlefield. Care must be continuous and seamless from the POI or wounding through a phased system providing essential care in theater, with evacuation for definitive and rehabilitative care in CONUS.

2-110. The MEDCOM is a dedicated, regionally focused command with a basis of allocation of one per theater.

2-111. The MEDCOM is a versatile, modular medical C2 structure composed of a main command post (MCP) and an operational command post (OCP) which are 100 percent mobile. The MEDCOM is comprised of two standard requirements code (SRC) identified modules to provide a scalable medical C2 capability to the corps commander. Further, the MEDCOM is capable of—

- Providing subordinate medical organizations to operate under the subordinate MEDBDE and/or MMB and providing FSTs and/or other augmentation required by BCTs, and echelons above brigade (EAB) medical companies.
- Operating as the medical force provider for provisioning of required assets to the MEDBDEs.
- Advising the corps and other senior level commanders on the medical aspects of their operations.
- Providing staff planning, supervision of operations, and administration of assigned and attached medical units.
- Providing reach to the MHS.

- Providing medical Army support to other Services (ASOS)/Title 10 responsibilities of the ASCC commander.
- Coordinating with the USAF theater patient movement requirements center (TPMRC) for medical regulating and movement of patients from MEDCOM MTFs.
- Providing consultation services, technical advice, and clinical policy development in all aspects of medical and surgical services.
- Containing the functional staff to coordinate hospitalization, nursing, PVNTMED (entomology, epidemiology, OEH surveillance, potable water inspection, pest management, food facility inspection, and control of medical and nonmedical waste), medical evacuation, medical regulating, MEDLOG to include blood management, dental services, veterinary services (zoonotic disease control, investigation and inspection of subsistence, and animal medicine), pharmacy, optometry, nutrition care, and COSC, neuropsychiatric (NP), and BH services, medical laboratory services, and area medical support to supported units.
- Providing technical advice and consultation on medical automated information systems and programs such as the Theater Medical Information Program (TMIP) and Medical Communications for Combat Casualty Care (MC4).
- Functioning as the senior theater joint medical C2 element when augmented with joint augmentation package.
- Peacetime planning, training for and with the MEDBDEs, and participation in theater engagement plans.
- Providing advice and assistance in facility selection and preparation.
- Coordinating and orchestrating MEDLOG operations to include Class VIII supply, distribution, medical maintenance and repair support, optical fabrication, and blood management.
- Planning for the single integrated medical logistics manager (SIMLM), mission when designated.
- Planning and coordination of religious support for MEDCOM staff, assigned and attached medical units, and casualties in MEDCOM subordinate unit MTFs.

Operational Command Post

2-112. The MEDCOM OCP is a deployable, versatile module, which provides medical C2, policy development, and technical guidance to subordinate medical units and provides interface and liaison with supported theater forces in EAC, corps, and subordinate BCTs. The MEDCOM OCP has an assigned SRC to facilitate the placement of personnel in the CP and to integrate the CP into the time-phased force deployment list (TPFDL). The MEDCOM OCP is depicted in Figure 2-9.

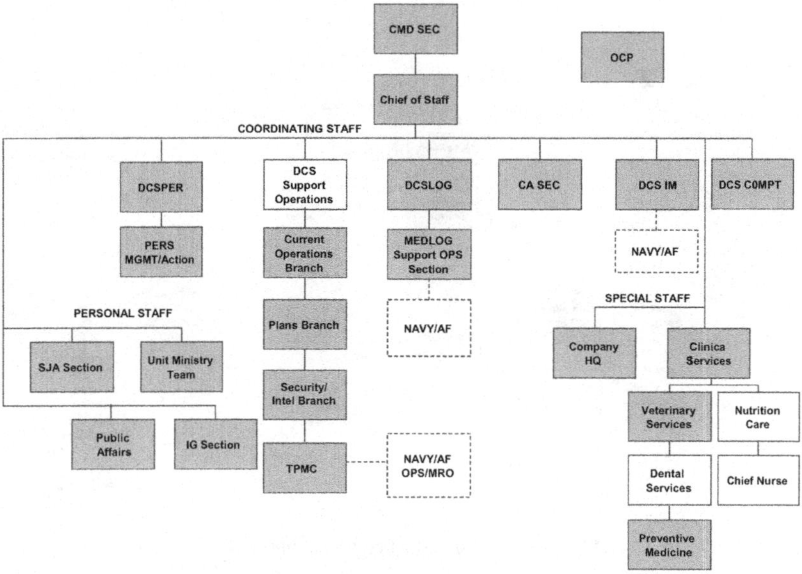

Figure 2-9. Operational command post, medical command

2-113. As the force continues to build within the theater, the MEDCOM's OCP can be incrementally expanded or augmented with additional planning and professional expertise. The deployment of personnel from the MCP model can then be built into the medical C2 required structure in order to provide campaign-quality health care to the deployed force.

Main Command Post

2-114. The MEDCOM MCP is a deployable augmentation module, which completes the staffing of the MEDCOM headquarters to enhance the delivery of campaign-quality health care to deployed forces in EAC, corps, and supported BCTs. It facilitates medical C2, policy development, and technical guidance to subordinate MEDBDEs and provides interface and liaison with supported theater forces in EAC, corps, and subordinate BCTs. If the OCP is deployed and needs additional personnel or clinical skills, the MEDCOM MCP can deploy personnel to the deployed OCP. This will provide a more robust medical C2 element and additional operational and planning capabilities; increased medical logistics staff; automated medical information systems; a more diverse clinical staff; and more depth in the command judge advocate (CJA), inspector general (IG), civil affairs (CA), and public affairs sections. The MEDCOM MCP has an assigned SRC to facilitate the placement on and the integration into the TPFDL. The MCP is depicted in Figure 2-10.

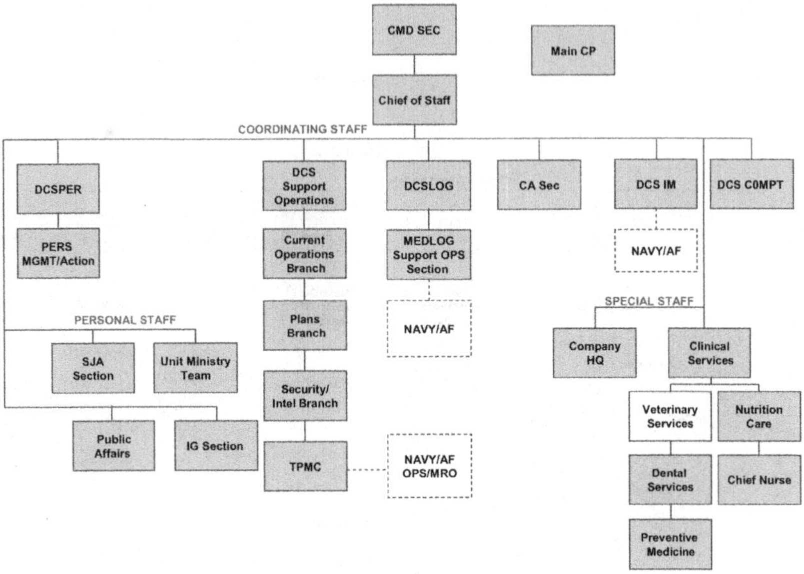

Figure 2-10. Main command post, medical command

MEDICAL BRIGADE

2-115. The MEDBDE provides a scalable expeditionary medical C2 capability for assigned and attached medical functional organizations task-organized in support of the deployed division/corps forces. The MEDBDE brings all requisite medical C2 and planning capabilities to provide responsive and effective HSS/FHP throughout the corps AO. The MEDBDE is depicted in Figure 2-11.

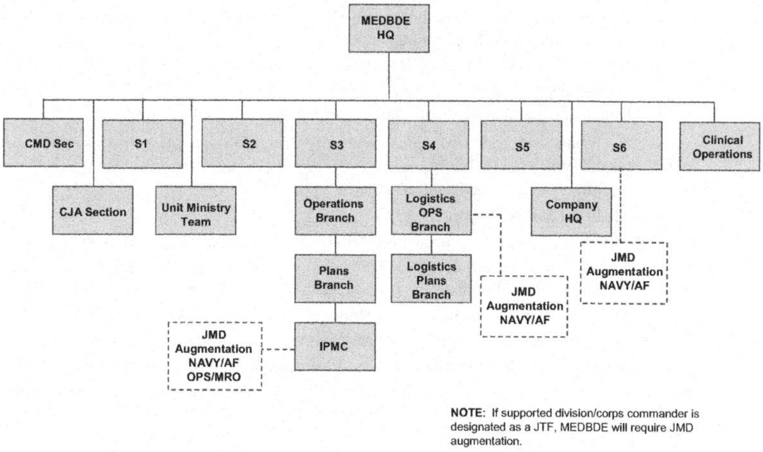

Figure 2-11. Medical brigade

2-116. The MEDBDE ensures the division/corps commander has the right mixture of medical professional (operational, technical, and clinical) expertise to synchronize the complex system of AHS functions required to maintain the health of his command by promoting fitness, preventing casualties from DNBI, and promptly treating and evacuating those injured on the battlefield.

2-117. The design and flexibility of the MEDBDE permits the division/corps commander the ability to meet expeditionary medical support requirements in support of early entry forces. As the supported corps commander's forces grow in both size and complexity, the MEDBDE can deploy additional modules that build upon one another to support full spectrum operations. The MEDBDE also provides the corps commander the appropriate medical C2 to continue to build his medical force capabilities through the integration of Army, joint, and multinational medical forces to ensure he can identify and counter any health threats in his supported areas. This will permit him to transition from expeditionary medical support operations to providing campaign-quality health care in his AO.

2-118. The MEDBDE has the capability to provide SRC elements for an early entry module (EEM), an expansion module, and the campaign module (CM), thus enabling its capability to be tailored to METT-TC factors of a specific division/corps operation.

2-119. The MEDBDE provides medical C2, administrative assistance, and technical supervision of assigned and attached medical units providing AHS in support of a division/corps. Further, it is capable of—

- Providing a rapidly responsive early entry medical C2 module that can quickly integrate into the early entry deployment sequence for crisis management.
- Providing full spectrum continuous medical C2 in support of all Army corps and joint (when augmented) forces.
- Providing operational medical augmentation to Role 2 BCT medical companies.
- Advising corps and BCT commanders on the medical aspects of their operations.
- Providing medical staff planning, operational and technical supervision, and administrative assistance for MMBs and CSHs task-organized under the MEDBDE.
- Coordinating with supporting TPMRC for medical regulating and medical evacuation from MMBs and hospitals to supporting EAC MTFs and CONUS.

- Providing medical consultation services and technical advice in—
  - Preventive medicine (medical and OEH surveillance, environmental health, sanitary engineering, and medical entomology).
  - Nursing services.
  - Dental services.
  - Behavioral health (to include COSC) and NP care.
  - Veterinary services (including food safety and inspection, animal medical care, and veterinary PVNTMED services).
  - Nutrition care and medical food service.
  - Medical laboratory support.
- Providing advice and recommendations for conduct of civil-military operations (CMO).
- Control and supervision of Class VIII supply and resupply movement to include blood management. When designated by the combatant command, serves as the SIMLM.
- Providing joint capable medical C2 capability when augmented with appropriate joint assets.
- Serves as the executive agent for veterinary services according to DODD 5200.31E.
- Coordinates ASOS for ship-to-shore/shore-to-ship medical evacuation mission.
- Plans and coordinates religious support for MEDBDE staff, assigned and attached medical units, and casualties in MEDBDE subordinate unit MTFs.

Chapter 3

Operational and Tactical Evacuation Planning

A comprehensive medical evacuation plan is essential to ensure effective, efficient, and responsive medical evacuation is provided to all wounded, injured, and ill Soldiers in the JOA. The Army medical evacuation plan flows from the GCC's guidance and intent and incorporates all missions and tasks directed by the GCC to be accomplished and is synchronized with supporting and supported units. In some scenarios, Army air and ground evacuation resources may be directed to provide support to sister Services, allied and coalition partners, and HN forces. Examples of a medical evacuation OPLAN and OPORD are contained in Appendix D.

SECTION I — THEATER MEDICAL EVACUATION PLANNING RESPONSIBILITIES

JOINT PLANNING

3-1. All medical evacuation planning within the theater reflects the GCC's guidance and intent. Planning at the joint level is governed by Joint Publications (JPs) 5-0, 5-00.1, and 5-00.2. Guidance on medical evacuation in joint operations is addressed in JP 4-02. When directed by the GCC, Army medical evacuation assets may be tasked to support other than Army forces engaged in the execution of the joint mission. These additional support missions will be clearly articulated in the GCC's OPLAN and OPORD. The ASCC surgeon, with the advice of the senior medical evacuation planner, will coordinate and synchronize these support operations with the combatant command surgeon, joint task force (JTF) surgeon, and the other Services and/or multinational partners as required to ensure that a comprehensive and effective, efficient, and responsive plan is developed and implemented.

MEDICAL COMMAND AND CONTROL ORGANIZATIONS

3-2. Army evacuation assets support the full spectrum of all Army operations and are capable of supporting Army forces engaged in the execution of the joint mission, as directed. Major combat operations supported by an ASCC will have a deployed MEDCOM. The C2 coordination and orders flow for AHS is illustrated in Figure 3-1. Evacuation planning should occur at each major level of command in order to effectively execute command, control, and influence over evacuation planning.

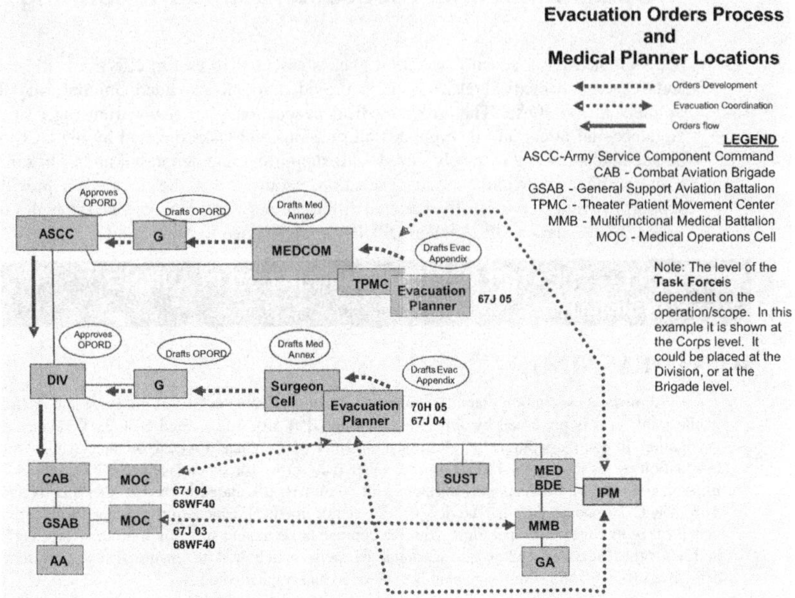

Figure 3-1. Command and control organizations for coordination and orders flow

3-3. Under the new Army modular design, some early entry operations may not require the deployment of a theater or a brigade C2 structure. In these cases, medical evacuation planning, coordination, and synchronization are initiated by the most senior medical command/staff level (such as the division surgeon, medical brigade, GSAB, and/or MMB). The senior leaders responsible for air and ground medical evacuation provide estimates and recommendations for the plan. This plan is normally provided as an appendix to the HSS annex and is disseminated down through the chain of command.

3-4. Dedicated Army rotary-wing aircraft (air ambulances) are under the C2 of the GSAB within the transformed Army aviation structure. Ground ambulances remain under the C2 of medical units/elements that are organic to movement and maneuver, fires, and sustainment units. The medical company, ground ambulance is under the C2 of the MMB.

3-5. To ensure an integrated medical evacuation system, synchronization of air with ground ambulance assets is essential. To affect this synchronization, the medical C2 headquarters must maintain influence over the execution of the medical evacuation plan. This influence is accomplished through the orders process.

Theater Patient Movement Center

3-6. In the design of the modular MEDCOM, the medical reengineering initiative (MRI) MEDCOM's medical regulating office (MRO) staff was reevaluated, realigned, and augmented with medical evacuators to more effectively coordinate and synchronize the various ground and air evacuation aspects of the operation and to supervise and facilitate the medical regulating functions. This staff section was renamed as the theater patient movement center (TPMC). The addition of medical evacuators enables this section to develop, coordinate, and synchronize the medical evacuation portion of the MEDCOM AHS

hour coordination and management responsibility for patient regulating and administration within the MEDCOM AO and directs patient evacuation between facilities within its AO as required to avoid patient overflow. This staff is responsible for preparing patient statistical reports for the command. This section also coordinates with the supporting TPMRC for the transfer of patients out of the MEDCOM/theater. The TPMRC regulates the flow of patients out of the theater to an MTF capable of providing the required specialty care and arranges for the transport of these patients by USAF aeromedical evacuation assets.

INTRATHEATER PATIENT MOVEMENT CENTER

3-7. As with the modular MEDCOM, the MRO of the modular MEDBDE was also redesigned in a similar fashion to enhance the capabilities for medical evacuation planning within the MEDBDE AO. This staff section was renamed the intratheater patient movement center (IPMC). This office is responsible for coordinating and synchronizing medical evacuation operations with the TPMC and the division surgeon cell. It regulates patients within its subordinate hospitals and arranges for transportation to transfer patients between facilities to relieve surgical backlogs, to obtain specialty care, and to ensure bed availability prior to start of operations, as appropriate. It also coordinates with the MEDCOM TPMC and the TPMRC for the evacuation of patients out of theater. This office maintains a 24-hour coordination and management responsibility for patient regulating and administration with the MEDBDE AO. It also prepares patient statistical reports for the MEDBDE commander. In the absence of the TPMC, IPMC assumes duties of the TPMC.

SECTION II — PLANNING PROCESS

EVACUATION PLANS AND ORDERS

CHARACTERISTICS

3-8. Operation plans and orders are the means by which commanders express their visualization, commander's intent, and decisions. They focus on results the commander expects to achieve. These OPLANs and OPORDs provide the basis for ensuring that medical evacuation operations are comprehensively planned, synchronized, responsive, and ensure a seamless continuum of care.

3-9. Evacuation plans and orders help form the basis commanders use to synchronize military operations. They encourage initiative by providing the *what* and *why* of a mission, and leave the *how to accomplish the mission* to subordinates. They give subordinates the operational and tactical freedom to accomplish the mission by providing the minimum restrictions and details necessary for synchronization and coordination. As medical evacuation operations are complex and may cross command and service boundaries, it is essential that the medical evacuation annex to the medical OPLAN/OPORD clearly delineates responsibilities, communications interoperability and procedures, and coordination requirements.

PLANS AND ORDERS

3-10. Publishing the commander's OPLAN/OPORD permits subordinate commanders to prepare supporting plans and orders. They can discern and implement instructions derived from a higher commander's plan and order. Appendix D provides an example of operational planning and a sample OPORD. Additionally, the commander's OPLAN/OPORD—

- Focuses subordinates' activities.
- Provides tasks and activities, constraints, and coordinating instructions necessary for mission accomplishment.
- Encourages agility, speed, and initiative during execution.
- Conveys instructions in a standard, recognizable, clear, and simple format.

3-11. The amount of detail provided in an evacuation plan or order depends on several factors, to include the experience and competence of subordinate commanders, cohesion and tactical experience of subordinate units, and the complexity of the operation.

3-12. Commanders balance these factors with their guidance and commander's intent, and determine the type of plan or order to issue. To maintain clarity and simplicity, plans and orders include annexes only when necessary and only when they pertain to the entire command. Annexes contain the details of support and synchronization necessary to accomplish the mission.

3-13. For a complete discussion of the military decision-making process and development of OPLANs/OPORDs refer to FM 5-0 and FMI 5-0.1.

Medical Evacuation Planning

3-14. Medical evacuation planning which supports the HSS/FHP plan is an ongoing process and is accomplished by medical operations element at battalion, brigade, division, corps, and theater levels. The HSS/FHP operations planning process must take under consideration all issues that could affect or influence HSS/FHP support for tactical operations. These issues should be considered in the initial developmental stages of theater, corps, division, or brigade plan as the supporting HSS/FHP annex is developed.

3-15. Planning is a dynamic and continuous process. The framework established in a robust initial plan can be built on in subsequent phases of an operation. The evacuation plan for a given AO will require adjustment and change during different phases of an operation. For example, transitioning from offensive operations to stability operations will generally dictate a significant change in evacuation coverage priorities.

3-16. Medical evacuation planning considerations, although essentially the same at all levels of command, may entail a broader scope, complexity, and detail at the higher levels of command due to the coordination and synchronization requirements which exist across command lines and service or multinational boundaries. Table 3-1 provides medical evacuation planning considerations at various levels of command based on METT-TC.

Table 3-1. Medical evacuation planning considerations

Mission	1. Define the protocol for a valid evacuation mission (mission statement). This protocol will serve as the authority to execute the mission by each medical evacuation unit/element/crew.
	2. Use the HSS/FHP annex and mission analysis to: • Task evacuation units where needed. This may involve tasking part or all of DS units to perform a GS or area support mission.
	3. Based on the type of operation: • Define trigger points for changes in the evacuation support plan. • Develop a concept of operations for all of the corps evacuation resources.
	4. Define task organization of units assigned to the corps (and when directed to the JTF) and support relationships, if possible. Lines of coordination and mutual support relationships should be clearly delineated.
	5. Area support and GS coverage areas will be

	defined with flexibility built into the plan. Coverage areas along division boundaries, for example, should not restrict or hinder evacuation of neighboring units that are geographically close. 6. Define pilot certifications for shore-to-ship medical air evacuation. 7. Develop standing operating procedure (SOP) requirements for single versus dual ship medical air evacuation. Address theater-specific procedures for in-flight link-up between medical evacuation aircraft and nonmedical evacuation aircraft for the purpose of dual ship requirements and/or security of pick-up zone. 8. Delineate any nonstandard air evacuation missions and allocate resources accordingly, such as support to Army special operations, joint and coalition forces. 9. Coordinate with any units that will have shared responsibilities for evacuation in a specific area and/or a specific mission. 10. Evaluate medical evacuation specific A2C2 considerations. Specifically, coordinating the crossing of brigade/division/corps boundaries/ area of operations. Medical evacuation must be afforded special expedited procedures for medical evacuation missions. Additionally, METT-TC dependent air corridors may have to be developed for the medical evacuation mission.
Enemy	1. Define the level or air-to-air threat and the threat faced by ground evacuation resources and the impact of this threat on the overall evacuation plan. This will be a major consideration in dictating which means of evacuation will be the primary means. 2. Based on the estimate of the number of enemy prisoners of war (EPW) or other categories of detainees involved in the operation— • Define procedures for evacuating wounded, injured, or ill persons through medical channels to include responsibilities of the echelon commander for designating guards for those evacuated directly from the point of capture to an MTF. • Allocate assets as necessary to provide this support and to support detainee collection points and holding areas.
Terrain and Weather	Define the types of terrain in the theater and use information as a major consideration in dictating the primary means of evacuation (ground or air in

	each sector) and any factors that may impair evacuation efforts.
Troops and Support Available	1. Disseminate locations of facilities used in evacuation, MTFs, MASFs, and aerial ports of debarkation (APOD). 2. Disseminate the capabilities and availability of joint evacuation assets and how to request their assistance. 3. Collect contact information from all evacuation units, assemble and publish for theater use. 4. Standardize a mandatory evacuation mission data collection format and collection method and schedule.
Time Available	1. Define the acceptable limits of evacuation time based on the distance from evacuation units. 2. Determine the proper allocation of resources to support the entire theater requirements. 3. Plan for trigger points (in time or distance) for changes in the evacuation plan to occur. 4. Define phases of the operation that reflect the commander's intent. Major changes in evacuation coverage usually occur when changes occur in the operational tempo (OPTEMPO). 5. Define briefing levels for launch authority. Consider developing a uniform theaterwide aviation risk assessment sheet.
Civil Considerations	1. Based on the HSS/FHP estimate and plan, determine requirements for the evacuation of HN civilians and others. 2. Define and disseminate protocols and procedures for evacuating civilians. 3. Allocate resources to support this mission, when directed.

3-17. For an effective HSS/FHP plan, medical staffs must coordinate and share operations information in order to synchronize support operations for mission accomplishment. See FMs 4-02, 4-02.1, 4-02.6, 4-02.12, 4-02.17, 4-02.25, 8-42 (4-02.42), and 8-55 (4-02.55), for doctrinal guidance on HSS/FHP operations. After the HSS/FHP plan is completed, it is incorporated into the C2 headquarters plan. After all annexes and appendixes of the plan are approved by the commander, it is incorporated into the OPORD.

SURGEON'S RESPONSIBILITIES

3-18. The surgeon is responsible for supervision and development of AHS input for the OPORD. The HSS plan serves as the base document for this input. The HSS plan is revised or updated based on mission analysis or changes in HSS requirements. The surgeon is tasked by the Assistant Chief of Staff (Operations and Plans) (G3)/S3 for HSS/FHP input to the OPORD for support of operations. The G3/S3 indicates timeline requirements. The surgeon and his staff are involved in all stages of the planning process, to identify all HSS requirements. Information on development of running estimates and the OPLAN/OPORD are discussed below. The medical plan/operations cell develops a HSS/FHP plan based on guidance received from the commander and the surgeon and provides HSS/FHP operational planning updates to the surgeon. The plan is briefed to the commander for approval, as required and provided to the G3/S3 according to the format of the in FM 8-55 (4-02.55).

MEDICAL EVACUATION TOOLS

3-19. It is essential that the evacuation plan for all combat operations be well planned, coordinated, and disseminated. In designing the medical evacuation plan, the medical planner should consider the following:

- Casualty collection points normally are predesignated along the axis of advance or evacuation routes.
- Forward of the BAS, the trauma specialist, CLS, and combat troops take casualties to the CCPs. These points facilitate acquisition by supporting ambulance teams and reduce evacuation time.
- When used by the BAS, CCPs help preserve BAS mobility, preclude carrying casualties forward, and reduce evacuation time to the sustainment area. When designating a CCP, the designating authority makes a decision whether or not to provide medical staff at the location. This decision is based upon the assessment of risk versus the availability of personnel.
- Normally, the role of care designating the point is responsible for staffing. Medical personnel may not be available to staff these points, and CLSs and ambulatory patients may be required to perform self-aid, buddy aid, or enhanced first aid. Casualty collecting points should be identified on operational overlays.

Note. A CCP staffed by a trauma treatment team is designated as a BAS (minus) rather than as a CCP.

3-20. Ambulance exchange points are a predetermined position where patients are exchanged from one evacuation platform to another.

3-21. These points are normally preplanned and are a part of the HSS/FHP annex to the OPLAN. In the forward area, the threat of enemy ground activities, large concentrations of lethal weapons systems, and effective use of antiaircraft weapons may dictate that the AXP be a predetermined rendezvous point for the rapid transfer of patients from one evacuation platform to another. The location of AXPs should be frequently changed to preclude attracting enemy fires.

3-22. Ambulance exchange points are established for many different reasons:

For example, the ambulance platoon of the heavy BSMC possesses a mixture of wheeled and tracked ambulances.

- The brigade tracked ambulances are provided so that they may keep up with maneuver elements. These vehicles carry the patients from the BAS to an AXP where the brigade wheeled ambulances take over for the relatively longer trip to the sustainment area.
- Ambulance exchange points are not limited to ground evacuation assets. Another example is a situation where the threat air defense artillery capability is such that air ambulances cannot fly as far forward as the BASs. However, an AXP could be established a few kilometers to the rear, still well forward of the BSA.
- The brigade combat team (BCT) tracked or wheeled ambulances could then transfer the patients to the air assets, thereby facilitating the rapid evacuation of patients and realizing a significant time savings.

3-23. By using AXPs evacuation assets are returned to their supporting positions faster. This facilitates evacuation as the returning crews are familiar with the road network and the supported unit's tactical situation. In the case of air ambulance (AA) assets, it is important because of the requirements for integration into the A2C2 system at each role and the enhancement to survivability provided by current threat and friendly air defense information.

Table 4-1. Categories of evacuation precedence

Priority I—URGENT	Is assigned to emergency cases that should be evacuated as soon as possible and within a maximum of 2 hours in order to save life, limb, or eyesight, to prevent complications of serious illness, or to avoid permanent disability.
Priority IA—URGENT-SURG	Is assigned to patients who must receive far forward surgical intervention to save life and to stabilize them for further evacuation.
Priority II—PRIORITY	Is assigned to sick and wounded personnel requiring prompt medical care. This precedence is used when the individual should be evacuated within 4 hours or his medical condition could deteriorate to such a degree that he will become an URGENT precedence, or whose requirements for special treatment are not available locally, or who will suffer unnecessary pain or disability.
Priority III—ROUTINE	Is assigned to sick and wounded personnel requiring evacuation but whose condition is not expected to deteriorate significantly. The sick and wounded in this category should be evacuated within 24 hours.
Priority IV—CONVENIENCE	Is assigned to patients for whom evacuation by medical platform is a matter of medical convenience rather than necessity.
The NATO STANAG 3204 has deleted the category of Priority IV—CONVENIENCE; however, it will still be included in the US Army evacuation priorities as there is a requirement for it on the battlefield.	

SECTION II — MEDICAL EVACUATION REQUESTS

4-5. This section discusses requests for medical evacuation. Specific procedures, frequencies, and security requirements for transmittal of medical evacuation requests are delineated through the orders process and are made a part of the unit/command SOPs. Each sector based on the METT-TC may be designated with a different method of evacuation as the primary means to effect evacuation. In sectors which have a high ground-to-air or air-to-air threat may rely on ground evacuation assets to move the majority of patients. In other sectors where the ground threat is high and comprised of small arms, improvised explosive devices, and bombs, medical evacuation operations may be more efficiently and effectively executed by AAs. An additional consideration in planning medical evacuation operations is to determine whether armed escorts are required for either the ground or AA mission. Those missions that require armed escort must be thoroughly coordinated and synchronized between the medical assets and force protection assets that will accompany them.

4-6. Medical evacuation requests often are sent from the POI, through intermediaries, such as higher headquarters, who then transmit the request up to the nearest medical evacuation unit. The unit relaying the request must ensure that it relays the exact information originally received. The radio call sign and frequency relayed (Line 2 of the request) should be that of the requesting unit and not that of the relaying unit. However, the intermediaries contact information can be given as additional information if a callback is necessary to clarify details of the mission. Figure 4-1 depicts the communication flow for evacuation requests.

4-7. The necessity for secure communication in transmitting a 9-line request is METT-TC dependent. A technically capable enemy may be able to intercept a nonsecure 9-line request and strike at the pickup zone causing additional harm to the patients and the medical evacuation crew. In combat and when directed by higher headquarters, always use secure communications when requesting medical evacuation.

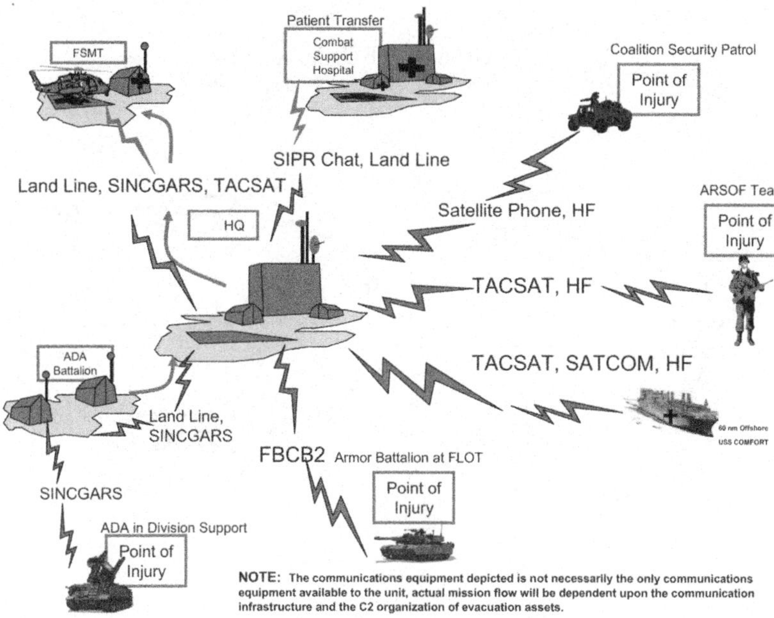

Figure 4-1. Flow of communication for evacuation request

GROUND EVACUATION REQUEST

4-8. Figure 4-2 provides an example of a request from the POI for ground evacuation. The patient is a ROUTINE precedence patient requiring evacuation from the POI to the supporting BAS (Role 1 MTF) with follow-on evacuation to an AXP for further evacuation to the rear. Should the patient require further treatment, either an air or ground evacuation could be used.

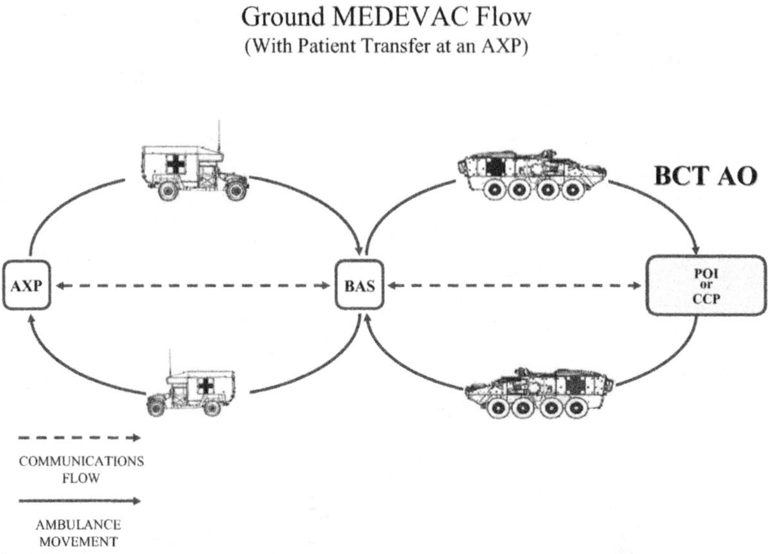

Figure 4-2. Ground evacuation request in a maneuver unit

4-9. Ground medical evacuation assets also provide DS and GS to supported units in the EAC. Figure 4-3 and Figure 4-4 depict the processes of providing DS to an ASMC and the GS support role of transferring a patient from a corps hospital to a MASF for further evacuation out of the theater.

4-10. In the corps area, the ASMC provides Roles 1 and 2 medical support on an area basis. The ASMC has organic evacuation assets and receives augmentation when required from the medical company ground ambulance assigned to the MMB.

4-11. Medical evacuation resources are also used to transfer a patient between hospitals within the theater and from a hospital to MASF for further evacuation out of theater. In Figure 4-3, this mission is being accomplished by a ground ambulance. However, depending upon the distance from the hospital to the MASF and/or APOD, this mission can also be accomplished by AA assets.

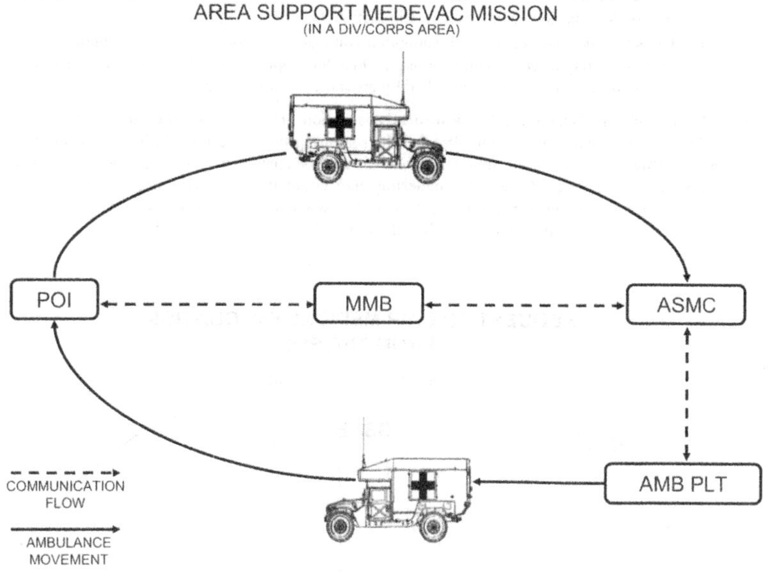

Figure 4-3. Ground evacuation request in the corps

AIR EVACUATION REQUEST

4-12. Air ambulances may fly as far forward as possible on the battlefield. Although evacuation by AA is the preferred means for all casualties, when high evacuation workloads exist, evacuation by AA should be the primary means used for URGENT and URGENT-SURG patients. Figure 4-5 depicts the zones of evacuation. The GSAB/CAB will place AA assets where they can best support the tactical commander's plan and facilitate the timely and responsive evacuation. To coordinate and synchronize the ground and air medical evacuation plans, a staff element was designed within the GSAB headquarters. This cell consists of the medical planners on the GSAB and CAB staff. The medical operations cell (MOC) will:

- Establish flight procedures specific to air evacuation missions within the CAB. This may include special routes or corridors as well as procedures for escort aircraft linkup.
- Ensure lines of communication (LOC) to supported units and higher levels of medical command are available. Further the MOC ensures supported units understand medical evacuation procedures and capability.
- Establish medical evacuation briefing and launch procedures. Ensure there is 24-hour access to those able to launch high and very high-risk missions.
- Maintain awareness of tactical and medical situation. Coordinate with MROs at higher levels to efficiently conduct the general support mission and to work in concert with adjacent units.
- Assist the AA CO and GSAB/CAB staff conduct medical evacuation operations.

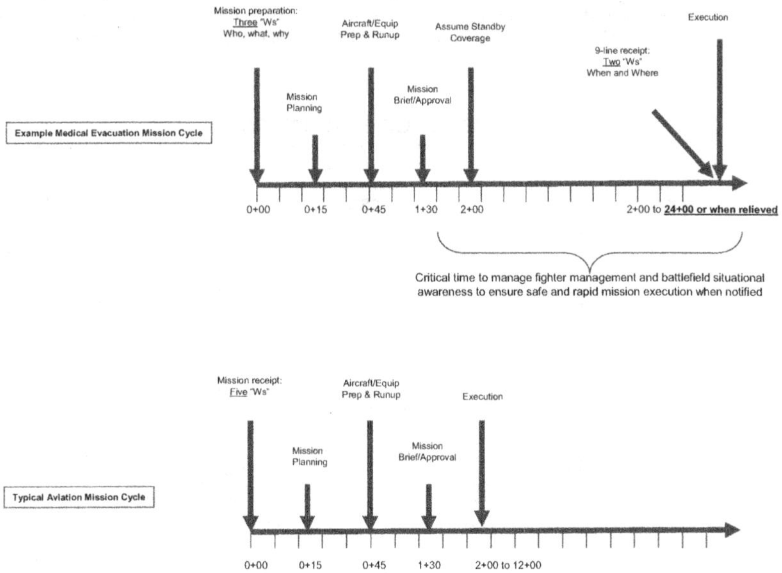

Figure 4-6. Aviation mission planning/execution cycle

JOINT INTERCONNECTIVITY

4-16. When directed by the GCC, Army evacuation assets may be used to evacuate patients from the other services or may be placed in direct support of other service, allied, or coalition units participating in the operation. Prior to the initiation of the operation, communications, and procedural issues should be synchronized to ensure the seamless execution of the operation. Interoperability issues can result in slower response times and may adversely effect medical evacuation operations. Figure 4-7 depicts the communications and patient flow in a joint scenario.

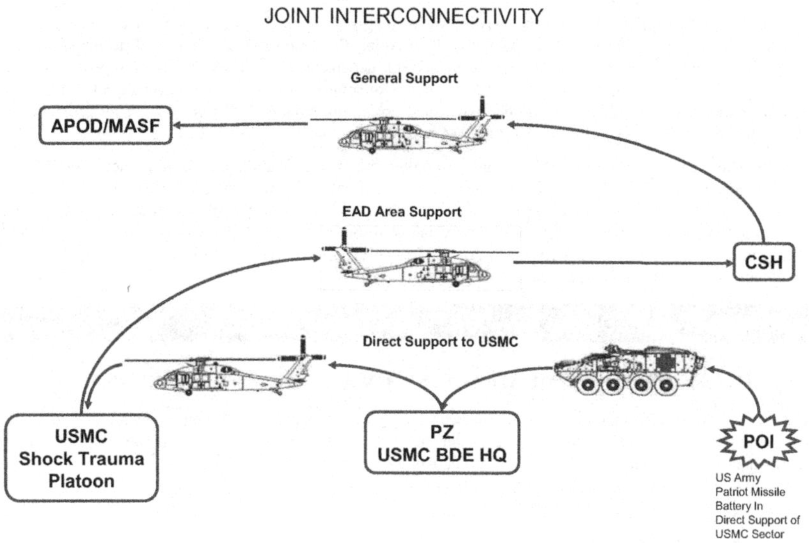

Figure 4-7. Air and ground ambulance evacuation in a joint environment

SECTION III — MEDICAL EVACUATION UNITS, ELEMENTS, AND PLATFORMS CONSIDERATIONS

MEDICAL EVACUATION SUPPORT PROTOCOL

4-17. Medical evacuation is performed by the higher role of medical care supporting forward and evacuating from the lower role. Medical evacuation support for theater, corps, and division is provided by organic ambulance teams, squads, and platoons. In addition, it is provided by the MMB medical company, ground ambulance medical company (GAMC), ASMC, and the GSAB medical company, AA.

4-18. The MMB maximizes the effective employment of ground ambulances resources whether in support of a division, corps, or at theater level. It coordinates through the combat aviation brigade's GSAB for effective use of AA support of its ASMCs. The medical brigade, MMBs, GSAB, sustainment brigade, BCTs and division surgeon section (DSS) coordinate and synchronize medical evacuation operations for the division AOs. The DSS coordinates with the MEDBDE for the evacuation of patients from division MTFs to supporting CSHs. The MEDBDE coordinates and arranges the required intratheater medical evacuation of patients to and from supporting CSHs. The MEDCOM coordinates for evacuation of patients out of the theater. To ensure that patients are evacuated to the appropriate MTFs, MROs are organic to the MEDCOM, and MEDBDE.

Medical Evacuation Platforms Factors

4-19. Evacuation platforms must be capable of keeping pace with the troops supported. The patient's medical condition is the overriding factor in determining the evacuation platform and destination MTF. The AA operates wherever needed on the battlefield, dependent on risk and METT-TC factors. Use of hardened armored MEV may be the vehicle of choice for some missions for short evacuation to an MTF or to a secure AXP for transfer to an air or wheeled ground ambulance. The criteria for an efficient and effective evacuation system were identified in Chapter 1.

4-20. The crew of the AA, assisted by onboard patient monitoring and diagnostic equipment, is trained in aeromedical procedures to provide optimum en route patient care. It is the platform of choice for most categories of patients. However, insufficient numbers of AAs are available to evacuate all patients expected in a battle zone. To conserve these valuable resources, medical planners should plan to use AAs to primarily move Priority I, URGENT, Priority IA, URGENT-SURG, and Priority II patients with other categories on a space-available basis.

SECTION IV — MEDICAL EVACUATION AT UNIT LEVEL

CONSIDERATIONS FOR MEDICAL EVACUATION MISSIONS

4-21. As METT-TC factors affect all medical evacuation missions and the employment of ambulance assets. The medical commander must consider the basic tenets that influence the employment of medical evacuation assets. These factors include the patient's medical condition and the:

- Tactical commander's plan for employment of combat forces.
- Enemy's most likely course of action.
- Anticipated patient load.
- Expected areas of patient density.
- Availability of medical evacuation resources to include ground and air crews.
- Availability, location, and type of supporting MTFs.
- Protection afforded medical personnel, patients, and medical units, vehicles, and aircraft under the provisions of the Geneva Conventions.
- Army airspace command and control plan.
- Engineer obstacle plans.
- Fire support plan (to ensure medical evacuation assets are not dispatched onto routes and at the times affected by the fire support mission).
- Road network/dedicated medical evacuation routes (contaminated and clean).
- Weather conditions.

Patient Acquisition

4-22. Units with organic medical evacuation assets have the primary responsibility for patient acquisition. Methods of employment and evacuation techniques differ depending upon the nature of the operation.

4-23. Units without organic ambulance assets are provided medical evacuation support on an area support basis. Units must develop techniques which facilitate the effective employment of their combat medics/company/platoon medics, enhance the ability to acquire patients in forward areas, and rapidly request medical evacuation support. The techniques developed should be included in the unit TSOP. As a minimum, the TSOP should include the—

- Vehicle assignment for the organic medical personnel.
- Vehicles designated to be used for casualty transport and/or patient evacuation.
- Procedures for requesting medical evacuation support (during routine operations or during mass casualty situations).
- Role of the first sergeant (1SG), platoon sergeants, and CLSs in medical evacuation.

Maneuver Battalion Medical Platoons

4-24. The medical platoon leader (a physician) along with the medical platoon field medical assistant should be included in all battalion tactical planning. This officer—

- Maintains knowledge of the concept of operations, commander's intent, and the anticipated medical requirements.
- Develops the HSS/FHP plan (FM 8-55 [FM 4-02.55] and FM 4-02.4) and provides HSS/FHP overlays with preplanned evacuation routes and CCPs, to ambulance squads or teams for inclusion in the battalion OPLAN.
- Ensures that the platoon sergeant provides strip maps or other navigational tools to the ambulance drivers, if needed.
- Requests augmentation support from the supporting medical company in advance of the operation, if required. When elements of other maneuver battalions are attached as part of a task force, then ensures that adequate medical elements are included in the support package.
- Ensures that orientation and support are provided for the medical personnel. This precludes taxing the medical elements of the receiving unit. These responsibilities are normally delegated to the medical operations officer (field medical assistant).

4-25. The ambulance/evacuation section squad leader/NCO ensures that the ambulance teams have a working knowledge of the terrain features in the AO. Whenever possible, he familiarizes with primary and secondary medical evacuation routes through route reconnaissance conducted by the squad leaders. This platoon sergeant and squad leader manages the employment of the ambulance teams and monitors the communications net to remain abreast of the tactical situation.

4-26. The following factors should be considered when selecting ambulance routes:

- Tactical mission.
- Coordinating evacuation plans and operations with the unit movement officer.
- Security of routes and security escort.
- Availability of routes.
- Physical characteristics of roads and cross-country routes (to include natural obstacles).
- Requirements to traverse roads in urban areas and potential obstructions from rubble and debris.
- Traffic density.
- Time and distance factors.
- Proximity of possible routes to areas that may be subject to enemy fire.
- Lines of patient drift.
- Cover, concealment, and available defilade for moving and stationary vehicles.
- Engineer obstacle plans.
- Fire support plan (to ensure medical evacuation assets are not dispatched onto routes and at the times affected by the fire support mission).

4-27. Depending upon the combat situation, the modes of evacuation may include walking soldiers who are wounded, manual and litter carries, nonmedical transportation assets, or dedicated medical evacuation platforms. Evacuation in the battalion area normally depends on the organic ambulances assigned. Evacuation by AA is dependent upon the patient's medical condition, availability of air assets, tactical situation, air superiority, and weather conditions.

4-28. The ambulance team is normally deployed to each maneuver company trains and one remains with the Role 1 MTF/BAS in the combat trains. Ambulances operate as far forward as the tactical situation permits. An ambulance team operating in a maneuver company AO, is normally under the tactical control of the maneuver company executive officer or 1SG. This ambulance team, however, remains under the OPCON of the medical platoon.

4-29. The medical operations officer ensures that the ambulances are located close to the anticipated patient workload. An ambulance team consists of one ambulance and two combat medics (on track vehicles, a third emergency care specialist is required to provide en route medical care). One or two of these teams serve in DS of a maneuver company. To become familiar with the specific terrain and battlefield situation, the team maintains contact with the supported maneuver company during most combat operations. Maneuver medical platoon ambulances not in DS are positioned strategically throughout the battalion area or are sited at the BAS to—

- Evacuate patients from the company aid posts, CCPs to the BAS.
- Reinforce the forward positioned ambulance teams.
- Support the combat forces held in reserve and/or scout, mortar and forward observer platoons.

4-30. Another employment option is to forward site the additional ambulance teams at company CCPs, as well as at the BASs. Many times the ambulance team finds battlefield casualties who have not been seen by a trauma specialist. In these cases, the team members dismount and then find, treat, and evacuate the patients. Ambulance teams not specifically dedicated to support combat elements can be used as messengers in medical channels and to provide transport of emergency medical personnel, equipment, and supplies.

4-31. During static situations where the maneuver company is not in enemy contact or is in reserve, the ambulance team returns to the BAS to serve as reinforcement to other elements in contact. However, during movement to contact, the ambulance team immediately deploys with its supported unit. In moving patients back to the CCP, the team may be assisted by nonmedical personnel. Specific duties of the ambulance team are to—

- Maintain contact with supported elements.
- Find and collect the wounded.
- Administer EMT.
- Initiate or complete the FMC.
- Evacuate patients to the BAS.
- Direct or guide ambulatory patients to the BAS.
- Resupply company/platoon medics.
- Serve as messengers in medical channels.
- Route recons from Role 1 MTF to Role 2 MTF or the higher role of medical care.

Casualty Collection Points

4-32. During the offense, CCPs may be used to avoid hampering the movement of the maneuver elements. In fast-moving situations, preplanned CCPs are included in the HSS plan and activated based on the crossing of phase lines, upon the occurrence of predetermined events, or on the execution of other control measures. It may be necessary to set up multiple CCPs for each phase of an operation. Rotating the use of these points precludes the enemy from using them to pinpoint maneuver elements or from attracting enemy fires. When the situation permits, medical evacuation of patients from CCPs, or AXPs may be accomplished by AAs.

4-33. Ambulance teams move using available terrain features for cover and concealment. They avoid prominent terrain features and likely targets. When stationary, the ambulance crew should conceal the vehicle as much as possible. When a casualty occurs in a tank, Bradley infantry fighting vehicle (BIFV), a Stryker wheeled armored vehicles, the ambulance team moves as close to the armored vehicle as possible. Assisted by the armored crew, if possible, the casualty is extracted from the vehicle and then administered EMT. The ambulance team moves the patient to the Role 1 MTF/BAS or to a CCP point to await further evacuation.

4-34. The senior company medic/trauma specialist normally remains with the company combat trains, but may be used anywhere in the company area, even assisting the ambulance teams in some situations. He may be used to direct ambulance teams to locations where vehicle crews need assistance or where injured or wounded crew members have been left. In some situations, crew members may have to rely on self-aid or buddy aid until the CLS or the platoon medic/trauma specialist arrives.

Area Medical Evacuation Support

4-35. Medical evacuation on an area basis is required at all roles in the AHS system. The division units without organic medical evacuation resources, such as combat engineers, will require evacuation support on an area basis. To ensure that these elements receive adequate support, the medical planner must include their requirements into the OPLAN. Prior coordination is essential to ensure that the locations of CCPs, AXPs, and BASs are disseminated to these elements and that any unique support requirements are included in the OPLAN. Further, the triggers for relocating these assets and their next locations should also be included.

MEDICAL EVACUATION SUPPORT FROM ROLE 1 AND ROLE 2 MEDICAL TREATMENT FACILITIES

4-36. Evacuation from Role 1 MTFs operated by treatment squad or team is normally provided by the BSMC/FSMC ambulance platoon and the GSAB AA company. Further, these ambulance assets provide medical evacuation support on an area basis to other units in the brigade area, as required.

Ground Ambulances

4-37. The ambulance teams from the ambulance platoon are normally collocated with the BSMC/FSMC treatment platoon for mutual support. They establish contact and locate one ambulance team with the medical platoon of each maneuver battalion. The remaining ambulances are used for BCT operations and area support. The ambulances are positioned at AXPs or CCPs, or are field-sited with Role 1 MTFs operated by the BSMC/FSMC treatment teams.

4-38. A FSMT from the supporting GSAB medical company, AAs provides DS AA support for the BSMC/FSMC Role 2 MTF. The teams normally remain under the operational control of the GSAB. The DS relationship provides authority to the BSMC/FSMC in coordination with the brigade support battalion/forward support battalion (BSB/FSB) distribution support section, brigade surgeons section to direct the integrated air and ground medical evacuation support for the BCT. The section leader of the FSMT should be included in the brigade tactical planning process.

Air Ambulances

4-39. The AA team evacuates Priority I, URGENT patients from as far forward as possible to the BSMC/FSMC. Further, when a FST is colocated with a BSMC/FSMC, AAs evacuate Priority IA, URGENT-SURG to this facility if the patient can not be flown to a Role 3 MTF condition due to distance or patient condition. External lift capabilities of aeromedical evacuation helicopters add an important dimension to its role on the battlefield in moving medical supplies and equipment.

4-40. The GSAB AAs may operate from the sustainment area and BSAs providing around-the-clock, immediate response, evacuation aircraft, as well as provide the BSMC/FSMC commander flexibility and agility in the movement of treatment teams and equipment to the forward battle area. It also provides the capability to rapidly resupply Class VIII supplies to combat units.

4-41. The medical elements within the BSB/FSB will request AA support through the brigade aviation element (BAE). The FSMT leader will provide coordination between the BSB/FSB, BAE, or the BCT S3.

4-42. In the BSMC/FSMC, the executive officer (XO) is the principal assistant to the commander for the tactical employment of the company assets. The XO should be included in all brigade tactical planning. Officer preparation will be required to reinforce or reconstitute forward HSS elements and to request augmentation through the brigade surgeon section (BSS), if required. The FSMT leader keeps the BSMC/FSMC XO apprised of the operational capability. This enables the XO to effect timely reinforcement or augmentation. The BSMC/FSMC XO must be familiar with the specific terrain and battlefield situation. Further, the XO should have a thorough understanding of the division and brigade commanders' ground tactical plan.

Brigade Medical Evacuation Plan

4-43. The BSS is responsible for developing the BCT medical evacuation plan. The BSMC/FSMC commander is responsible for the execution of brigade medical evacuation plan, to include the use of both ground and air assets. The BSS medical planner should include the medical company commander, XO, brigade S1, brigade XO, medical platoon leaders, FSMT leader, the BSB/FSB medical plans and operations officer, and the BAE company commander (if available) in the planning process.

Ground Ambulance Evacuation Brigade Role 2 Medical Treatment Facilities

4-44. Evacuation from the BSMC/FSMC is provided by ground ambulances from the GAMC or from AAs from the GSAB AA company. The GAMC provides support to the BSMCs and medical elements in the division AO. They usually evacuate only those patients who will not RTD within 72 hours.

Ground Ambulance Support for Echelons Above Brigade/Division Units

4-45. The ASMC ambulance platoon normally collocates with the treatment platoon for mutual support and area taskings. It performs ground evacuation and en route patient care for supported units in the sustainment area. The ASMC may also evacuate patients from the supported Role 2 MTFs in the sustainment area to the ASMC Role 2 MTF. The ambulance platoon is mobile in operations as its assets may be totally deployed at one time. The platoon normally forward stations a portion of its teams in support of those units in the sustainment area.

4-46. The remaining teams are used for GS support to division sustainment, movement and maneuver and fires units, and for reinforcing support or ambulance shuttles. Platoons or squads from the GAMC will be in DS, or OPCON to, and collocated with supported division units.

SECTION V — EXCHANGE OF PROPERTY

PROPERTY EXCHANGE AND PATIENT MOVEMENT ITEMS

4-47. United States Army medical evacuation operations require that whenever a patient is evacuated from one MTF to another, or is transferred from one ambulance to another, medical items of equipment (casualty evacuation bags [cold weather-type bags], blankets, litters, and splints) remain with the patient. To prevent rapid and unnecessary depletion of supplies and equipment, the receiving Army element exchanges like property with the transferring element. This reciprocal procedure will be practiced to the fullest extent possible through all phases of evacuation from the most forward element through the most rearward hospital.

4-48. With USAF aeromedical evacuation operations, a major factor in the evacuation of patients is that specific medical equipment and durable supplies designated as patient movement items (PMI) must be available to support the patient during the evacuation. Examples of PMI include—

- Ventilators.
- Litters.
- Patient monitors.
- Pulse oximeters.

These items will be available for exchange at the supporting ASFs and MASFs. Refer to JP 4-02 and FM 4-02.1 for additional information on PMI.

This subparagraph implements STANAG 2128 and QSTAG 436.

4-49. Medical property of allied nations (NATO and ABCA armies) accompanying patients of allied nations will be returned to the parent nation at once, if possible. If it is not possible, like items will be exchanged as in the paragraph above.

4-50. When medical property of coalition forces without ratified standardization agreements accompanies patients of coalition forces, it is returned as soon as practical. Commanders should consult with their staff judge advocate early in the planning process to ensure appropriate policy and procedures are developed and disseminated.

SECTION VI — MEDICAL EVACUATION SUPPORT FOR THE OFFENSE, THE DEFENSE, AND STABILITY OPERATIONS

MEDICAL EVACUATION SUPPORT FOR THE OFFENSE

4-51. The offense is the decisive form of war, the commander's only means of attaining a positive goal or of completely destroying an enemy force (FM 3-0). The offense is characterized by rapid movement, deep penetrations, aggressive action, and the ability to sustain momentum regardless of counterfires and countermeasures.

4-52. When considering the evacuation plans to support an offensive action, the medical planner must consider many factors (FM 8-55 [FM 4-02.55]). The forms of maneuver, as well as the enemy's capabilities, influence the character of the patient workload and its time and space distribution. The analysis of this workload determines the allocation of AHS resources and the location or relocation of MTFs. Evacuation support of offensive operations must be responsive to several essential characteristics. As operations achieve success, the areas of casualty density move away from the supporting facilities. This causes the routes of medical evacuation to lengthen. Heaviest patient workloads occur during disruption of enemy main defenses, at terrain or tactical barriers, during the assault on final objectives, and during enemy counterattacks. The accurate prediction of these workload points by the medical planner is essential if medical evacuation operations are to be successful.

4-53. In traditional combat operations, the major casualty area of the division BCTs is normally the zone of the main attack. As the main attack accomplishes the primary task of the division, it receives first priority in the allocation of combat power. The allocation of combat forces dictates roughly the areas which are likely to have the greatest casualty density. As a general rule, all BCT MTFs are located initially as far forward as combat operations permit. This allows the maximum use of these facilities before lengthening evacuation lines force their displacement forward. In operations that feature deep battles with CBRN weapons targeted at supporting logistical bases, MASCAL operations may be conducted in sustainment area.

4-54. As advancing combat formations extend control of the battle area forward, supporting medical elements overtake patients. This facilitates the acquisition of the battle wounded and reduces the vital time elapsed between wounding and treatment. In offensive operations, two basic problems confront the supporting evacuation units.

- First, contact with the supported unit must be maintained. Responsibility for the contact follows the normal HSS pattern of rear to front. The contact is maintained by forward deployed air and ground evacuation resources.
- Secondly, the mobility of the MTFs supporting the combat formations must be maintained. Periodically, BCT medical companies and collocated FSTs, are cleared of their patients so that they may move forward. In some scenarios, CSHs may also have to evacuate patients to reduce patient overflow, surgical backlogs, or to increase bed availability.

4-55. This requirement for prompt evacuation of patients from forward MTFs requires available ambulances to be leveled well forward from the outset. The requirement for periodic movement of large numbers of patients from divisional and corps facilities further stresses the evacuation system.

TYPES OF OFFENSIVE OPERATIONS

Movement to Contact/Attack

4-56. Medical evacuation support in movement to contact/attack is key to the tactical plan. Prior deployment of evacuation resources with parent and supported units permits uninterrupted and effective evacuation support.

Exploitation and Pursuit

4-57. Evacuation support of exploitation and pursuit operations resembles those discussed for the envelopment (paragraph 4-59). Since exploitation and pursuit operations can rarely be planned in detail, evacuation operations must adhere to TSOPs and innovative C2. These actions are often characterized by—

- Decentralized operations.
- Unsecured ground evacuation routes.
- Exceptionally long distances for evacuation.
- Increased reliance on convoys and AAs.

MEDICAL EVACUATION SUPPORT FOR DIFFERENT FORMS OF OFFENSIVE MANEUVER

Penetration and Attack

4-58. In this tactic, the attack passes through the enemy's principal defensive position, ruptures it, and neutralizes or destroys the enemy forces. Of all forms of offensive maneuver, the penetration of main enemy defenses normally produces the heaviest medical evacuation workload. Patient acquisition starts slowly, but becomes more rapid as the attack progresses. The evacuation routes lengthen as the operation progresses. The penetration maneuver is often preceded by heavy preparatory fires which may evoke heavy return fire. These enemy fires may modify the decision to place evacuation assets as far forward as possible. Patient evacuation may be slow and difficult due to damage to roads or the inaccessibility of patients. Evacuation support problems multiply when some combat units remain near the point of original penetration. This is done to hold or widen the gap in enemy defenses while the bulk of division forces exploit or pursue the enemy. Treatment elements are placed near each shoulder of the penetration; ground evacuation cannot take place across an avenue of heavy combat traffic. Besides the heavy traffic, the area of the penetration is normally a target for both conventional and nonconventional CBRN weapons.

Envelopment

4-59. In the envelopment, the main or enveloping attack passes around or over the enemy's principal defensive positions. The purpose is to seize objectives which cut the enemy's escape routes and subject him to destruction in place from flank to rear. Since the envelopment maneuver involves no direct breach of the enemy's principal defensive positions, the medical evacuation system may not be confronted with a heavy workload in the opening phase—

- Ambulances are echeloned well forward in all levels of AHS to quickly evacuate the patients generated by suddenly occurring contact.
- Medical treatment facilities moving with their respective formations overtake patients during evacuation and reduce delays in treatment. After triage and treatment, the patients are evacuated to division/corps-level facilities by accompanying division/corps ground ambulance assets or AAs.

- When the isolated nature of the envelopment maneuver precludes prompt evacuation, the patients are carried forward with the treatment element. Again, nonmedical vehicles may be pressed into emergency use for this purpose. When possible, nonmedical vehicles should be staffed by a combat medic or CLS.
- When patients must be carried forward with the enveloping forces, AHS commanders use halts at assembly areas and phase lines to arrange combat protection for ground ambulance convoys through unsecured areas.
- Further, the commander may take advantage of friendly fires and suppression of enemy air defenses to call for prearranged AA support missions, or emergency use of medium-lift helicopter backhaul capabilities.

Infiltration

4-60. Infiltration is a form of maneuver used during offensive operations.

- The division can attack after infiltration or use it as a means of obtaining intelligence and harassing the enemy.
- Though it is not restricted to small units or dismounted actions, the division employs these techniques with a portion of its units, in conjunction with offensive operations conducted by the remainder of its units.

Army Health System of Infiltration/Maneuver

4-61. The AHS of infiltration/maneuver is restricted by the amount of medical equipment, supplies, and transportation assets that can be introduced into the attack area. No deployment of division-level medical units without their organic transportation should be attempted.

4-62. Elements of unit-level AHS should be accompanied by their organic vehicles, and ambulances should receive priority for deployment. It may be necessary to man-carry enough BAS equipment into the attack area to provide EMT and ATM; however, this results in degrading mobility. When the element is committed without its ambulances, patients are evacuated to the BAS by litter bearer teams. This requires reinforcement of the medical platoon by division or corps medical personnel or improvisation of litter teams using combat troops (if available and approved by the tactical commander). Patient evacuation from the BAS and medical resupply of the force may be provided by litter bearers, depending upon distances and degree of secrecy required.

4-63. When airborne and air assault forces are used, infiltrating elements may land at various points within the enemy's rear area and proceed on foot to designated attack positions. As in surface movement, the amount of medical equipment taken may be limited. In airborne operations, the evacuation of patients will be by litter bearers or frontline ambulances to CCPs or the BAS and then by division-level ambulances to the Role 2 MTF. In air assault operations, the evacuation is by litter bearers to CCPs or to the BAS and then by AAs to a Role 2 MTF. Once the combat element begins the assault on the objective, secrecy is no longer important and its isolated location requires HSS characteristic to airborne and air assault operations until ground linkup.

Turning Movement

4-64. The turning movement is a variant to the envelopment in which the attacker attempts to avoid the defense entirely; rather, the attacker seeks to secure key terrain deep in the enemy's rear and along his LOC. Faced with a major threat to his rear, the enemy is thus "turned" out of his defensive positions and forced to attack rearward at a disadvantage.

> *Note.* General MacArthur's invasion at Inchon during the Korean War is an example of a classic turning movement. Casualties were initially light as the main defenses were avoided; however, as the invasion developed, resistance stiffened and higher casualty rates were experienced. Further, as fighting occurred in a populated area (Seoul), significant civilian casualties resulted. The lack of Korean health care providers caused many of these civilians to seek medical aid from US field medical units.

Medical Evacuation for Turning Movement

4-65. Medical evacuation support to the turning movement is provided basically in the same manner as to the envelopment. As the operation is conducted in the enemy's rear area, LOC and evacuation routes may be unsecured resulting in delays in resupply and evacuation. In the Inchon example, a hospital ship was located off the coast to accept patients evacuated from the fighting. However, due to the precarious tides, evacuation and resupply were often delayed for hours and sometimes days since the harbor could not be navigated by small vessels. It was not until Kimpo Airfield fell that timely evacuation could occur. The deployed HSS units must be able to quickly clear the battlefield of patients, evacuate them from the forward areas, and sustain the patients in rear areas until evacuation routes are established.

MEDICAL EVACUATION SUPPORT FOR THE DEFENSE

TYPES OF DEFENSE

4-66. There are three forms of the defense:

- **Area defense** concentrates on denying enemy access to designated terrain for a specific period of time, rather than on the outright destruction of the enemy.
- **Mobile defense** focuses on denying the enemy force by allowing him to advance to a point where he is exposed to a decisive counterattack by the striking force. The primary defeat mechanism, the counterattack, is supplemented by the fires of the fixing force.
- **Retrograde** is an organized movement to the rear and away from the enemy. The enemy may force these operations or a commander may execute them voluntarily. Within the retrograde operation there are three forms: delay, withdrawal, and retirement.

4-67. Support is generally more difficult to provide in the defense. The patient load reflects lower casualty rates, but forward area patient acquisition is complicated by enemy actions and the maneuver of combat forces.

4-68. Medical personnel generally are permitted much less time to reach the patient, complete vital EMT, and remove him from the battle site. Increased casualties among exposed medical personnel further reduce the medical treatment and evacuation capabilities. Heaviest patient workloads, including those produced by enemy artillery and CBRN weapons, may be expected during the preparation or initial phase of the enemy attack and in the counterattack phase.

4-69. The enemy attack may disrupt ground and air routes and delay evacuation of patients to and from treatment elements. The depth and dispersion of the defense create significant time and distance problems for evacuation assets. Combat elements may be forced to withdraw while carrying their remaining patients to the rear. The enemy exercises the initiative early in the operation which may preclude accurate prediction of initial areas of casualty density. This makes the effective integration of air assets into the evacuation plan essential. The use of AAs must be coordinated and synchronized with the supporting GSAB to ensure the synchronized execution of evacuation operations occurs.

4-70. The support requirements for retrogrades may vary widely depending upon the tactical plan, the enemy reaction, and the METT-TC factors. Firm rules that apply equally to all types of retrograde operations are not feasible, but considerations include:

- Requirement for maximum security and secrecy in movement.

- Influence of refugee movement that may impede medical evacuation missions conducted in friendly territory.
- Integration of evacuation routes and obstacle plans should be accomplished.
- Difficulties in controlling and coordinating movements of the force which may produce lucrative targets for the enemy.
- Movements at night or during periods of limited visibility.
- Time and means available to remove patients from the battlefield. In stable situations and in the advance, time is important only as it affects the physical well-being of the wounded. In retrograde operations, time is more important. As available time decreases, medical managers at all levels closely evaluate the capability to collect, treat, and evacuate all patients.
- Medical evacuation routes will also be required for the movement of troops and materiel. This causes patient evacuation in retrograde movements to be more difficult than in any other type of operation.
  - Command, control, and communications may be disrupted by the enemy.
  - Successful medical evacuation requires including ambulances on the priority list for movement.
  - Providing for the transportation of the slightly wounded in cargo vehicles and providing guidance to subordinate commanders defining their responsibilities in collecting and evacuating patients.
  - Special emphasis must be placed on the triage of patients and consideration given to the type of transportation assets available for evacuation.
- When the patient load exceeds the means to move them, the tactical commander must make the decision as to whether patients are to be left behind. The medical staff officer keeps the tactical commander informed in order that he may make a timely decision. Medical personnel and supplies must be left with patients who cannot be evacuated (refer to FM 4-02 for additional information).

Air Ambulance Evacuation Support to Covering Forces in Defensive Operations

4-71. The division covering force uses combat power to screen the FLOT and force the early deployment of the enemy's main force. Organic AHS elements of the covering forces establish aid stations of minimal size or operate in the split-team mode (FM 4-02.4). Ambulances are deployed well forward to promptly clear patients from combat units. Medical care is limited to EMT and ATM, followed by rapid evacuation whenever possible.

4-72. The covering force surgeon maintains communications with attached aviation elements and uses these assets (augmented by medical personnel to provide en route medical care when feasible) to provide backhaul of casualties, rather than use ground ambulances. The early evacuation of patients from BASs ensures their mobility for rearward displacement. The BCT or armor cavalry regiment (ACR) medical company/troop is equipped to receive patients generated in the covering force area. Depending upon the fluidity of the defensive situation, the FSMT (if available) may have to rapidly evacuate patient's further rearward to a BSMC/FSMC and collocated FST or a CSH.

4-73. The wide dispersion of units and the manner in which they withdraw make patient acquisition difficult. When covering forces withdraw, patients are transported to the rear by the medical element that acquired them. Seriously wounded or injured patients are given priority for evacuation by air. There are usually few COSRs during covering force operations, although delayed symptoms may occur once the element is withdrawn to a safe area. Ground ambulances augmented by nonmedical transportation assets evacuate the remaining patients.

MEDICAL EVACUATION SUPPORT FOR STABILITY OPERATIONS

4-74. Stability operations promote and protect US national interests by influencing the threat, political, and information dimensions of the operational environment. They include developmental, cooperative activities during peacetime and coercive actions in response to crisis.

4-75. Some of the purposes for which the US Forces are employed to conduct stability operations are as follows:

- Civil control.
- Civil security.
- Provide essential services.
- Support economic and infrastructure development.
- Governance.

4-76. Stability operations may complement the main effort or they may themselves constitute the main effort. Table 4-2 identifies the types of stability operations and the types of missions the BCT may conduct.

Table 4-2. Types of stability operations

Types Of Operations	***Types Of Missions***
Civil Control	Establish police forces training.
	Integrate trained police into operations.
	Counter organized crime.
	Establish judicial system.
	Transition to indigenous police forces.
Civil Security	Initiate security forces training.
	Establish National Security Forces.
	Integrate security forces into operations.
	Isolate population from insurgency.
	Defeat insurgency.
	Transition to National Security Forces.
Essential Services	Sewage treatment plants operating.
	Establish trash disposal.
	Water treatment plants operating.

Table 4-2. Types of stability operations

	Restore electrical power. Reopen hospitals and clinics.
	Reopen schools and universities.
Economic/Infrastructure Development	Implement Employment Programs.
	Secure vital natural resources.
	Repair/rebuild distribution infrastructure.
	Prioritize reconstruction projects.
	Essential banking services available.
	Implement public works projects.
Governance	Identify and recruit local leaders.
	Facilitate establishment of sector representation.
	Facilitate establishment of neighborhood councils.
	Facilitate establishment of district councils.
	Support and secure elections.

4-77. Medical evacuation support to forces deployed in stability operations is dependent upon the specific type of operation, anticipated duration of the operation, number of forces deployed, theater evacuation policy, medical troop ceiling, and anticipated level of violence. In most situations, medical evacuation support follows the traditional support provided to combat forces. If there is a shortened theater evacuation policy, a limited medical troop ceiling, and limited hospitalization assets within the AO, organic and DS ambulance support is provided from the POI to the supporting Roles 1 or 2 MTF and, once the patient is stabilized for further evacuation, from the treatment element to an airfield for evacuation out of the theater.

4-78. Reconstruction operations use Army forces to assist civil authorities, foreign or domestic, as they prepare for or respond to crises and relieve suffering. In reconstruction operations, Army forces provide essential support, services, assets, or specialized resources to help civil authorities deal with situations beyond their capabilities. The purpose of reconstruction operations is to meet the immediate needs of designated groups for a limited time, until civil authorities can do without Army assistance.

4-79. Humanitarian and civic assistance operations can include a number of activities such as disaster relief, civil support, refugee assistance, the provision of medical care to isolated populations, and refeeding programs resulting from famines or natural disasters. Medical evacuation assets may be used to evacuate the injured from disaster sites, to provide the emergency transport of critically needed medical supplies and personnel to remote locations, or to perform emergency rescues during times of flooding, wildfires, or other natural disasters.

SECTION VII — MEDICAL EVACUATION SUPPORT FOR ENABLING/SHAPING OPERATIONS

PASSAGE OF LINES

4-80. Passage of lines presents a challenge for the medical planner. There will be a number of medical evacuation units using the same air and road networks. Coordination and synchronization are essential if confusion and over evacuation are to be avoided. The information required to operate in the BCT or division AO includes:

- Radio frequencies and call signs.
- Operations plans and TSOPs.
- Location of MTFs.
- Location of CCPs and AXPs.
- Main supply route, forward arming and refueling points (FARPs), and A2C2 data.

SECURITY OPERATIONS

4-81. In security operations, the covering forces are dependent upon organic resources found in the maneuver battalion medical platoon for initial support. The level of command for the covering force (division or corps) determines the responsibility for the subsequent evacuation plan. In a corps covering force, for example, the corps AHS structure has the responsibility for establishing and operating the medical evacuation system to support the forward deployed corps forces. This is done to prevent the divisions following the covering forces from becoming overloaded with patients prior to the hand off and passage of lines. The use of CCPs, AXPs, and nonmedical transportation assets (CASEVAC) to move the wounded is essential. The covering force battle may be extremely violent. Patient loads will be high and the distance to MTFs may be much longer than usual. The effectiveness of the medical evacuation system depends upon the forward positioning of a number of ground ambulances and the effective integration of AAs into the evacuation plan.

ADVANCE, FLANK, AND REAR GUARDS

4-82. In advance, flank, and rear guards, these forces normally receive medical evacuation support through the attachment of evacuation teams. The teams evacuate patients to predesignated CCPs along a main axis of advance or to the nearest treatment element providing area support. Employment of AAs provides a measure of agility and flexibility.

RIVER CROSSING OPERATIONS

4-83. The river barrier itself exerts decisive influence on the use of division and BCT medical units. Attack across a river line creates an AHS delivery problem comparable to that of the amphibious assault. Army health system protection elements cross as soon as combat operations permit.

4-84. Early crossing of treatment elements reduces turnaround time for all crossing equipment that is used to load patients on the far shore. Maximum use of AA assets is made to prevent excessive patient buildup in far shore treatment facilities. Near shore MTFs are placed as far forward as assault operations and protective considerations permit to reduce ambulance shuttle distances from off-loading points. For detailed information on river crossing operations, refer to FM 90-13.

RECONNAISSANCE OPERATIONS

4-85. The reconnaissance in force is an attack to discover and test the enemy's position and strength or to develop other intelligence. The division usually probes with multiple combat units of limited size, retaining sufficient reserves to quickly exploit known enemy weaknesses.

4-86. Ambulances are positioned well forward at both unit and division levels. Ambulances are moved at night to enhance secrecy. The echeloning of ambulances is an indication to the enemy that an attack is imminent due to the forward placement of HSS. Role 2 MTFs are not established until a significant patient workload develops.

4-87. Patients received at BASs of reconnoitering units are evacuated to Role 2 MTFs as early as practical or are carried forward with the force until a suitable opportunity for evacuation presents itself. Maximum possible use of AA assets is made to cover extended distances and to overcome potentially unsecured ground evacuation routes.

RETROGRADE OPERATIONS

4-88. A retrograde operation is a movement to the rear or away from the enemy. This type of operation may be forced by enemy action or may be executed voluntarily.

AEROMEDICAL EVACUATION SUPPORT IN RETROGRADE OPERATIONS

4-89. Aeromedical evacuation support in retrograde operations may vary widely depending upon the enemy reaction and the situation. Firm rules that apply equally to all types of retrograde operations are impossible to establish. Factors to consider in planning aeromedical evacuation support for retrograde operations include—

- Patient's condition/status.
- Mission, enemy, terrain and weather, troops and support available, time available and civil considerations.
- Requirement for maximum efforts in secrecy and employment of aircraft survivability equipment.
- Influence of refugee movement (which may impede military medical movements conducted in friendly territory).
- Integration of evacuation routes and other Army aviation routes.
- Difficulty in controlling and coordinating movements of the force which may produce lucrative targets for the enemy.

4-90. During retrograde operations, battalion, brigade, and division surgeons must evaluate their capability to collect, treat, and evacuate patients. Three factors influence the number of patients evacuated from the battlefield and the time and place of patient treatment:

- Time available prior to MTF displacement.
- Location of the enemy advance.
- Means of evacuation.

4-91. Due to congestion on main supply routes and general evacuation routes, the AAs may provide the best means of evacuation during retrograde operations. Seriously wounded patients should be evacuated by the fastest and most comfortable means and should receive medical care en route. Proper sorting (triage) and rapid evacuation of patients lessen the need for establishing a complete Role 2 MTF operation. Therefore, the FSMT (if available) leader should be prepared to surge his efforts in the retrograde. Combat and operational stress reaction casualty rates are usually low as compared to wounded in action (WIA) rates during controlled retrograde operations, but increase after safety is reached.

4-92. Commanders must decide to displace medical evacuation units before they hamper the maneuver forces conducting the retrograde movement. During mission planning, the medical evacuation units/elements integrate with the supported unit's displacement plan. The medical evacuation element leaders must coordinate with the supported unit commander and the BSMC for displacement decision points and locations.

WITHDRAWAL OPERATIONS

4-93. A withdrawal operation is one in which a deployed force disengages from an enemy force. It may be forced by enemy pressure or conducted voluntarily.

4-94. Although the deployed force disengages from the enemy, contact is maintained by security elements while the main force moves to the sustainment area, forms march columns, and moves to a predesignated location. Aeromedical evacuation elements must be prepared to support the withdrawing force as well as any forces left in contact. Time available to acquire, treat, and evacuate patients from the line of contact is limited.

AEROMEDICAL EVACUATION SUPPORT IN WITHDRAWAL OPERATIONS

4-95. The AA medical company element supporting the withdrawing force must ensure it maintains sufficient station capability at its present location to support the patients while taking action to move to a new location.

4-96. In addition to the evacuation support provided to the force left in contact, there is a requirement to support the force moving to the sustainment area. If the main force infiltrates to the sustainment area, patients are carried by their parent units to the assembly area or MTFs in the sustainment areas. Medical assets are positioned in the assembly area to receive patients.

4-97. The division establishes Role 2 MTFs at a minimal size and well to the rear of the first line of alternate positions. If combat and environmental conditions indicate a light patient workload and the road network permits rapid ambulance movement, the division may use another Role 2 MTF to provide GS to all withdrawing brigades.

4-98. Role 2 MTFs conduct leapfrog rearward, occupying successive positions placed along the withdrawal route to minimize the requirement for multiple displacements by any one MTF. This avoids unnecessary interference with combat operations while providing continuous medical support. With the exception of an ambulance squad to support the covering force the division normally does not employ ambulances further forward than the BAS. AA elements will normally evacuate to sustainment area and corps positions to avoid impeding the withdrawal of brigades.

4-99. Preparation for the withdrawal operation includes distribution of extra medical consumable supplies and nonexpendable exchange items to each medical element. This is done to overcome the potential effects of isolated MTFs or the possible intermittent operation of the medical evacuation system. The AA medical company may receive the mission to assist with this distribution.

4-100. If withdrawal is made under enemy pressure, the provision of aeromedical evacuation is modified. Since the time available is critical, aeromedical resources cannot remain in the forward areas. The assets are usually withdrawn as a unit. Patients occurring during the withdrawal are carried to the sustainment areas by the parent unit, normally using nonmedical vehicles.

DELAYING OPERATIONS

4-101. Delaying operations occur when forces are insufficient to attack or to defend and when the defensive plan calls for drawing the attacker into an unfavorable situation. The delay maneuver may include friendly maneuver to successive positions across a broad front.

4-102. Delaying brigades split their combat power, moving their less mobile forces directly to the next defensive position while the elements remaining in contact fight to the sustainment area.

4-103. Aeromedical support for a delaying operation is conducted like withdrawal operations.

RETIREMENT OPERATIONS

4-104. A retirement operation is a rearward movement of a force not in contact with the enemy. It is conducted according to the force's OPLAN and without pressure by enemy forces.

4-105. Because the division is no longer in contact with the enemy, it can march (in multiple columns) directly to the rear. The AHS requirements for this type of operation are similar to those in a movement to contact. The patient workload is light. The FSMT (if available) leader will coordinate with the BSMC commander, and the AA medical company CP will coordinate with the DSS for guidance in the positioning of company assets to support this operation.

SECTION VIII — SPECIAL FORCES OPERATIONS

4-106. Army special operations forces (ARSOF) have limited organic AHS resources. They are, therefore, dependent upon the theater AHS for the majority of their health care requirements. To ensure that ARSOF receive comprehensive and timely AHS and logistics, medical planners must consider each medical function and determine the support organic and theater AHS assets will provide. The significant challenge for ARSOF medical planners is to ensure the efficient delivery and use of the conventional Class VIII (medical materiel) system for support of ARSOF missions.

4-107. Army special operations forces consist of special forces (SF), rangers, special operations aviation (SOA), psychological operations (PSYOP), and CA. The SOA, CA, and PSYOP have limited medical capabilities and are very dependent upon the theater HSS assets and the organizations they are supporting. Special forces and rangers often operate far removed from conventional AHS and must be more self-reliant and sustaining than conventional forces. Accordingly, SF and ranger medical personnel receive enhanced medical training above that provided for a trauma specialist. The Army educates the SOF medic to be an independent care practitioner qualified to provide ATM to combat casualties. When deployed on independent operations, the two SF medics are the sole source of medical care for their operational detachment, the indigenous forces, and their families that the detachment supports. They can train the indigenous populace in basic medical skills and establish an austere AHS system. Nonmedical ARSOF personnel receive medical training at the combat lifesaver level. Special operation combat medics (SOCMs) are trained in emergency medical treatment and in combat trauma management, but they do not possess the invasive surgical skills of the SF medical sergeant. The SOCMs provide medical care in combat environments where resuscitative surgery is more rapidly accessible than in the small unit missions of SF. Physicians and physician assistants assigned to units with SOCMs can provide ATM. Army health system planning for missions where the risk of penetrating trauma is high and further complicated by time and extended distance requires extensive planning and coordination. The capability for medical personnel and units that can perform resuscitative surgery and invasive stabilization must be provided regardless of the ARSOF unit involved.

4-108. Although the ARSOF health care provider receives enhanced medical training exceeding the level and scope found in conventional forces, he depends heavily on the conventional AHS system to conserve the combat strength of the ARSOF (particularly in the area of medical evacuation where the ARSOF does not have a dedicated system).

4-109. Ideally, medical evacuation for ARSOF personnel should follow the doctrinal flow sequence. The ARSOF medical planner must be innovative and follow the tenets of immediate far forward stabilization. He directs evacuation to the appropriate MTF when the condition of the patient warrants it, with whatever means of transportation are available.

4-110. Medical evacuation of ARSOF casualties is an operational matter. That is, it must reflect the commander's concept of the operation. It can only succeed when the medical planner integrates the medical evacuation plan with the tactical plan and logistics airflow.

4-111. Conventional medical evacuation assets are not normally used when operations occur in hostile and denied areas. Use the conventional medical evacuation system once the casualty is extracted from the hostile or denied area. For additional information on AHS to ARSOF, refer to FM 3-05, FM 4-02.6. FM 8-42 (FM 4-02.42), and FM 4-02.43.

SECTION IX — URBAN OPERATIONS

TERRAIN AND ENVIRONMENT

4-112. Throughout history, battles have been fought on urbanized terrain. Some recent examples urban operations (UO) include Hue, Beirut, Panama City, and Baghdad. Military operations on urbanized terrain are those military actions planned and conducted on a terrain where man-made structures impact on the tactical options available to the commander. This terrain is characterized by a three-dimensional battlefield, having considerable rubble, ready-made fortified fighting positions, and an isolating effect on all movement and maneuvers and fires, and sustainment units. In this environment, the requirement for a sound and understandable evacuation plan cannot be overstated. Of concern to medical and tactical planners is the need to plan, train, prepare, and equip for evacuation from under, above, and at ground level.

TERRAIN CONSIDERATIONS

4-113. Medical evacuation in the urban environment is a labor-intensive effort. Due to rubble, debris, barricades, and destroyed roadways, much of the evacuation effort will be by manual litter teams. When this occurs, establish a litter or ambulance shuttle system. The shuttle system reduces the distance that litter teams have to carry the wounded or injured Soldiers. This enhances the litter teams' effectiveness by providing brief respites, reducing fatigue. Further, the litter teams stay in the forward areas. They are familiar with the geography of the area and what areas have or have not been searched for casualties. In moving patients by litter, you should:

- Use covered evacuation routes such as subways, whenever possible.
- Use easily identifiable points for navigation CCPs.
- Rest frequently by alternating litter teams.

EVACUATION PLANNING CONSIDERATIONS FOR URBAN OPERATIONS

4-114. Conducting medical evacuation operations in the urban environment challenges the medical planner. He must ensure that the HSS/FHP plan includes special or unique materiel requirements or improvised use of standard equipment. The plan must be sufficiently flexible to support unanticipated situations.

4-115. Special equipment requirements include, but are not be limited to:

- Axes, crowbars, and other tools used to break through barriers.
- Special harnesses; portable block and tackle equipment; grappling hooks; collapsible litters; lightweight collapsible ladders; heavy gloves; and casualty blankets with shielding. This equipment, using pulleys, is for lowering casualties from buildings or moving them from one building to another at some distance above the ground.
- Equipment for the safe and quick retrieval from craters, basements, sewers, and subways. Casualties may have to be extracted from beneath rubble and debris.

4-116. Air ambulances equipped with a rescue hoist may be able to evacuate casualties from the roofs of buildings or may be able to insert needed medical personnel and supplies.

COMMUNICATIONS IN THE URBAN ENVIRONMENT

4-117. The urban environment degrades communications. Line of sight radios will be ineffective and individual Soldiers will not have access to radio equipment. The medical evacuation teams will have difficulty in locating injured or wounded Soldiers due to their isolation within buildings, or by being hidden in rubble and debris. Injured Soldiers can display alternate forms of communications, such as markers, panels, or field expedients (field/uniform jackets or T-shirts) indicating where they may be found.

Establishing Casualty Collection Points

4-118. Preplan and establish CCPs at relatively secure areas accessible to both ground and AAs. The location of these points are depicted on the HSS/FHP overlay to the OPLAN. Casualty collection points should—

- Offer cover from enemy fires.
- Be as far forward as the tactical situation permits.
- Be identifiable by an unmistakable feature (natural or man-made).
- Allow rapid turnaround of ambulances.

4-119. The tactical commander must approve route markings to the MTF and if to display of the Geneva Red Cross at the facility. (Camouflaging or not displaying the Geneva Red Cross can forfeit the protections, for both medical personnel and their patients, afforded under the Geneva Conventions. Refer to Appendix A and FM 4-02 for additional information.) The location of the MTF must be as accessible as possible, but well separated from fuel and ammunition depots, motor pools, reserve forces, or other lucrative enemy targets, as well as civilian hazards such as gas stations or chemical factories.

Ground Evacuation For Urban Operations

4-120. When using ground evacuation assets in support of UO, the medical planner must be aware that built-up areas may have significant obstacles to vehicular movement. Factors requiring consideration include the following:

- Rubble and other battle damage complicate and canalize transportation operations within the urban terrain.
- Bypassed pockets of resistance and ambushes pose a constant threat along evacuation routes.
- Land navigation with most tactical maps proves to be difficult. Using commercial city maps when available can aid in establishing evacuation routes.
- Ambulance teams must dismount from the ambulance, search for, and rescue casualties.
- Movement of patients becomes a personnel intensive effort. There are insufficient medical personnel to search for, collect, and treat the wounded. Medical units may require assistance in the form of litter bearers and search teams from supported units.
- Refugees may hamper movement into and around urban areas.
- Dislocated civilians, EPW, and detainees receive medical treatment according to the command policy and the Geneva Conventions.

Air Evacuation for Urban Operations

4-121. When using AA assets in support of UO, the medical planner must consider enemy air defense capabilities and terrain features, both natural and man-made, within and adjacent to the built-up areas. Air ambulances are the preferred means of evacuation in UO. Considerations in the use of AAs include the following:

- Movement is highly restricted and is canalized over secured areas, down wide roads, and open areas.
- Telephone and electrical wire and communication antennas hinder aircraft movement.
- Secure landing zones (LZs) must be available.
- Landing zones may include buildings with helipads on their roofs or sturdy buildings, such as parking garages.
- Snipers with air defense capabilities may occupy upper stories of the urban area's taller buildings.

Specialty Training

4-122. Medical personnel require special training in the tactics, techniques, and procedures required to operate in an UO environment. If they are to survive in this environment, they must know how to:

- Cross open areas safely.
- Avoid barricades and mines.
- Enter and depart safely from buildings.
- Recognize situations where booby traps or ambushes are likely and would be advantageous to the enemy. Detailed information on the conduct of combat operations in the urban environment is contained in FM 3-06 and FM 3-06.11.

Note. Medical personnel do not engage in offensive-type actions. They must rely on the supported unit to provide covering fires and to clear rooms and buildings prior to entry.

4-123. Units can modify and apply many of the extraction and patient evacuation techniques used in a mountainous terrain for evacuation in UO.

SECTION X — HOMELAND SECURITY

4-124. To preserve the freedoms guaranteed by the Constitution of the US, the Nation must have a homeland that is secure from threats and violence, including terrorism. Homeland security (HS) is the Nation's first priority, and it requires a national effort; in which the DOD has a key role in that effort. The National Strategy for Homeland Security (NSHS) complements the National Security Strategy of the US by providing a comprehensive framework for organizing the efforts of federal, state, local, and private organizations whose primary functions are often unrelated to national security. Critical to understanding the overall relationship is an understanding of the distinction between the role that DOD plays with respect to securing the Nation and HS, and the policy in the NSHS, which has the Department of Homeland Security (DHS) as the lead. Homeland security at the national level has a specific focus on terrorist threats. The DOD focus in supporting HS is broader. Military application of the NSHS calls for preparation, detection, deterrence, prevention, defending, and responding to threats and aggression aimed at the homeland. The DOD also provides military assistance to civil authorities (MACA), including consequence management activities. The Armed Forces of the US support the NSHS through two distinct but interrelated mission areas–homeland defense (HD) and civil support. The United States Northern Command (USNORTHCOM) is the unified command that has the DOD responsibility for HD and civil support.

HOMELAND DEFENSE

4-125. Homeland defense is the protection of US territory, sovereignty, domestic population, and critical infrastructure against external threats and aggression. The purpose of HD is to protect against and mitigate the impact of incursions or attacks on sovereign territory, the domestic population, and defense critical infrastructure. For HD missions, as directed by the President of the US and/or the Secretary of Defense, DOD serves as the lead federal agency.

Homeland Defense Characteristics

4-126. The DOD is the lead, supported by other agencies, in defending against traditional external threats/aggression (such as air and missile attack). However, against internal asymmetric, nontraditional threats (such as terrorism), DOD may be in support of DHS. When ordered to conduct HD operations within US territory, DOD will coordinate closely with other federal agencies or departments.

TYPES OF HOMELAND DEFENSE OPERATIONS

4-127. The types of homeland defense operations that the future brigade combat teams is expected to perform include—

- Offensive operations.
- Defensive operations.
- Humanitarian and civic assistance.
- Support to counterdrug operations.
- Show of force.
- Domestic support operations.
- Security operations (screen, guard, area security, and local security).
- Reconnaissance operations (zone, route, and area reconnaissance).

CIVIL SUPPORT

4-128. Civil support missions consist of military support to US civil authorities for domestic emergencies, and for designated law enforcement within the scope of restrictions required by the Posse Comitatus Act and other support approved by the Secretary of Defense (see FM 3-07).

4-129. Employment of military forces within the US, its territories, and possessions, under the auspices of civil support, typically falls under the broad mission of MACA. The MACA missions consist of three mission subsets. These mission subsets consist of military support to civil authorities, military support to civilian law enforcement agencies, and military assistance for civil disturbances.

MILITARY SUPPORT TO CIVIL AUTHORITIES

4-130. Military support to civil authorities (MSCA) is the most widely recognized form of DOD civil support because it usually consists of support for high-profile emergencies such as natural or man-made disasters that often invoke Presidential or state emergency/disaster declarations:

- **Natural Disasters.** In the event of a natural disaster or emergency, there will be a concerted US Government effort to support the affected areas. The DOD may be asked to provide assistance to DHS/Federal Emergency Management Agency (FEMA) in an attempt to save lives, protect property, and lessen the threat of a catastrophe in the US. Examples of natural disasters include, but are not limited to severe weather, wildland fire fighting, and animal disease eradication. When natural disasters occur and military assistance is anticipated, DHS/FEMA will request a defense coordinating officer (DCO) who serves as the single DOD point of contact within the disaster area. The DCO will be under the OPCON of the designated supported combatant command. Units supporting the event are normally OPCON to the supported combatant command.
- **Special Events.** The DOD provides support to a special event, a planned program of athletic competition or related activities involving participants from the US and/or other nations. Historic examples of these events are the Olympic Games and the Pan American Games.
- **Man-made Disasters.** Manmade disasters can be accidental or intentional. An example of an accidental event is an oil spill. The implications of a deliberate or unintentional large release of a CBRNE are severe. A catastrophic CBRNE event or attack may occur with little or no warning, resulting (either immediately or over time) in mass casualties and producing a mass exodus of evacuees. The response capabilities and resources of the local jurisdiction (to include mutual aid from surrounding jurisdictions and response support from the state) may be insufficient and quickly, if not immediately, overwhelmed.

Domestic Support Operations

4-131. Domestic disasters occur due to various causes and many may require the assistance of DOD and more specifically, the Army medical evacuation assets to fill response shortfalls. In the post-9/11 environment, terrorism or other intentional acts are recognized as a more likely cause of contingencies within the US than many had acknowledged since the civil defense era of the forties and fifties. This threat increases the likelihood that a large-scale disaster will occur that requires the augmentation of civil authorities' resources by DOD. The US DHS and USNORTHCOM have been established to coordinate the preparedness for and response to such events for civil authorities and DOD respectively.

4-132. Domestic support operations are typically conducted to prevent loss of life and property. Certain operations may be conducted in the form of immediate and automatic response by US military commanders as defined in DOD Directive 3025.15, to save lives, prevent human suffering, or to mitigate great property damage under imminently serious conditions. Other requests for support for civil disasters are made by civil authorities or a DCO (in a Stafford Act declared emergency) through the Joint Directorate of Military Support per the directive and work down through command channels from the Secretary of Defense.

4-133. Homeland Security Presidential Directive 5 recognized that there currently exists an artificial separation of local, state, and federal response to an incident and the role of DOD is not well streamlined or defined. This recognition led the White House and the DHS to initiate the transition from the Federal Response Plan to the National Response Plan, which serves as the doctrine for how the US responds to disasters no matter the cause.

4-134. Medical evacuation support plans to augment civil authorities are necessary for a variety of contingencies. Domestic support planning must address a range of problems such as—

- Preidentifying medical evacuation capabilities, units and personnel available to support various contingencies large enough to require civil support.
- Command and control relationships between civil authorities and DOD forces especially when DOD units are divided.
- Support for deployed DOD forces when no DOD logistics operations are deployed, including medical support.
- Cost capture and reimbursement from civil authorities to DOD in non-Stafford Act emergencies.

SECTION XI — OTHER TYPES OF MEDICAL EVACUATION SUPPORT MISSIONS

EVACUATION OF MILITARY WORKING DOGS

4-135. Military working dogs (MWDs) when injured or ill may be evacuated on any transportation means available. The using unit is responsible for the evacuation of the animal. Use of dedicated medical evacuation assets (air or ground ambulances) is authorized based on mission priority and availability. When possible, the handler should accompany the animal during the evacuation. Using units should include the location of veterinary treatment facility/support units on operational overlays. See FM 4-02.18 for information on veterinary support for MWDs.

PERSONNEL RECOVERY OPERATIONS

4-136. Air ambulances do participate in PR operations in a support role and are not protected from attack while engaged in a PR mission over contested or denied territory. If the involvement in these operations consists solely of evacuating wounded crewmembers from a crash site in friendly territory, AAs retain the protection accorded to them under the provisions of the Geneva Conventions. Air ambulances flying in contested or denied areas are not protected from attack unless flying at times and routes agreed to by the enemy and may be summoned to land. Air ambulances must obey a summons to land. Personnel retain their protections under the Conventions. However, if AAs participate in the actual search and rescue phases of the operation, they are not solely engaged in the provision of AHS and are, therefore, not afforded the protections. Further, because the mission does not fall within the protected activities, the AAs participating in the PR operation must remove the Geneva Conventions emblem from the aircraft.

SHORE-TO-SHIP EVACUATION OPERATIONS

4-137. It is imperative that Army AA units be able to interface on the first day of battle with US Navy (USN) air-capable ships. Lessons learned from past operations, have shown that US Army helicopters should be able to operate to and from USN air-capable ships. An interservice agreement between the Army and the Navy allows for deck-landing qualification of Army pilots (refer to FM 1-564 for additional information). It is important that units having contingency missions requiring Navy support establish training requirements to obtain naval-operations orientation, water egress training, water survival, and deck-landing qualification. This enhances the successful accomplishment of the aeromedical evacuation mission to naval vessels.

COMMUNICATIONS

4-138. In past joint operations, communications have been burdensome for both Army and Navy elements. It is essential to establish commonality of communication requirements during training exercises and to establish communication equipment and frequencies for medical evacuation to Navy vessels. This will provide smooth integration of Army helicopters into the Navy airspace management system during actual operations.

NAVIGATION

4-139. As Navy vessels may operate relatively long distances from the ground combat operations, Army AA units need to be proficient in over water navigation. The use of navigational aids (NAVAIDS) from the Navy element in support of the operation is the first priority for over water navigation. Basic dead-reckoning remains a secondary measure.

MEDICAL EVACUATION OF ENEMY PRISONERS OF WAR AND DETAINEES

4-140. Sick, injured, and wounded EPW and detainees are treated and evacuated in military police (MP) channels when possible. They must be physically segregated from US, allied, and coalition patients. Guards for these prisoners are provided according to the BCT, division or corps TSOP and are from other than medical resources. The echelon commander is normally responsible for this support. The US provides the same standard of medical care for wounded, sick, and injured EPW and detainees as that given to US, allied, and coalition Soldiers. Wounded, sick, or injured EPW and detainees in the CZ may be treated and returned to MP channels for evacuation; or the EPW and detainees may be stabilized and moved through medical channels to theater hospitals for treatment. Enemy prisoners of war and detainees are not evacuated from the theater for medical treatment.

4-141. When EPW are evacuated through medical channels, medical personnel—

- Report this action through medical channels to the next higher headquarters.
- Request disposition instructions from the MEDBDE IPMC.
- The MEDBDE IPMC is responsible for—
  - Coordinating the transportation means.
  - Identifying the MTF to which the EPW/detainees will be taken.
  - Coordinating, in conjunction with the hospital commander, with the Detainee Reporting System to account for EPW within medical channels.

SECTION XII — MEDICAL EVACUATION REQUEST

4-142. Procedures for requesting medical evacuation support must be institutionalized down to the unit level. Procedural guidance and standardization of request procedures are provided below. The same format used to request aeromedical evacuation is also used for requesting ground evacuation.

Table 4-3. Procedures for information collection and medical evacuation request preparation

Line	*Item*	*Explanation*	*Where/How Obtained*	*Who Normally Provides*	*Reason*
1	Location of pickup site	Encrypt the grid coordinates of the pickup site. When using the DRYAD Numeral Cipher, the same "SET" line will be used to encrypt the grid zone letters and the coordinates. To preclude misunderstanding, a statement is made that the grid zone letters are included in the message (unless unit SOP specifies it's use at all times)	From map	Unit leader(s)	Required so evacuation vehicle knows where to pickup patient. Also, so that the unit coordinating the evacuation mission can plan the route for the evacuation vehicle (if the evacuation vehicle must pick up from more than one location).
2	Radio frequency, call sign and suffix	Encrypt the frequency of the radio at the pickup site, not a relay frequency. The call sign (and suffix if used) of person to be contacted at the pickup site may be transmitted in the clear.	From SOI	RTO	Required so that evacuation vehicle can contact requesting unit while enroute (obtain additional information or change in situation or directions).
3	Number of patients by precedence	Report only applicable information and encrypt the brevity codes. A—URGENT B—URGENT-SURG C—PRIORITY D—ROUTINE E—CONVENIENCE If two or more categories must be reported in the same request, insert the word "BREAK" between each category.	From evaluation of patient(s)	Medic or senior person present	Required by unit controlling vehicles to assist in prioritizing missions.

4	Special equipment required	Encrypt the application brevity codes. A—None B—Hoist C—Extraction equipment D—Ventilator	From evaluation of patient/situation	Medic or senior person present	Required so that the equipment can be placed on board the evacuation vehicle prior to the start of the mission.
5	Number of patients by type	Report only applicable information and encrypt the brevity code. If requesting medical evacuation for both types, insert the word "BREAK" between the litter entry and ambulatory entry. L+# of patients–Litter A+# of patients–Ambulatory (sitting)	From evaluation of patient(s)	Medic or senior person present	Required so that the appropriate number of evacuation vehicles may be dispatched to the pickup site. They should be configured to carry the patients requiring evacuation.
6	Security of pickup site (wartime)	N—No enemy troops in area P—Possibly enemy troops in area (approach with caution). E—Enemy troops in area (approach with caution). X—Enemy troops in area (armed escort required).	From evaluation of situation	Unit leader	Required to assist the evacuation crew in assessing the situation and determining if assistance is required. More definitive guidance can be furnished to the evacuation vehicle while it is enroute (specific location of enemy to assist an aircraft in planning its approach).
6	Number and type of wound, injury or illness (peacetime)	Specific information regarding patient wounds by type (gunshot or shrapnel). Report serious bleeding, along with patient's blood type if known.	From evaluation of patient(s)	Medic or senior person present	Required to assist evacuation personnel in determining treatment and special equipment needed.
7	Method of marking pickup site	Encrypt the brevity codes. A—Panels B—Pyrotechnic signal C—Smoke signal D—None E—Other	Based on situation and availability of materials	Medic or senior person present	Required to assist the evacuation crew in identifying the specific location of the pick up. Note that the color of the panel or smoke should not be transmitted until the evacuation vehicle contacts the unit (just prior to its arrival). For security, the crew should identify the color and the unit verifies it.
8	Patient	The number of patients in each category need not	From evacuation	Medic or	Required to assist in

	nationality and status	be transmitted. Encrypt only the applicable brevity codes. A—US military B—US citizen C—Non-US military D—Non-US citizen E—enemy prisoner of war (EPW)	platform	senior person present	planning for destination facilities and need for guards. Unit requesting support should ensure that there is an English speaking representative at the pickup site.
8	CBRN contamination (wartime)	Include this line only when applicable. Encrypt the applicable brevity codes. C—Chemical B—Biological R—Radiological N—Nuclear	From situation	Medic or senior person present	Required to assist in planning for the mission. (determine which evacuation vehicle will accomplish the mission and when it will be accomplished)
9	Terrain description (peacetime)	Includes details of terrain features in and around proposed landing site. If possible, describe relationship of site to prominent terrain feature (lake, mountain, tower).	From area survey	Personnel present	Required to allow evacuation personnel to assess route/avenue of approach into area. Of particular importance if hoist operation is required.

Chapter 5
Medical Evacuation in Specific Environments

This chapter addresses medical evacuation in specific environments or under special circumstances. The medical evacuation effort must be well-planned and its execution synchronized to be effective. Further, medical evacuation personnel must be flexible and ready to improvise, if needed, to meet the demands of unique situations.

SECTION I — MOUNTAIN OPERATIONS

5-1. In the past, armies have experienced great difficulty in evacuating patients from mountainous areas because they are extremely diverse in nature. Some mountains are dry and barren with temperatures ranging from extreme heat in the summer to extreme cold in the winter. In tropical regions, mountains are frequently covered by lush jungles and heavy seasonal rains occur. Many areas display high rocky crags with glaciated peaks and year-round snow cover. Elevations can also vary from as little as 1,000 feet to over 16,000 feet with drastic and rapidly occurring weather changes.

5-2. Operations in mountainous terrain require some procedure modifications. This is due to the environmental impact on personnel and equipment. Important physical characteristics and considerations that influence medical evacuation are—

- Rugged peaks, steep ridges, and deep valleys.
- Limited number of trafficable roads.
- Reduced communications ranges.
- Unpredictability of and severe changes in weather.
- Decreased partial pressure of oxygen.
- Limited availability of LZs.

5-3. In order to effectively support the tactical plan, the HSS/FHP plan must provide maximum flexibility. The HSS planner should consider using all means of evacuation. Due to the length of evacuation times and the limited means of ground evacuation, it is important to triage and prioritize patients prior to movement.

- The availability of improved, hard-surfaced roads is extremely limited, if they exist at all. Usually, improved roads are only found in valley corridors. Such roads are often dependent upon a system of narrow bridges spanning mountain streams and ravines. They may also twist along ridgelines and cling to steep shoulders.
- Secondary roads and trails may be primitive and scarce. However, they may provide the only routes capable of vehicular traffic. Cross-compartment travel between adjacent valleys may be impossible by ground vehicle. Off-road travel requires detailed planning, even for short distances.
- Because of rough terrain, the Role 2 MTF may not be able to reach the BAS by ground vehicle. An ambulance shuttle system is established with an AXP for air and ground evacuation vehicles to meet litter bearers. Litter bearers and beasts of burden may be the only means of evacuation available. Any available personnel may be used as litter bearers (nonmedical personnel from supported units may be required to augment the litter bearer teams). Close coordination between Role 2 MTFs and BASs in establishing CCPs or AXPs is necessary to—
  - Reduce distance traveled by litter bearers.
  - Reduce evacuation time.

5-11. Jungle combat operations are characterized by ambushes and other unconventional warfare operations. The security threat caused by infiltrators requires that LOC be patrolled often and that convoys be escorted. It is, therefore, essential that HSS be performed as far forward as the tactical situation permits. Positioning assets forward—

- Improves response time.
- Reduces road movement.
- Allows the HSS elements to take advantage of the security offered by combat units.

5-12. The thick foliage often makes evacuation by ground more difficult than in other types of terrain. Factors such as the threat, limited road network, and reliance on nonmedical personnel for convoy security make air evacuation the preferred means. By using the ambulance shuttle system, patients can be transferred from forward operating ground ambulances to either ground or AAs operating further to the rear. In situations where evacuation assets are delayed by various factors (weather or terrain), patients are held for longer periods of time at forward locations. This will dictate the need for additional medical supplies. Health service support planners must try to anticipate these delays whenever possible. The increased disease and infection incidences associated with the jungle environment may worsen the patient's condition; therefore, timely evacuation is essential.

5-13. In some remote and densely foliaged jungles, the only means of evacuation may be by litter. Ambulances may not be practical on trails, unimproved muddy roads, or in swamps. As in mountain operations, there is a higher proportion of litter cases than usual. In the jungle even a slightly wounded Soldier may find it impossible to walk through dense undergrowth. At best, litter teams can carry patients only a few hundred meters over rough jungle terrain before needing rest or relief. Litter carries should be kept as short as possible and medical elements prepositioned and retained forward.

5-14. Other special planning considerations in jungle operations include—

- Water is vital in the jungle and is plentiful. Water from natural sources, however, should be considered contaminated. Water purification procedures must be taught to all Soldiers. (Refer to FMs 21-10 and 4-25.12 for additional information.)
- Due to the tropical climate, units should pack hot weather clothing when deploying to jungle areas. Insect (mosquito) nets, insect repellent, permethrin treated uniforms and sunscreen should be issued to all Soldiers operating in this environment.
- The jungle environment is ideal for the transmission of a large number of diseases. The rate of DNBI casualties is potentially the highest in this climate. The heat, humidity, and terrain places the troops at high risk for dehydration, heat injury, skin diseases, endemic diseases, and immersion foot. Small wounds can rapidly become infected and lead to loss of effectiveness and possibly require evacuation. High standards of personal hygiene must be taught, encouraged, and maintained by the command. Mosquitoes and other arthropods that carry disease flourish under jungle conditions. Use of all possible personal protective measures must be ensured. Food- and waterborne diseases leading to diarrhea or other symptoms will abound. Food service sanitation measures must be strictly followed. The potential for contamination of food and water increases with each time they are handled, stored, or transported. Soldiers must be encouraged to consume adequate amounts of water that has been purified and to eat only approved foods. In the jungle it is necessary for the commander to pay meticulous attention to the details of PVNTMED measures to maintain an effective fighting force. For additional information on PVNTMED measures, refer to FM 21-10 and FM 4-25.12.
- Health service support personnel should be trained in survival and support techniques in jungle environments. For example, training should be conducted in—
    - Hot weather acclimatization and survival.
    - Prevention, early detection, and treatment of arthropod-, food-, and waterborne diseases.
    - Land navigation in a jungle environment.
    - Field sanitation and other PVNTMED measures.
    - Care and maintenance of equipment and supplies.

- Due to the increased heat and humidity, vehicles and equipment require additional maintenance. Equipment tends to rust quickly and must be cleaned and oiled more frequently. Canvas items rot and rubber deteriorates much faster than in more temperate climates.
- The range of FM communications in the jungle is significantly reduced due to the dense undergrowth, heavy rains, and hilly terrain. The range of a radio set operated in the jungle may be reduced by 10 to 25 percent. The heavy rain and high humidity of the tropics also reduce the range (about 20 percent) and reliability of wire communications. The transmission range can be extended by using additional radio relays and field expedient antennas.
- Utility helicopters are not able to lift the same size loads that can be lifted in more temperate areas. This results in a reduced patient load in some evacuation aircraft. Again, frequent and accurate performance planning is essential for mission accomplishment.
- There may be few suitable LZs. Many LZs will only be large enough to support one or two helicopters at a time.
- Hoist operations may be required more frequently in the thick jungle vegetation where LZs are not available. The forest penetrator should be carried on all operations.
- For aviation-specific information, refer to training circular (TC) 1-400.

SECTION III — DESERT OPERATIONS

5-15. Deserts are arid, barren regions of the earth incapable of supporting normal life due to a lack of fresh water. Although deserts are often thought of as hot climates, it is important to note that temperatures range from over 136 degrees (°) Fahrenheit (F) in some deserts, to bitter cold in others. Day to night fluctuations in temperature can exceed 70° F. Desert terrain can have mountains, rocky plateaus, or sandy dunes; some desert areas may contain all of these characteristics. Rain, when it falls, often causes flooding in low-lying areas. Winds can have a devastating effect upon AHS operations by destroying equipment and supplies and causing dust storms. Dust storms make navigation and patient treatment difficult. Since deserts vary considerably in the type of terrain and temperature, and in their cultural makeup, current medical intelligence should be obtained prior to deployment on operations conducted on desert terrain.

5-16. People have lived and fought in desert areas for thousands of years. However, the environmental effects on personnel can be extreme, especially for Soldiers not prepared for these operations.

- To be effective, Soldiers must be properly acclimatized to the desert. Two weeks are usually required to satisfactorily acclimatize troops to hot environments, using progressive degrees of heat exposure and physical exertion. Other potential acclimatization problems that may be encountered are the effects of dry air and altitude on the respiratory system. Since many desert areas are located in mountainous terrain, Soldiers must be acclimatized for both the altitude and the temperature. In some areas of the world, such as the Gobi Desert in East Asia, people must be acclimatized to the cold in addition to the dryness. (For additional information, refer to FM 21-10, FM 8-250, and TB MED 507.)
- Units deployed in desert areas typically have long LOC and are widely dispersed. As unit elements become more dispersed, commanders must rely more heavily on junior officers and NCOs to ensure that C2 is maintained and that discipline and PVNTMED programs are enforced. For a unit to be effective, a high level of discipline must exist at all levels of the organization.
- Water is the most basic need in a desert. Without it, Soldiers cannot function effectively for more than a few hours. Thirst is not an adequate indicator of the need for water. It is necessary for each commander to establish and enforce a supervised drinking program. Experience has shown many times that Soldiers do not drink enough fluids unless forced to do so. It is important to cool the water, if at all possible, to make it more appealing. Water supplies should be carefully guarded against accidental loss, sabotage, or contamination. Extra water must be carried by AHS vehicles for patients to drink and to cool heat casualties.
- Soldiers deployed in the desert are susceptible to endemic diseases and environmental injuries. Proper water discipline, vaccines, prophylactic measures, field sanitation measures, personal

hygiene, and other PVNTMED measures can reduce these risks. Cold weather injuries, heat injuries, and respiratory disease can also be prevalent. Proper clothing, equipment, and a water discipline program must have command emphasis in desert operations.

- Winds may very easily damage materiel such as aircraft, antennas, and tents. Equipment is protected by using covers, tie-downs, and shelters. Terrain helps shield equipment from the wind if site selection is done carefully. In some cases, special tools, such as extra long metal tent stakes, are necessary.
- The effects of wind and sand are interrelated. Desert sand starts to become airborne when the wind reaches about 20 knots. Sandstorms—
  - Restrict visibility.
  - Pose a hazard to eyes (especially for Soldiers wearing contact lenses).
  - Can contaminate water supplies (if they are not protected).
  - Make navigation difficult.
- The sun may cause sunburn of the skin and eyes (if protection is not used).

5-17. Seven characteristics of the desert environment that may adversely effect equipment are—

- Terrain. Trafficability varies with the type of terrain covered. Open, flat, and rocky terrain affords higher trafficability than do mountainous areas, lava beds, or salt marshes. Drivers must be well-trained in judging the terrain over which they are driving to select the best alternative routes.
  - Tracked vehicles are best suited for desert operations. However, they can throw tracks when traversing a rocky area. Their use is also limited in rough terrain with steep slopes.
  - Wheeled vehicles may be used in desert operations; however, they normally have a lower average speed than tracked vehicles and a higher incidence of damage and malfunction. Wheeled vehicles often bog down in sandy areas and cannot traverse many of the rougher areas.
  - In planning for desert operations, vehicles should carry extra repair parts (fan belts, tires, and other items apt to malfunction).
- Excessive heat causes vehicles to overheat, leading to greater than normal wear. The frequency of leaks on vehicles and aircraft is greater than in some other environments. Engine and transmission seals tend to dry out and crack; fuel lines wear out quickly; and water requirements for cooling vehicle engines are greater. Loss of water, through evaporation, must be included in logistical planning. Using vehicle and aircraft covers reduce the effects of heat while vehicles and aircraft are not in use. Refer to FM 1-202.
  - Batteries do not hold their charge efficiently in intense heat. Dry battery supplies should be increased to compensate for a higher usage rate.
  - Communications equipment must be protected from the heat in the desert. Dust covers are used on this type of equipment. If the equipment has ventilating ports, these should be cleaned regularly to avoid clogging.
  - Medical supplies must be protected from the heat to prevent deterioration. The shelf life of some medical supplies decreases when stored in hot climates.
  - Medical equipment must be protected from the effects of heat. It may be protected using the same techniques as those used to protect communications equipment.
- The sun burns unprotected skin and it may damage unprotected eyes. Soldiers should dress in loosely fitting clothing, use sunburn cream or oils to protect exposed skin, and wear sunglasses or goggles to protect their eyes. Soldiers should remain fully clothed. Removing clothing increases direct exposure of the skin to the sun and eliminates the beneficial cooling effects of the moisture trapped in clothing. Radiant light or its heat effects may be detrimental to plastics, lubricants, pressurized gasses, rubber, and some fluids. All vehicles and aircraft should be kept well-ventilated, and windshields should be covered to reduce heat buildup inside. Supplies of all types should be stored in a well-ventilated, shady area. Placing supplies in covered holes in the ground may reduce adverse heat effects.

- Dust and sand present one of the greatest dangers to the proper functioning of equipment. Sand mixed with lubricants forms an abrasive paste. Lubrication fittings, bearings, and filters should be inspected frequently and changed when required.
  - Aircraft should not be exposed to dust and sand any more than is absolutely necessary. Ground handling instead of hovering reduces sand ingestion. Dust and sand increase failure of microphone switches, signal distribution panels, and circuit breakers, and cause electrical motors and generators to burn out. Wheel and flight control bearings require more frequent cleaning; engines should be flushed frequently.
  - Medical and communications equipment may be adversely affected by dust and sand. Over a period of time, electrical insulation is damaged by windblown sand. When combined with the effects of lubricants on the insulation, dust and sand can become a major communications problem. Special care should be taken to brush dust off radio equipment and to keep ventilating ports and channels clear.
  - Sand can accumulate in airframes, on the bottom of armored vehicles, and in bearings on all types of equipment. This accumulation, combined with oil and condensation, adds extra weight to aircraft and vehicles as well as jamming their control linkages. Sand and grease buildups must be removed from bearings to ensure safe operation and control of aircraft and vehicles.
  - Dust trails created by hovering aircraft or ground vehicles can be seen in excess of 10 miles on a relatively flat desert. This exposes these assets to direct and indirect enemy fires. Ground vehicles should reduce their speed to the point that they do not create a dust signature.
- Humidity is a factor in some desert areas of the world, especially in the Middle East. Humidity can become a problem for short periods of time in other desert areas. Light coats of lubrication can help prevent rust; however, these benefits should be weighed against the dust-gathering qualities of oil. Demisting equipment is used on optics and night vision equipment to combat the effects of humidity.
- Temperature variation can cause condensation in humid desert areas affecting optics, fuel lines, air tanks, and weapons. Expansion and contraction of air and fluids cause tires to over inflate during the day and under inflate at night. Fuel tanks may overflow during the day causing a fire hazard. Oil fluid levels become overfull and cause leaks during the day, or insufficient lubrication occurs when the oil cools. Vehicle operators and crew chiefs must ensure that the effects of temperature variations do not become a significant problem.
- Static electricity is prevalent in the desert. This is important to remember during refueling operations and when oxygen is being used onboard vehicles or aircraft. Proper refueling procedures must be followed. Static electricity also causes severe shock to ground personnel in sling load and hoist operations.

CAUTION

During hoist and sling load operations, the load must touch the ground before the ground crew can handle it.

5-18. To ensure success in desert operations, detailed planning is required. Factors to consider include—

- Additional quantities of water are required for HSS/FHP operations for the survival of both medical personnel and their patients. Load plans for all vehicles and aircraft must include water. Water is as mission essential as any piece of unit equipment. It should be a priority item when load plans are developed.
- Prescribed load lists are expanded to carry sufficient quantities of repair parts easily degraded by the environmental factors. For example, rubber and plastic fittings and tubes, or spare parts for communications equipment.

- Plan for the effects of wind, sand, and sun on all equipment. All plastic and glass surfaces on vehicles, aircraft, and other equipment should be covered when not in use. Covers should be ordered or made prior to deployment.
- Fuel planning is critical due to power limitations, extended range requirements, and increased vulnerability of refueling sites in the relatively open desert terrain. Careful planning of FARPs is essential for mission accomplishment.
- Units should plan to pack both hot and cold weather clothing when deploying due to varying temperatures between day and night.
- Petroleum, oils, and lubricants (POL) should be of the proper viscosity for desert operations. Maintenance services are also performed more frequently on ground vehicles and aircraft, thus requiring a larger amount of POL than normal.
- Extra filters of all types are planned for due to a higher consumption rate.

5-19. Training for desert operations is not significantly different than training for operations in other areas except for the following:

- Because many desert areas are in mountainous terrain and because high temperatures increase density altitude, aeromedical evacuation units should conduct mountain training to prepare for contingencies in desert areas. Further, procedures and techniques for evacuation in mountainous terrain must be practiced by all AHS personnel. Special equipment requirements (paragraph 5-7) must also be considered.
- Navigation in desert terrain varies from simple to extremely difficult. Factors affecting navigation are the type of desert and the scale and quality of the available navigational charts. At times, aircraft may have to use dead-reckoning navigational techniques (time, distance, and heading). Ground vehicles must have compasses available, as they have to rely on compass headings and odometer readings to navigate. Ground and AA crews should be able to interpret navigational charts and maps of all types and scales. Use of convoys is a viable technique to improve security and to ensure that ground vehicles do not get lost. Aircraft may be used to assist in navigation by convoys in those areas in which there are poor road networks and the terrain offers no distinctive features by which to navigate.

5-20. In principle, medical evacuation operations in the desert do not differ greatly from these operations in other environments.

5-21. Helicopter landing sites should be chosen with care. Common mistakes made by many units when establishing the LZ are—

- Locating the pad relative to the patient and tents, vehicles, and other obstacles. A common tendency is to locate the helipad downwind of MTFs so that approaches may be made into the wind towards the facility. In mountainous deserts, winds normally channel down the valleys and are more predictable along valley floors. A better site selection for an LZ is with the MTF, alongside the approach and takeoff zone. Thus, the landing direction is up or down the valley, depending on the airflow, and the MTF is not over flown.
- Situating landing sites in washes, small confined areas between large rocks, or close to moving tracked vehicles. Map coordinates are rarely accurate unless the site is beside a major terrain feature. Therefore, LZs should be located next to major terrain features or on higher ground where they can be seen from the air at a distance of 2 to 3 kilometers, if possible. Lack of distinctive features in the open desert and on large-scale maps makes pinpoint navigation difficult, especially at night.
- Marking of helicopter LZs is done so that the pad can be seen from the air, but the markings should not be a hazard in themselves. If engineer tape is used, it should be firmly secured to prevent it from blowing loose. Panel markers are not a good tool to use as they are difficult to see. If panel markers are used, they need to be secured. If used, flares or marker smoke should not be deployed on or directly upwind from the pad. Smoke grenades or flares should not be thrown under the aircraft as it lands. Avoid using white smoke to mark the LZ. Colored smoke is probably the best daylight marking method. It is difficult to detect a smoke grenade more

than 2 to 3 kilometers away, but an aircraft in the general vicinity can normally see it. Radios are used to guide aircraft to the LZs, but this creates an electronic signature. Units requesting medical evacuation must be prepared to signal the evacuation aircraft upon its arrival. Normally, map coordinates will guide the aircraft to within 2 to 3 kilometers of the LZ. Even from nap-of-the-earth (NOE) altitudes, the aircrew may be able to see several units in the area. The requesting unit must signal the aircraft to ensure the designated LZ is used.

5-22. When planning for night flights, considerations include the following—

- Moonlight aids the medical evacuation pilot by providing him with the light to see with either unaided vision or night vision goggles (NVGs). When adequate ambient light exists, medical evacuation crews function almost as effectively at night as they do during daylight. The small arms threat is somewhat reduced at night, although it still exists from radar-guided weapons, infrared-sited weapons, and passive night vision device-equipped weapons systems. Flying into a bright moon with NVGs on can be compared to flying into the sun during the day. The goggles darken and visibility becomes extremely poor. Flight routes should not be planned to fly directly into a bright moon if NVGs are to be used.
- The lack of visual cues over sand is similar to that over water. It is very easy for pilots to become disoriented and fly into the ground. Reliance on radar altimeters is a must over flat sandy areas of the desert.
- Frequently, desert areas do not have sufficient ambient light to allow adequate night vision, even with the aid of NVGs. A pilot wearing NVGs is often unable to see the ground at an altitude of 100 feet using a landing light equipped with a pink light filter. Under these conditions, dead reckoning is the only effective navigation method unless global positioning satellites, Doppler equipment, or NAVAIDS are available. Unfiltered light can be used with or without NVGs; however, this increases the risk of exposing the aircraft's position to the enemy.
- Desert warfare is usually characterized by extended battle zones which increase evacuation distance and time. Combat health support units are located further to the sustainment area in the desert. Establishing an ambulance shuttle system or CCPs is useful. Combat health support units require a greater number of vehicles for operating in deserts than in other environments. Air evacuation by fixed and rotary-wing aircraft is the preferred method due to their speed and range. Further, using aircraft reduces the load on ground vehicles. Augmentation from higher echelon HSS may also be required to meet the extended evacuation needs.
- Smoke is used extensively on the modern battlefield by both sides. It can be effectively used to mask friendly actions to include medical evacuation. (Refer to Appendix E for further information.)
  - Smoke can be a major hazard, especially to medical evacuation helicopters. Smoke reduces visibility and forces an aircraft higher where it can be acquired by threat weapons systems. The phenomenon of inversion occurs often in the desert. When this happens, medical evacuation vehicles and aircraft may be able to work underneath the smoke using the smoke layer for overhead concealment.
  - Medical units must coordinate closely with supported organizations on smoke operations. Smoke can either help or hinder the evacuation mission, depending upon how it is used.
- Communications in the desert are affected by a number of factors. Atmospheric interference and the skip of signals occur frequently. Mineral deposits in the desert may unexpectedly disrupt communications. Many of these problems can be overcome by using additional radio relays, pre-established control measures, and visual signals.
- Artificial lights may be used at times in the desert. They are very easily detected. Even with blackout lights, vehicles using lights can be detected for miles with NVG. Serious consideration should be given to driving without using lights when the tactical situation dictates. Ground guides are used to help vehicles navigate through areas that are not clearly marked or through area where troops are present.
- Wind is one of the most significant environmental factors affecting medical evacuation in the desert. Wind can be destructive to both structures and equipment; tents, antennas, and aircraft

can be easily damaged. Wind direction and speed vary greatly within the space of a few miles. Velocity is substantially increased when wind channels between hills and direction changes due to interference of terrain features. The wind frequently makes aeromedical evacuation impossible by exceeding the operating limitations of the aircraft. At other times, it may limit the use of some potential LZs. Blowing sand and dust can slow down the evacuation system by making navigation by either ground or AA difficult, if not impossible. High winds are predictable to a certain extent. For example, at certain times of the year in the Mojave Desert high winds occur every day at dusk and last for 3 to 4 hours. At other times, high winds, based on frontal weather patterns, can remain for several days at a time. These factors should be considered by HSS planners, and medical evacuation assets should be massed or relocated accordingly.

- The desert provides little or no protection from enemy air defenses except in mountainous terrain. Aircraft may have to be flown above NOE altitudes to prevent a dust signature. These factors cause increased exposure and vulnerability of AAs to enemy air defenses and may limit their employment.

5-23. Further Information. Refer to FM 90-3 for additional information on desert operations.

SECTION IV — EXTREME COLD WEATHER OPERATIONS

5-24. Operations in the extreme cold have many of the limiting factors found in desert operations. The tundra and glacial areas are harsh, arid, and barren. Temperatures may reach lows of -80° F to -100° F which, combined with gale force winds, make exposure unsurvivable.

- The greatest environmental detriment to operations is blowing snow. This results in a loss of depth perception from total white conditions. Blowing snow is caused by the wind or by the rotor wash of helicopters; its effect reduces visibility to zero.
- Other environmental considerations are as extreme but easier to circumvent. Solid footing is suspect in both the dead of winter and in the summer. Snow and ice cover crevasses, holes, and otherwise unstable ground. In traversing suspect ground situations, consider linking Soldiers by rope. During the summer, ground transportation is more restricted than in any other environment due to the marsh and muskeg composition of the arctic tundra. Patients must be sustained for a longer duration due to terrain delays and the lack of direct lines of evacuation.
- Greater responsibility has to be placed on each Soldier, especially for maintenance of nutrition and water consumption. It is imperative to stress that leadership and training are important in the prevention of cold weather injury. Strict adherence to the guidelines found in FMs 21-10 and 31-70 assures an effective fighting force. Water conservation is essential; however, adequate consumption by the individual should be enforced.
- Factors to consider for conducting evacuation in arctic operations include the following:
  - Arctic warfare is usually characterized by extended battle zones that increase evacuation distance and time. Establishing an ambulance shuttle system or CCPs is useful. Augmentation from higher echelon AHS may also be required to meet the extended evacuation needs.
  - Additional supplies of water should be carried by ambulances and maintained at CCPs, if possible.
  - Due to the decreased temperature and frozen environment, ambulance maintenance requirements are increased. Lubricants must be of the correct viscosity for the temperature. In extreme cold, batteries perform less efficiently. Consult the specific aircraft maintenance manual for the recommended battery and procedures for extreme cold weather operations.

Note. All ambulances are considered deadlined without a functional heater for the patient compartment.

  - The proper storage of medical supplies is essential to prevent loss from freezing.

  - There are few terrain features or road networks; therefore, evacuation routes must be surveyed and marked over open terrain. At extreme latitudes, operations during the winter months are conducted in extended hours of darkness. The use of NVG may be required. Compass accuracy is inconsistent due to a geomagnetic phenomenon. Beacons and homing devices are essential for air navigation.
  - Weather is extremely unpredictable. There are too few observers to allow for accurate assessment of weather patterns. Unfavorable weather conditions cause unexpected delays; therefore, medical personnel must be prepared to provide survival measures for their patients and themselves.
  - Landing zones must be chosen with extreme care in both winter and summer. Blowing snow mandates instrument-assisted takeoffs and running landings. Landing areas must be correspondingly larger. The full weight of the aircraft cannot be allowed to settle on the skis until after firm ground conditions are established. Movement of patients to and from the aircraft is difficult. Where an aircraft lands is where it stays. A rocking motion, to free the skis prior to lift off, is performed using the cyclic and antitorque controls.
- Thorough planning and strict preparation are the keys to survival. Factors to consider include the following:
  - Mud obstacles at noon may become an avenue of approach at midnight.
  - Snow complicates all work. Snow-covered terrain hampers reinforcements, muffles noise, makes cross-country driving hazardous, and creates different camouflage requirements.
  - Because of thermal sights, a complete reappraisal of concealment is required.
  - Tracks in the snow destroy concealment.
  - No Soldier is assigned to any job alone. The buddy system is used at all times.
  - Anticipate that all maintenance tasks will take twice as long.
  - Bare metal can stick to skin or wet garments in subfreezing temperatures.
  - Fuel spilled on skin or garments increases the freezing factor; it is one of the greatest causes of injury in winter operations.
  - When operating in the cold, anticipate increased POL needs. Fuel consumption can rise as much as 25 percent for vehicles operating in deep snow, slush, or mud.
  - The recommended fuel for Yukon stoves is diesel.
  - Make every effort to warm gearboxes and engines before starting.
  - The first consideration in the AO is heat; followed by shelter for sustained work.
  - Soldiers need to stand clear of taut cables; steel tends to be brittle and breaks in extremely cold temperature.
  - Fire extinguishers are winterized by adding 15 percent nitrogen to the carbon dioxide.
  - Degradation of battery life requires changes as much as six times more frequently than in a more temperate environment.
  - Radio sets are warmed up prior to transmission. The sets may be turned on but should not transmit for at least one-half hour.

Note. Single-channel ground and airborne radio system (SINCGARS) radios do not require a warm up period.

  - Frost shields (such as using the plastic bag in which the batteries are packed) should be placed over microphones.
  - Grounding rods have to be buried horizontally instead of pounded in vertically. Recovery of stakes and rods placed in the ground is significantly more difficult.
  - Flooring is needed in heated areas because of the thawing of the tundra.
  - Soldiers must take breaks for water and warmth.
  - Static electricity presents a serious safety hazard especially around flammable materials.

5-25. For additional information, refer to FM 31-70.

SECTION V — CHEMICAL, BIOLOGICAL, RADIOLOGICAL, AND NUCLEAR ENVIRONMENTS

5-26. Medical evacuation and treatment are conducted continuously throughout operations conducted in a CBRN contaminated environment. The AHS commander must have a comprehensive plan which is rehearsed on a periodic basis to ensure the timely evacuation and treatment of casualties in a CBRN environment. Techniques and procedures which are essential for operating in a contaminated environment should be contained in the unit TSOP. The number of casualties and their medical condition, type of contaminant, the size of the land area contaminated, the expected duration of operation, risk assessment and acceptable level of risk, and the number of AHS assets (medical personnel, medical units, and evacuation vehicles and aircraft) initially contaminated will determine the quantity and type of uncontaminated AHS resources, if any, which will be introduced into the contaminated environment to ensure timely medical treatment and evacuation occur.

5-27. The Commander must take into consideration the number of assets he is willing to commit during evacuation operations in a CBRN contaminated environment. Since the combinations of evacuation methods are nearly endless, the commander has greater flexibility in tailoring an evacuation plan to meet his particular tactical situation in a CBRN contaminated environment.

5-28. On the modern battlefield there are three basic modes of evacuating patients (personnel, ground vehicles, and aircraft).

- In using personnel to physically carry the casualties, the commander must realize the inherent stress involved. Cumbersome mission-oriented protective posture (MOPP) gear needed in a contaminated environment (added to climate, increased workloads, and the fatigue of battle) greatly reduces the effectiveness of unit personnel.
- If the commander must send evacuation personnel into a radiological contaminated area, he must establish operational exposure guidance (OEG) for the medical evacuation operation. Radiation exposure records are maintained by the unit CBRN NCO and are made available to the commander, staff, and surgeon. Based on OEG, the commander decides which medical evacuation assets to send into the contaminated environment.

5-29. Commanders should make every effort to limit the number of evacuation assets which are contaminated while still maintaining a timely and effective medical treatment and evacuation operation.

- It is expected that a certain number of both ground and AAs will become contaminated in the course of battle. The commander can, therefore, segregate the contaminated ones. This results in the smallest impact on his available assets and the greatest possibility for continuing the patient evacuation mission. Optimize the use of resources, medical or nonmedical, which are already contaminated before employing uncontaminated resources.
- Once a vehicle or aircraft has entered a contaminated area, it is highly unlikely that it will be able to be spared long enough to undergo a complete decontamination. This depends upon the contaminant, the tempo of the battle, and the resources available. Normally, contaminated vehicles (air and ground) have restricted use and are confined to dirty environments.
- Introducing uncontaminated aircraft into a contaminated area should be avoided, whenever possible. Ground ambulances should be used instead of AAs as long as their use does not adversely affect the patient's medical condition. Ground ambulances are more plentiful and are easier to decontaminate. This does not, however, preclude using aircraft in a contaminated environment or in the evacuation of contaminated patients.
- The relative positions of the contaminated area, FLOT, and threat air defense systems determine if and where helicopters are to be used. The commander may choose to restrict one or more helicopters to the contaminated areas and use ground vehicles to cross the line separating contaminated and clean areas. The ground ambulance can proceed to the receiving MTF with a

patient decontamination station. The patient can then be transferred to a clean ground or AA if further evacuation is required. The routes used by ground vehicles to cross between contaminated and clean areas are considered dirty routes and should not be crossed by clean vehicles. The effects of wind and time upon the contaminants must also be considered.

- The rotor wash of the helicopters must always be kept in mind when evacuating contaminated casualties. The intense winds disturb the contaminants in the area and further aggravate the condition by additionally spreading the contaminants. Ideally, the aircraft must be allowed to land and reduce to a flat pitch prior to bringing any patients near. This will be dictated by the tactical situation, but allows some reduction in the effects of the downwash. A helicopter must not land too close to a decontamination station (especially upwind) because any trace of contaminants in the rotor wash will compromise the decontamination procedure.
- Immediate decontamination of aircraft and ground vehicles should be accomplished to minimize crew exposure. Units should develop their own procedures for thorough decontamination and document them in their TSOPs. A sample aircraft decontamination station that may be tailored to a particular unit's needs is provided in FM 3-11.5.
- Evacuation of patients must continue even in a contaminated environment. The commander must recognize the constraints placed upon him by resources and plan and train to overcome deficiencies.
- Refer to FM 4-02.7 for additional information on AHS operations in a CBRN environment.

SECTION VI — SHORE-TO-SHIP EVACUATION OPERATIONS

5-30. It is imperative that Army aeromedical evacuation units be able to interface with USN air-capable ships. Lessons learned from past operations, such as Vietnam and Grenada, have shown that US Army helicopters should be able to operate to and from USN air-capable ships. An interservice agreement between the Army and the Navy allows for deck-landing qualification of Army pilots. (Refer to FM 1-564 for additional information.)

- It is important that units having contingency missions requiring Navy support establish training requirements to obtain naval-operations orientation, water egress training, water survival, and deck-landing qualification. This enhances the successful accomplishment of the aeromedical evacuation mission to naval vessels.
- In past joint operations, communications have been burdensome for both Army and Navy elements. Commonality of communication requirements should be established during training exercises. Communication equipment and frequencies for medical evacuation to Navy vessels must be established. This will provide smooth integration of Army helicopters into the Navy airspace management system during actual operations.
- As the Navy vessels may operate relatively long distances from the ground combat operations, Army aeromedical evacuation units need to be proficient in over water navigation. The use of NAVAIDS from the Navy element in support of the operation is the first priority for over water navigation. Basic dead-reckoning remains a secondary measure.

5-31. An important aspect of joint operations is the medical capabilities of Navy vessels servicing the CZ. Knowledge of ship's medical capabilities assists the MRO to direct patients to proper treatment sites. There are many classes of ships which can meet the medical needs of ground forces. Destroyer tenders and aircraft carriers have helicopter landing areas, one operating room (OR) and, at a minimum, one medical officer. Amphibious ships have the most extensive medical facilities of any Navy combat ship. The Navy has 59 amphibious ships in active commission plus two tank landing ships, which are operated by the Navy Reserve Force. The primary mission of the amphibious ships is to transport and support the Fleet Marine Force. The ships have the additional duty of casualty receiving and treatment ships (CRTS). During normal operations, the medical staff is kept to a minimum. The medical staff is augmented when expanded capabilities are needed. Current information regarding landing requirements and medical capabilities should be obtained during training periods with the Navy. Casualty care is secondary to the combat mission of all US combat ships.

- Amphibious assault ships "WASP" class are designated by the Navy as amphibious helicopter assault carrier dock (LHD) (followed by a number) and have the largest patient care facilities on any US combat ship. The WASP class ships have six main ORs, four dental ORs, bed capacity that can be expanded to 600, and it carries 1,500 pints of frozen blood. This ship can receive casualties from helicopters or landing craft.
- Amphibious assault ships "TARAWA" class are designated by the Navy as amphibious helicopter assault carrier (LHA) (followed by a number). The TARAWA class ships have three main ORs, two dental ORs, an overflow bed capacity of 300, and carries 1,500 pints of frozen blood. The ship can receive casualties from helicopters or landing craft.
- Amphibious assault ships "IWO JIMA" class are designated by the Navy as amphibious assault helicopter carrier—aviation (LPH) (followed by a number). These ships were specifically designed to operate helicopters. The IWO JIMA class ships have two ORs and an overflow bed capacity of 200.
- The amphibious transport dock (LPD) (followed by a number) has less medical capabilities than the LHD, LHA, or LPH ships. It can be designated as secondary casualty receiving ship.
- The older dock landing ship (LSD) (followed by a number) can be used as a secondary CRTS when augmented. The newer class of LSD currently under construction can be used as a casualty receiving ship with a capacity for 50 wounded.
- The tank landing ship (LST) (followed by a number) is another type of ship used in amphibious operations. It is designed with a helicopter platform and a stern ramp. Patients can be delivered by air or boat when required by tactical or mass casualty situations. When the LST is augmented with medical personnel and materiel, it can be used for the emergency treatment and evacuation of patients.
- The troop transport (AP) (followed by a number) is not in active service. When available, the troop transport can be outfitted with special medical facilities and carry sick, injured, and wounded personnel.

5-32. The Military Sealift Command operates two hospital ships. The United States Navy Ship (USNS) MERCY T-AH 19 and the USNS COMFORT T-AH 20. One ship is based on each coast and, when needed, will be assigned medical staffs from military hospitals, getting underway within 5 days. The hospital ships MTFs were designed for a total capacity of 1,000 casualties, including 500 acute care beds and 500 recuperation beds. The hospital ships have 50 trauma stations in the casualty receiving area; 12 operating rooms; a 20-bed recovery room; 80 intensive care beds; and 16 intermediate, light, limited care wards. The maximum patient flow rate, for which the helicopter facility and the casualty reception area were designed, is 300 patients per 24 hours. There is a limited capability to receive casualties from boats.

5-33. The US Army has the shore-to-ship medical evacuation mission on an area support basis for Marine forces deployed on land.

SECTION VII — AIRBORNE AND AIR ASSAULT OPERATIONS

5-34. Airborne and air assault operations are some of the most complicated military operations to undertake, not only from an operational perspective, but also from a sustainment and conservation of the fighting strength perspective. A number of personnel within the medical chain that need to coordinate, not only amongst the AMEDD, but also include the aviation community as what proponent own the aircraft (for example, BAE, BSS, medical platoon, and such). Manual evacuation techniques may have to be used and planners should plan for redundant communications, only ground medical evacuation may or may not be available.

5-35. The airborne and air assault operational forces are specialized forces employed to maximize their design characteristics. Airborne units are a flexible force that can be strategically or tactically deployed. They can be inserted rapidly anywhere in the world as either a deterrent or strike force. Air assault units are flexible and lethal fighting organizations. They are ideally suited for rapid employment to critical areas beyond the reach of ground forces.

5-36. After airborne forces have landed in the objective area, they reorganize and maneuver to seize objectives. When it is necessary for assault aircraft to land in the drop zone, they are parked and unloaded rapidly. Then, they may be used to transport Soldiers injured during the parachute assault. It must be understood that organic medical units may experience an overload of patients during the early phases of an airborne assault. These units have to hold the patients until either ground link-up is made or evacuation can be established at airheads. Aeromedical evacuation from the airhead is accomplished using tactical and strategic USAF aircraft.

5-37. The air assault division's organic aircraft have the ability to attack from any direction, over fly obstacles, and bypass enemy positions. Evacuation of patients in the assault phase is accomplished by AAs. Air ambulances may accompany the air assault task force (AATF) or respond from laager sites once the initial assault has taken place. If AAs are providing on-call support, it will be necessary to fly secure air avenues of approach.

5-38. When both airborne and air assault divisions have been employed and become a part of other conventional forces, their operations are similar to that of light infantry forces. During initial deployment, division medical evacuation assets may be used to evacuate patients to the airhead for air evacuation directly to division/corps hospitals.

Chapter 6

Medical Regulating

GENERAL

6-1. Medical regulating is the coordination and control of moving patients to MTFs which are best able to provide the required specialty care. This system is designed to ensure the efficient and safe movement of patients.

PURPOSE OF MEDICAL REGULATING

6-2. Medical regulating entails identifying the patients awaiting evacuation, locating the available beds, and coordinating the transportation means for movement. Careful control of patient evacuation to appropriate hospitals is necessary to—

- Effect an even distribution of cases.
- Ensure adequate beds are available for current and anticipated needs.
- Route patients requiring specialized treatment to the appropriate MTF.

6-3. The factors that influence the scheduling of patient movement include the following:

- Patient's medical condition (stabilized to withstand evacuation).
- Tactical situation.
- Availability of evacuation means.
- Locations of MTFs with special capabilities or resources.
- Current bed status of MTFs.
- Surgical backlogs.
- Number and location of patients by diagnostic category.
- Location of airfields, seaports, and other transportation hubs.
- Communications capabilities (to include radio silence procedures).

MEDICAL REGULATING TERMINOLOGY

6-4. As medical regulating may include coordination with other services, it is necessary to use the correct terminology. These terms include—

- Intracorps Medical Regulating—This is the system by which patients are transferred or evacuated from an FSB or main support battalion (MSB) to a combat support hospital (CSH).
- Intratheater Medical Regulating—This is the system by which patients are transferred or evacuated from one hospital to another within the AO. This includes evacuations between CZ hospitals, between EAC hospitals, or from CZ hospitals to EAC hospitals.
- Intertheater Medical Regulating—Moving patients between, into, and out of the different theaters of the geographic combatant commands and into the CONUS or another supporting theater. (JP 4-02.)
- Patient Administrator—The patient administrator (PAD) accomplishes the medical regulating function at the hospital level in addition to his normal duties. His medical regulating functions include consolidating all evacuation requests within the hospital and forwarding an evacuation request to his next higher headquarters for action. The PAD is also responsible for keeping his next higher MRO apprised of the current beds available and the OR status.

- Medical Regulating Officer—The MRO functions as the responsible individual at C2 headquarters for receiving and consolidating evacuation requests. These requests are initiated by the DSS or subordinate hospitals. The MRO also maintains the current patient status, bed status, and the surgical backlog at subordinate hospitals. His duties include—
  - Managing what patient classes are regulated into his facility.
  - Determining what resources are available to move the patients and coordinating for the use of these assets.
  - Maintaining accountability of patients within the MTFs.
  - Preparing reports as required.
- Theater Patient Movement Requirements Center—The TPMRC is a joint agency normally located at or near the unified theater headquarters. The theater surgeon supervises the functions of this office. These functions include—
  - Maintaining direct liaison with the Global Patient Movement Requirements Center (GPMRC), the MROs of component services, and the transportation agencies which furnish the means for evacuation.
  - Obtaining periodic reports of available beds from the services MROs providing hospitalization.
  - Selecting hospitals based on the reported bed availability to receive patients within EAC.
- Global Patient Movement Requirements Center—The GPMRC is a joint agency located in CONUS and established by the USTRANSCOM. The GPMRC receives requests from the TPMRCs. The primary role of the GPMRC is to apportion intertheater assets to the TPMRCs, collaborate and integrate proposed TPMRC intertheater plans and schedules, and communicate lift and bed requirements. The destination hospital is determined based on the patient's medical needs, the available transportation resources, and MTF capabilities.
- Joint Patient Movement Requirements Center (JPMRC)—Joint patient movement requirements center provides theater patient movement requirements center-type domain, automatic information system support and operations for a joint task force operating within a unified command area of responsibility. The theater patient movement requirements center maintains overall responsibility for theater patient movement operations, but the joint patient movement requirements center is responsible for patient movement operations within its area scope of responsibility and coordinates with the theater patient movement requirements center for intra-theater patient movement and the GPMRC for inter-theater patient movement. Also called JPMRC.
- Theater Aeromedical Evacuation System (TAES)—The TAES is a functional organization which is provided by the USAF and performs the mission of theater AE. Significant components of the TAES include the following:
  - Aeromedical evacuation command squadron.
  - Aeromedical evacuation control team (AECT).
  - Mobile aeromedical staging facility.
  - Contingency aeromedical staging facility (CASF) (under the C2 of the medical group commander).
  - Aeromedical evacuation crews.
  - Critical care air transport teams (CCATTs) (not necessarily Air Force).
- Aeromedical Evacuation Command Squadron—The AE command squadron provides C2 of all assigned TAES forces and can deploy in advance of other AE components to arrange support requirements for AE forces. The command squadron advises commanders and other personnel/agencies on AE operations, capabilities, and requirements and provides procedural and technical guidance for attached and transiting AE elements.
- Aeromedical Evacuation Control Team—The AECT is a CASF (under the C2 of the medical group commander) and is an USAF element responsible for linking validated patient movement requests with the appropriate airlift platform. The AECT is normally located within the Air

Mobility Division of the theater air operations center (AOC) for intratheater and the tanker airlift control center (TACC) for intertheater coordination of AE operations using USAF aircraft. The CONUS and strategic AOC for the USAF is the TACC located at Scott Air Force Base, Illinois. The AECT coordinates with and receives requirements from the PMRC for execution.

- Aeromedical Evacuation Liaison Team (AELT)–The AELT provides a direct communications link and immediate coordination between the user service and the TAES. The AELT is located at any level where Air Force fixed-wing requests are initiated, verifies and coordinates with the AECT/patient movement requirements center (PMRC) the physiology of flight issues and patient flight/movement requirements, assists with patient preparation for flight, and directs patient on-load activities.
- Mobile Aeromedical Staging Facility—The MASF is a USAF staging facility employed at forward airfields in the CZ to provide a temporary staging capability for preparation of patients being evacuated from corps to EAC hospitals. The MASF is employed to ensure patients are prepared for aircraft loading with the main focus of reducing AE aircraft ground time.
- Contingency Aeromedical Staging Facility—The CASF, similar to the MASF, is a USAF staging facility. The CASF, which provides greater capability and longer holding periods than the MASF, is a fixed facility located at theater AE strategic hubs and interfaces with the TAES. As opposed to the TAES elements, which are under the C2 of the AE command squadron, the CASF reports to the medical group commander.
- Military Sealift Command (MSC)—The MSC is the USN element responsible for coordinating movement of supplies, equipment, and personnel into the JOA by Navy ships. Further, it coordinates, through the TPMRC, the medical evacuation of patients by ship from the JOA to the support base or hospital ship, as required.
- Corps Movement Control Center (CMCC)—The CMCC is the corps movement control organization. It provides centralized movement control and highway regulation for movement of personnel and material into, within, and out of the corps area. When USAF capabilities are exceeded, the CMCC coordinates requests for additional air and ground resources. It also obtains the necessary clearances to support the medical evacuation mission from the CZ.
- Movement Control Battalion (MCB)—The MCB mission is to provide movement management services and highway traffic regulations and to coordinate for personnel and material movements into, within, and out of the theater. The MCB coordinates with allied and HN movement control agencies. It also coordinates with the USTRANSCOM and its subordinate units (such as the Air Mobility Command and MSC) and prepares movement and port clearance plans and programs.
- Joint Military Transportation Board (JMTB)—The JMTB is a joint staff composed of members of the Army, USAF, and USN that coordinates transportation requirements for patients requiring intertheater evacuation.
- Transportation Command Regulating and Command and Control Evacuation System—The Transportation Command Regulating and Command and Control Evacuation System is the single DOD automated information system that supports medical regulating, patient movement, and patient in-transit visibility requirements.
- Automated Patient Evacuation System—The Automated Patient Evacuation System is the system that automates the patient movement portion of medical evacuation.

MEDICAL REGULATING FOR THE DIVISION

6-5. Medical regulating in and from the division is the responsibility of the DSS. Medical regulating in the division is not as formalized as the rest of the medical regulating system. It is usually operated procedurally so as not to depend solely on communications to effect rapid evacuation. The medical regulating function in the DSS is concerned primarily with.

- Tracking the movement of patients throughout the division MTFs and into the corps facilities.
- Monitoring the use of ambulance assets.

6-6. Air and ground ambulances placed in GS of the BCTs can be field sited in the sustainment area when the mission directs. When these assets go forward to the BSMC or the ASMC to evacuate patients to corps MTFs, they have corps MTF destinations predetermined (blocks of beds). The DSS, in coordination with the medical group/brigade MRO, establishes the number of patients a supporting corps hospital can accept during a particular period of time. These blocks of available beds are then provided to the GS ambulances prior to the call for missions.

> ***Note.*** Under the MRI force structure, the medical group functions were absorbed by the medical brigade.

- Once an evacuation mission is completed, the originating division MTF contacts the patient disposition section of the BSS and provides—
  - Patient numbers by category and precedence.
  - Departure times.
  - Modes of transportation.
  - Destination MTFs.
  - Any other information required by TSOP.
- The DSS, in turn, notifies the medical group/brigade MRO via the patient administration net, which is monitored by the corps MTFs. Since corps ground ambulances have no onboard communications ability and AAs have no amplitude-modulated high-frequency capability at present, all patient information is passed to the gaining MTFs via the patient administration nets. To reduce the turnaround time for ground ambulances and to move more serious patients to the CSHs in the sustainment area—
  - Air ambulances are given blocks of beds in the corps hospitals farther to the rear.
  - Ground ambulances are normally given blocks of beds in the more forward deployed CSHs.

6-7. Medical evacuation can be effected immediately, procedurally, and under conditions of communications silence without interrupting the continuum of care by—

- Preparing patient estimates.
- Prioritizing and task-organizing ambulance support.
- Assigning blocks of hospital bed designations prior to the start of the mission.
- The assistant chief of staff (personnel) (G1) is also involved in medical regulating as the G1 is responsible for personnel accountability and casualty reporting. Casualty liaison teams (CLTs) will normally be colocated at CSH to assist and coordinate patient tracking and casualty reporting.

MEDICAL REGULATING WITHIN THE COMBAT ZONE

6-8. The requirement to transfer patients from one hospital to another within the CZ occurs. This results from—

- Surgical backlogs.
- Mass casualty situations.
- Specialty care requirements.
- Planned movement of an MTF.

6-9. When it is necessary to transfer a patient, the attending physician notifies the hospital PAD. The PAD consolidates all such requests from the hospital and requests movement authority from the IPMC.

6-10. If the medical IPMC can transfer the patient or patients to its subordinate hospitals, he designates the hospitals to receive the patients and notifies both the requesting and receiving hospitals of the transfer. The IPMC also tasks subordinate medical evacuation units for the assets to transfer the patients.

6-11. If the MEDBDE cannot provide the needed hospitalization within its own resources, the MRO forwards the request to the medical TPMC for action. The TPMC then designates the receiving hospitals and notifies the subordinate IPMCs. The medical IPMCs disseminate the information to the hospital PAD and coordinate the evacuation resources for the transfer. The MRO also coordinates the regulation of patients to—

- Other US military service hospitals and naval hospital ships.
- Allied nations' military hospitals.
- Other authorized supporting facilities.

MEDICAL REGULATING FROM THE COMBAT ZONE TO ECHELONS ABOVE CORPS

6-12. Hospital attending physicians and oral and maxillofacial surgeons submit daily reports to the hospital PAD listing the patients requiring evacuation. The PAD assembles this information and transmits the report to the medical brigade headquarters. This report is a request for transportation, as well as a notification of the number of patients requiring evacuation. The report classifies the patients according to—

- Diagnostic category.
- Desired on-load points.
- When the patients will be available for evacuation.

6-13. The medical brigade TPMC consolidates these reports from each hospital attached to the medical brigade and forwards his report to the corps TPMC. The medical brigade MRO consolidates the reports and transmits the data to the theater TPMC.

6-14. If a JPMRC has been activated within the theater, the MEDCOM MRO consolidates all reports from the CZ medical brigades and forwards them to the JPMRC. The JPMRC designates hospitals in the EAC to receive the patients. The designation is based on the previously received bed status reports from all service components and available means of evacuation. The JPMRC then notifies the MEDCOM MRO of designated hospitals. The TPMC accomplishes this task if the JPMRC is not activated.

6-15. The primary means of moving patients from the CZ out of theater is USAF aircraft. With the elements of the TAES deployed, it is possible to find AELTs at each echelon and as far forward as Air Force fixed-wing requests are initiated. The PMRC monitors the MRO patient evacuation requests and passes requirements to the AELT. At the same time they pass airlift requirements to the AECT, seeking an aircraft to perform the evacuation mission. The AELT, through the MRO patient movement request, requests the PMRC/AECT to move patients. Included in the request are the originating medical facility (OMF) and the destination airfields. The airfields selected are those serving the hospitals designated to receive patients.

6-16. The AECT is a component of the TAES and performs the mission of coordinating the movement of and providing in-flight medical care to patients while under the USAF control. The AECT receives patient movement requirements from the TPMRC, then works with the airlift control team (ALCT) in the AOC to meet the evacuation requirements.

6-17. The AOC coordinates the forward movement of cargo and personnel aboard USAF aircraft with other USAF units, Army transportation representatives, and USN agencies. Certain of these aircraft are scheduled to evacuate patients on their return trips. These aircraft seldom go forward solely to evacuate patients; however, these missions may be used for retrograde airlift and mixed cargo/AE missions is the primary means of AE due to limited aircraft and the need to maximize the use of this asset.

6-18. After the schedules have been arranged, the AECT returns the detailed flight schedule to the JPMRC who passes the information to the TPMC/AELT and the parent AE element.

6-19. The MEDCOM MRO, in coordination with the AELT, issues these instructions to the IPMCs (with the authority to move patients in Army CZ facilities) and the receiving hospitals. The hospitals must prepare to receive the patients at the destination airfields. (This may be accomplished by collocating an ASMC at the airfield to receive the incoming patients.) The patients are sorted by destination hospital and moved by Army medical evacuation means. The instructions mentioned above include, as a minimum, the—

- Number of patients to be moved.
- On-load airfield.
- Destination airfield.
- United States Air Force aircraft mission number.
- Estimated time of arrival at the destination airfield.

6-20. The TPMC issues the flight and movement instructions to its subordinate medical IPMCs. The IPMCs then direct the evacuation units and hospitals within their AOs to move the patients to the on-load airfield according to the arrival time of the aircraft. This movement must be closely controlled, as a MASF can accommodate up to 50 patients. The patients cannot be delivered to the MASF earlier than 6 hours prior to arrival of the aircraft and no later than 1 hour prior to arrival.

MEDICAL REGULATING WITHIN ECHELONS ABOVE CORPS

6-21. Medical regulating within the EAC is similar to the system used within the CZ. Attending physicians within the Role 3 hospitals notify the hospital PAD of patients requiring evacuation to CSHs. The PAD then consolidates the requests from the hospital and forwards the consolidated requests to the IPMC. The IPMC, in turn, consolidates the requests and forwards them to the TPMC.

6-22. The TPMC, based on periodic bed status and availability reports from subordinate hospitals, designates specific hospitals to receive the patients. The hospitals are designated based on bed availability, to include specialty beds, to support the specific patient. The TPMC then notifies the requesting medical IPMC of the designated hospitals and, in turn, notifies the designated hospitals.

INTERTHEATER MEDICAL REGULATING

6-23. The patients who are evacuated to EAC are treated there and then further evacuated to CONUS. The attending physicians at the hospital notify the PAD. The PAD then consolidates these requests and forwards them to the IPMC. This IPMC forwards the consolidated request to the TPMC who, in turn, consolidates and forwards a request to the JPMRC (if established) or the TPMRC.

6-24. Upon request of the J/TPMRC for authority to evacuate patients to CONUS, the GPMRC directs the distribution of these patients into hospitals throughout the CONUS; advises the J/TPMRC of the destination hospital; and provides the authority for such movement. As a rule, the destination hospitals are military facilities. Civilian national disaster medical system member hospitals and other federal hospitals may also receive patients. The Department of Veterans Affairs hospitals, for example, may receive patients who are expected to be discharged from service. The GPMRC validates the TPMRC patient movement requirements and, if moving by air, tasks the TACC to plan, schedule, and execute the intertheater evacuation.

6-25. When the J/TPMRC receives the authorization to move patients, it notifies the TPMC of destination hospitals in CONUS. The TPMC coordinates with the JMTB to arrange movement of CONUS-bound patients. The TPMC then authorizes the movement to ASFs/aeromedical staging squadrons (ASTSs) that are located on or near air bases or airstrips capable of handling long-range aircraft. Transportation is arranged, within Army channels, to move patients from the hospitals to the staging facilities. The medical brigade, in coordination with the AELT, then notifies the subordinate CSHs of the flight schedule and the evacuation arrangements for movement to strategic airheads. At strategic airheads, there is an established ASF/ASTS. When the patients are delivered to the USAF, the responsibility for those patients is transferred from the Army hospital to the TAES. Upon arrival in CONUS further movement is the responsibility of the GPMRC.

6-26. All patients may not be able to be moved by air from the theater to CONUS. In that event, the MSC is used to move them by surface means. The movement authority also comes from the G/T/JPMRC or TPMC which has arranged with the Navy Service Component Command for the movement of patients by hospital ships. When the patients are moved by ships, the MEDCOM has to provide holding facilities at the port (collocating an ASMC can provide this support). Patients are delivered to these holding facilities and held there until loaded aboard the ships.

MOBILE AEROMEDICAL STAGING FACILITY

6-27. The MASF is a 13-person, mobile, tented, temporary staging facility deployed to provide supportive patient care and administration. Each MASF is capable of a throughput of 40 patients per 24 hours and should be able to sustain this tempo for 72 hours before augmentation is required and is not intended to hold patients overnight or for an extended period.

6-28. This theater system is used to evacuate patients from—

- United States Air Force operational locations within the CZ to hospital facilities outside the CZ.
- Airhead or airborne objective areas where airborne operations include USAF forward logistics support.

6-29. Mobile aeromedical staging facilities—

- Provide supportive medical care when not augmented by the CCATTs. (When augmented by CCATTs, continued stabilization of patients can be accomplished. Aeromedical evacuation crews and CCATTs fly airlift missions to provide in-flight patient care.)
- Confirm sending facility has prepared patients for evacuation.
- In coordination with the OMF, ensure patient evacuation manifests are completed. (If the OMF is an Role 2 facility [such as an FSMC augmented with an FST], the forms required to complete the manifest may not be available and the MASF will be required to complete the appropriate forms.)
- Identify patient baggage tags.

6-30. Upon deployment, the OMF provides an adequate quantity of medications for patients' transit time to the regulated destinations.

6-31. The MASF staff also establishes liaison with the OMF. The AELT is composed of one medical service corps officer for administrative assistance and a flight nurse. The team provides the initial interface between the user service and the TAES. The AELT is located at any level of the combat forces where Air Force fixed-wing requests are initiated.

LIMITATIONS OF THE UNITED STATES AIR FORCE THEATER AEROMEDICAL EVACUATION SYSTEM

6-32. There are a number of limitations that are inherent in the current system. These include the following:

- Absence of biological weapon and chemical weapon agent decontamination ability.
- The MASF does not have the capability to provide patient meals.
- The AECT ensures the initial 30-days medical resupply package arrives at the MASF. The MASF/AELT/aeromedical evacuation operations team rely on the user service for all other logistical support.
- It is the Army's responsibility to provide food and other logistical support required including moving patients back to Army facilities should USAF AE support be delayed.

ORIGINATING MEDICAL FACILITY'S RESPONSIBILITIES

6-33. Once the authorization to move the patient has been given, the OMF must complete the following administrative procedures prior to entering the patient into the TAES:

- The patient's baggage tag, patient evacuation manifest, and patient evacuation tag are the specified evacuation forms for all services and are completed as required by triservice regulation.

Note. If the OMF is a Role 2 facility (such as the BSMC), the forms may not be available; in which case, the forms will need to be completed by the MASF.

- All of the patient's medical records must be collected together and packaged. The dental records are forwarded separately in the event they are needed for identification.
- At the appropriate time, the OMF provides transportation to the MASF and assists in the offload.
- The OMF must provide the necessary medications, medical supplies, and equipment to support the patients' travel time to the regulated destination.
- Any requirements for armed guards must be met by the echelon commander.

Note. Medical units do not provide guards for prisoners or EPW in their care. When guards are required, they are provided by the echelon commander. The OMF will coordinate for this support when needed.

- A limited amount of personal baggage is authorized if each piece is properly tagged and delivered to the MASF with the patient. Patients will always be evacuated with CBRN-protective equipment, less the protective overgarment.
- Each patient must be clearly identified with a wristband or equivalent identification and properly classified as to his medical condition.
- The OMF must ensure that each patient is properly briefed and prepared for his evacuation prior to his arrival at the MASF.

MEDICAL REGULATING FOR ARMY SPECIAL OPERATIONS FORCES

6-34. As in medical evacuation, the medical regulating plan must be integrated with the ARSOF operational and logistic plan. Maximum use of opportune (operational and logistics) aircraft and command and logistics communications nets must be coordinated to expedite mission requests and ensure success.

6-35. The ARSOF medical planner must constantly coordinate with the battalion or group operations and logistics sections to obtain up-to-date information of opportune transportation assets to be used for evacuation. In an operation, or when the theater is not sufficiently developed to allow the TAES to be used effectively, the primary means of air evacuation will be those SOA or USAF SOF airframes conducting the clandestine mission. It is essential that coordination is made through the Theater Special Operations Command (TSOC) or the highest C2 element for flight medics or para rescue men to accompany the flight when backhauling the casualties. Otherwise, a medic from the SOF unit being supported may have to accompany the patient, leaving the mission without proper medical support, or the casualty may have to be transported without en route care.

6-36. For all other special operations, the supporting medical evacuation unit provides air and ground ambulances according to standard doctrinal procedures. United States Air Force MASFs or AELTs may be collocated at SOF support bases, or C2 bases, particularly during contingency operations where the build-up phase allows for prepositioning of assets.

6-37. During sustained special operations missions, the TSOC cannot afford to lose the services of low-density ARSOF skilled Soldiers who become casualties. Every effort must be made to preserve ARSOF capabilities and combat power in theater. A determination must be made as to who can be treated and returned to duty at hospitals within the EAC. As an exception to the theater evacuation policy, the combatant command may retain injured or wounded ARSOF in theater where they can be returned to limited duty without jeopardizing their recovery and health. There they can assume the support duties performed by other ARSOF Soldiers, freeing the latter for operational duties.

Appendix A

Geneva Conventions and Medical Evacuations

The conduct of armed hostilities on land is regulated by customary international law and lawmaking treaties such as The Hague and Geneva Conventions. The rights and duties set forth in the Conventions are part of the supreme Law of the Land. The US is obligated to adhere to these obligations even when an opponent does not. It is a DOD and Army policy to conduct operations in a manner consistent with these obligations. An in-depth discussion of the provisions applicable to medical units and personnel is provided in FM 4-02. This appendix discusses only those articles or actions which affect medical evacuation operations.

DISTINCTIVE MARKINGS AND CAMOUFLAGE OF MEDICAL FACILITIES AND EVACUATION PLATFORMS

This paragraph implements STANAG 2454.

A-1. All US medical facilities and units, except veterinary, display the distinctive flag of the Geneva Conventions. This flag consists of a Red Cross on a white background. It is displayed over the unit or facility and in other places as necessary to adequately identify the unit or facility as a medical facility.

AUTHORIZED EMBLEM

A-2. The Geneva Conventions authorizes the use of the following distinctive emblems on a white background: Red Cross; Red Crescent; and Red Lion and Sun. In operations conducted in countries using an emblem other than the Red Cross on a white background, US Soldiers must be made aware of the different official emblems. Although not specifically authorized as a symbol in lieu of the Red Cross, enemies of Israel in past wars have recognized the Red Star of David and have afforded it the same respect as the Red Cross. This showed compliance with the general rule that the wounded and sick must be respected and protected when they are recognized as such, even when not properly marked.

A-3. United States forces are legally entitled to only display the Red Cross. However, commanders have authorized the display of both the Red Cross and the Red Crescent to accommodate HN concerns and to ensure that confusion of emblems would not occur. Such use of the Red Crescent must be in a smaller size than the Red Cross.

This paragraph implements STANAG 2931.

Camouflage of the Distinctive Emblem

A-4. Camouflage of medical facilities (medical units, medical vehicles, and medical aircraft on the ground) is authorized when the lack of camouflage might compromise tactical operations. The marking of facilities and the use of camouflage are incompatible and should not be undertaken concurrently.

A-5. If failure to camouflage endangers or compromises tactical operations, the camouflage of medical facilities may be ordered by a NATO commander of at least brigade level or equivalent. Such an order is temporary and local in nature and is rescinded as soon as circumstances permit.

Note. There is no such thing as a *camouflaged* Red Cross. When camouflaging a medical unit or ambulance, either cover up the Red Cross or take it down. A *black cross* on an olive drab or any other background is not a symbol recognized under the Geneva Conventions.

Medical Aircraft

A-6. Medical aircraft exclusively employed for the removal of wounded and sick and for the transport of medical personnel and equipment shall not be attacked, but shall be respected by the belligerents, while flying at heights, times, and on routes specifically agreed upon between the belligerents concerned.

A-7. The medical aircraft shall bear, clearly marked, the distinctive emblem together with their national colors on their lower, upper, and lateral surfaces.

A-8. Unless agreed otherwise, flights over enemy or enemy-occupied territory are prohibited.

A-9. Medical aircraft shall obey every summons to land. In the event that a landing is thus imposed, the aircraft with its occupants may continue its flight after examination, if any.

A-10. In the event of involuntary landing in enemy or enemy-occupied territory, the wounded and sick, as well as the crew of the aircraft, shall be prisoners of war; medical personnel will be treated as prescribed in these Conventions.

SELF-DEFENSE AND DEFENSE OF PATIENTS

A-11. When engaging in medical evacuation operations, medical personnel are entitled to defend themselves and their patients. They are only permitted to use individual small arms.

A-12. The mounting or use of offensive weapons on dedicated medical evacuation vehicles and aircraft jeopardizes the protections afforded by the Geneva Conventions. These offensive weapons may include, but are not limited to machine guns, grenade launchers, hand grenades, and light antitank weapons.

A-13. Medical personnel are only permitted to fire in their personal defense and for the protection of the wounded and sick in their charge against marauders and other persons violating the Law of War.

ENEMY PRISONERS OF WAR/DETAINEES

A-14. Sick, injured, wounded enemy EPW, and detainees are treated and evacuated through normal medical channels but are physically segregated from US, allied, or coalition patients.

A-15. Personnel resources to guard EPW or detainee patients are provided by the echelon commander. Medical personnel do not guard EPW or detainee patients.

COMPLIANCE WITH THE GENEVA CONVENTIONS

A-16. The US is a party to the 1949 Geneva Conventions. Two of these Conventions afford protection for medical personnel, facilities, and evacuation platforms (to include aircraft on the ground). All HSS personnel should thoroughly understand the provisions of the Geneva Conventions that apply to medical activities.

CONSEQUENCES OF VIOLATIONS

A-17. Violation of these Conventions can result in the loss of protection afforded by them. Medical personnel should inform the tactical commander of the consequences of violating the provisions of these Conventions. The consequences may include—

- Medical evacuation assets subjected to attack and destruction by the enemy.
- Health service support capability degraded.
- Captured medical personnel becoming prisoners of war rather than retained persons. They may not be permitted to treat their fellow prisoners.
- Loss of protected status for medical unit, personnel, or evacuation platforms (to include aircraft on the ground).

PERCEPTION OF IMPROPRIETY

A-18. Because even the perception of impropriety can be detrimental to the mission and US interests, HSS commanders must ensure they do not give the impression of impropriety in the conduct of medical evacuation operations. For example, if a medical commander included in the unit's TSOP rules governing the use of automatic or crew-served weapons, it would give the impression that the unit possessed and intended to use these types of weapons. Under the provisions of the Geneva Conventions, medical units are only authorized individual small arms for use in the defense of the patients under their care and for themselves. Even though the unit does not possess these types of weapons, the entry in the TSOP could be misinterpreted and a case made that the commander intended to use these weapons in violation of the Geneva Conventions.

Appendix B
Legacy Units

Currently there are two units that are still active but will be transitioning out of the Army's inventory. They are the medical evacuation battalion and the medical company, AA (UH-1V).

SECTION I — MEDICAL EVACUATION BATTALION

B-1. The HHD, medical evacuation battalion, serves as the central manager of ground and air evacuation assets within the corps and EAC.

B-2. Assignment. The medical evacuation battalion is assigned to the MEDCOM in the EAC or to the medical brigade in the corps. It is normally further assigned to a medical group for C2.

Note. Under the MRI design, the medical group was replaced by the MEDBDE.

B-3. Air and ground ambulance companies assigned to the MEDCOM or MEDBDE are attached to the medical evacuation battalion for C2.

B-4. The basis of allocation is one medical evacuation battalion per a combination of three to seven of the following units:

- Medical companies, AA.
- Medical companies, ground ambulance.

B-5. Mission and capabilities.

- The mission of the medical evacuation battalion is to provide C2 of air and ground medical evacuation units within the AO. It tactically locates in the area where it can best control subordinate air and ground ambulance companies.
- The medical evacuation battalion is designed to focus on C2, planning, patient evacuation, subordinate unit support, and vehicle management. Specific capabilities are—
  - Command and control, planning and supervision of operations and training, and administration of a combination of air and ground ambulance companies.
  - Staff and technical supervision of aviation operations, safety, standardization, and aviation unit maintenance (AVUM)-level maintenance within the attached AA companies.
  - Coordination of medical evacuation operations and communications functions.
  - Coordination of logistics and service support to attached units.
  - Aviation medicine and unit-level AHS.
- This unit is dependent upon appropriate elements of the corps or ASCC for—
  - Personnel service support.
  - Health service support, to include hospitalization.
  - Mortuary affairs support.
  - Laundry, shower, and clothing repair.
  - Communications security equipment maintenance.
  - Military police support.

B-6. Organization and functions.

- Medical Evacuation Battalion (Figure B-1). The headquarters and headquarters detachment (HHD), medical evacuation battalion, is organized into a—
  - Battalion headquarters section.
  - Adjutant (US Army) section.
  - Intelligence Officer (US Army)/S3 section.
  - Supply Officer (US Army) section.
  - Detachment headquarters.
  - Treatment team.
- Battalion Headquarters Section. This section provides C2 of the assigned and attached air and ground ambulance companies. It also assists the commander on all military intelligence matters (to include the health threat), organization, training, operations, planning, personnel service support, and logistics support. Further, it provides information on the health of the command and aviation medicine expertise, as well as providing supervision over technical and flight aspects of administration, training, and safety within subordinate aviation units.
- Adjutant (US Army) Section. This section operates according to METT-TC and is the principle coordinating staff officer responsible for the delivery of human resources support. The S1 is responsible for the execution of all of the HR core competencies to include —PRM; PASR; PIM; R5 operations; casualty operations; EPS; and postal operations. It also includes MWR operations and HR planning and staff operations. The S1 also coordinates with elements of supporting agencies for finance, legal, religious, and administrative services.
- Intelligence Officer (US Army)/S3 Section. This section assists the S2/S3 officer in the execution of his duties and is capable of sustained 24-hour operations. This section remains abreast of the tactical situation and determines future medical evacuation requirements. It plans for ground and air evacuation operations, coordinates command post operations, and maintains the status of the AA units and plans for their employment. Further, this section maintains communications systems and nets, determines intelligence requirements, coordinates with movement control elements, and prepares orders and overlays.
- Supply Officer (US Army) Section. The S4 section assists the S4 officer in the execution of his duties. This section plans, coordinates, and supervises the requisitioning, receipt, storage, issue, and accounting for all classes of supply. Further, it monitors and keeps the commander informed on all matters pertaining to maintenance on assigned aircraft, ground vehicles, and medical equipment. This section also serves as the interface with the supporting medical battalion, logistics (forward/rear) for medical nonexpendable and durable item supply transactions.
- Detachment Headquarters Section. The detachment headquarters section provides, C2, administration, and logistics support for assigned personnel. It is also responsible for company supply and armament functions, food service operations, maintenance operations, and unit administration.
- Treatment Team. The treatment team provides unit-level AHS to assigned and attached elements collocated with the detachment headquarters and to adjacent units on an area support basis. The physician is a flight surgeon and provides staff assistance to the battalion commander on all matters pertaining to aviation medicine. The flight surgeon provides care and treatment for all assigned and attached aircrew members. This physician is dual-hatted as the battalion surgeon.

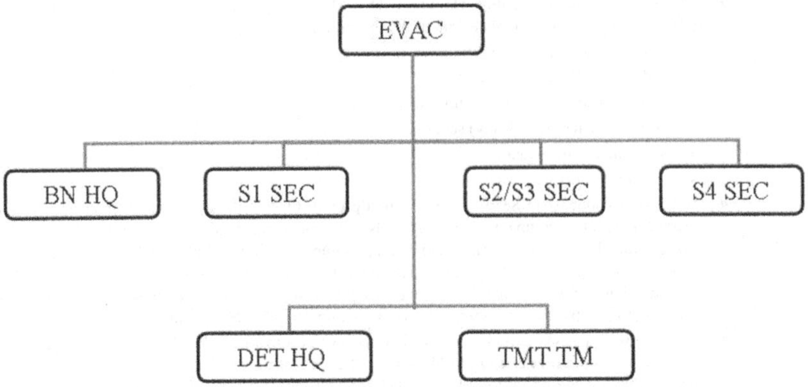

Figure B-1. Medical evacuation battalion

SECTION II — MEDICAL COMPANY, AIR AMBULANCE

B-7. The medical company, AA, provides aeromedical evacuation for all categories of patients consistent with evacuation precedence and other operational considerations. Medical evacuation is effected from as far forward as possible in the tactical AO to division- and corps-level MTFs.

B-8. Assignment.

- The medical company, AA, is normally assigned to the MEDCOM or MEDBDE and attached to the medical evacuation battalion for C2.
- The basis of allocation is one unit in support of each division or equivalent force supported. Further, one unit is in GS in the corps per two division or fraction thereof; or .333 units per separate brigades or ACRs.

B-9. Mission and capabilities.

- The mission of the medical company, AA, is to provide.
  - Aeromedical evacuation support within the theater, either DS to the divisions or GS to the corps.
  - Emergency movement of medical personnel, equipment, and supplies including whole blood, blood products, and biologicals.
- Specific capabilities of this unit are to operate on a 24-hour-a-day basis and evacuate patients based on operational capability (dependent on type of aircraft).
  - Operate 15 AAs (UH-60s). These ambulances are each capable of carrying six litter patients and one ambulatory patient, or seven ambulatory patients, or some combination thereof. Single patient lift capability is 90 litter patients, or 105 ambulatory patients, or some combination thereof. In-flight medical treatment and surveillance of patients is provided by a flight medic.
  - Operate 15 AAs (UH-1H/V). These ambulances are capable of carrying six litter, or nine ambulatory patients, or some combination thereof. Single patient lift capability is 90 litter, 135 ambulatory, or some combination thereof. In-flight medical treatment and patient surveillance are provided by a flight medic.
- Provide internal/external load capability for the movement of medical personnel and equipment.

- Perform AVUM on all organic aircraft and organizational maintenance on all organic avionics equipment. It also performs unit-level maintenance on all organic equipment less medical.
- Provide air crash rescue support, less fire suppression.
- Operate as an area support MEDEVAC section and three FSMTs to provide flexibility in supporting division, brigade, or brigade TF equivalent operations.

B-10. This unit is dependent upon:

- Support elements of corps or ASCC for—
  - Financial management, legal, and religious support.
  - Human resources support.
  - Logistics.
  - Health service support, to include medical supply and equipment.
  - Food service support.
  - Communications security equipment maintenance.
  - Mortuary affairs support.
  - Military police support.
  - Laundry, shower, and clothing repair.
  - Engineer support for heliport/landing strip construction and maintenance.
- The supporting aviation intermediate maintenance (AVIM) organization for AVIM support.

B-11. Organization and functions.

- The medical company, AA (Figure B-2), is organized into a—
  - Company headquarters.
  - Flight operations platoon consisting of a platoon headquarters, a flight operations section, and an airfield service section.
  - Aircraft maintenance platoon consisting of a platoon headquarters, a component repair section, and a maintenance section.
  - Air ambulance platoon consisting of a platoon headquarters, an area support MEDEVAC section, and three FSMTs.

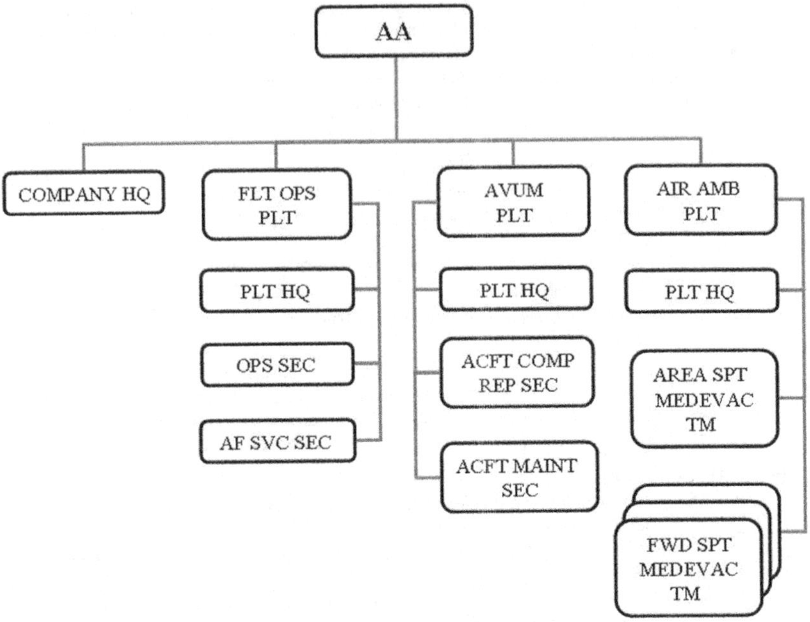

Figure B-2. Medical company, air ambulance

Appendix C

Tactical Standing Operating Procedures

This appendix provides a sample TSOP for an MMB engaged in medical evacuation missions. It should not be considered as all-inclusive. It may be supplemented with the information and procedures required for operating within a specific command or special operation.

PURPOSE

C-1. The TSOP prescribes policy, guidance, and procedures for routine support of tactical operations of a specific unit. It should cover broad areas of unit operations, but be sufficiently detailed to provide newly assigned personnel with the guidance required for them to perform their mission. A TSOP may be modified by the TSOP and OPLANs/OPORDs of higher headquarters. It applies to a specific unit and all subordinate units assigned and attached. Should a TSOP not be in conformity with the TSOP of the higher headquarters, the higher headquarters TSOP governs. The TSOP is periodically reviewed and updated as required.

FORMAT FOR THE TACTICAL STANDING OPERATING PROCEDURE

C-2. There is not a standard format for all TSOPs; however, it is recommended that a unit TSOP follow the format used by its higher headquarters. The TSOP can be divided into sections (specific functional areas or major operational areas) and further subdivided into annexes. An annex can be further subdivided into appendixes and then into tabs. Appendixes can be used to provide detailed information on major subdivisions of the annex, and tabs can be used to provide additional information (such as report formats or area layouts) addressed in the appendix.

C-3. Regardless of the format used, the TSOP follows a logical sequence in the presentation of material. It should discuss the chain of command, major functions and staff sections of the unit, operational requirements, required reports, necessary coordination with higher and subordinate elements for mission accomplishment, programs (such as command information, preventive medicine measures [PMM], and combat stress control) and other relevant topics.

C-4. Pagination of the TSOP can be accomplished by starting with page 1 and numbering the remaining pages sequentially. If the TSOP is subdivided into sections, annexes, appendixes, and tabs, a numbering system that clearly identifies the location of the page within the document can be used. Annexes are identified by letter and are listed alphabetically. Appendixes are identified by numbers and arranged sequentially within a specific annex. Tabs are identified by a letter and are listed alphabetically within a specific appendix. After numbering the initial sections using the standard numbering system (sequentially starting with page 1 through to the end of the sections), number the annexes and their subdivisions. They are numbered as the letter of the annex, the number of the appendix, the letter of the tab, and the page number. For example, page 4 of Annex D is written as "D-4"; page 2 of Appendix 3 to Annex D is written as "D-3-2"; page 5 of Tab A to Appendix 3 of Annex D is written as "D-3-A-5." This system of numbering makes the pages readily identifiable as to their place within the document as a whole.

C-5. In addition to using a numbering system to identify specific pages within the TSOP, descriptive headings should be used on all pages to identify the subordinate elements of the TSOP.

- The first page of the TSOP should be prepared on the unit's letterhead. The remaining pages of the major sections should include the unit identification in the upper right-hand corner of the page (for example: "Multifunctional Medical Battalion").
- A sample heading for an Annex is: "ANNEX B (Command Post) to Multifunctional Medical Battalion."

- A sample heading for an Appendix to Annex B is: "APPENDIX 3 (Command Post Security) to ANNEX B (Command Post) to Multifunctional Medical Battalion."
- A sample heading for a Tab to Appendix 3 to Annex B is: "Tab A (Tactical Operations Center [TOC] Security) to APPENDIX 3 (Command Post Security) to ANNEX B (Command Post) to Multifunctional Medical Battalion."

C-6. As the TSOP is developed there may be an overlap of material from one annex to another. This is due in part to similar functions that are common to two or more staff sections. Where overlaps occur, the material presented should not be contradictory. All discrepancies will be resolved prior to the authentication and publication of the TSOP. The TSOP is authenticated by the unit commander.

Sample Tactical Standing Operating Procedures (Sections)

C-7. The information contained in this paragraph can be supplemented. It is not intended to be an all-inclusive listing. Different commands will have unique requirements that need to be included. The first section of the TSOP identifies the specific unit/headquarters that developed it.

- Scope–This document establishes and prescribes procedures to be followed by the unit identified and its assigned, attached, or OPCON units/elements.
- Purpose–This document provides policy and guidance for routine tactical operations of this headquarters and its assigned, attached, or OPCON units.
- Applicability–Except when modified by policy guidance, TSOP, or OPLANs/OPORDs of the higher headquarters, this document applies to this unit and to all units assigned, attached, or under OPCON for combat operations. These orders, however, do not replace judgment and common sense. In cases of nonconformity, the document of the higher headquarters governs. Each subordinate element will prepare a unit TSOP, conforming to the guidance herein.
- References–This paragraph can include any pertinent regulations, policy letters, higher headquarters TSOP, or any other appropriate documents.
- General information–This paragraph discusses the required state of readiness of the unit; primary, secondary, and contingency missions; procedures for operating within another command's AO; and procedures for resolution of conflicts with governing regulations, policies, and procedures.

C-8. The second section of the TSOP discusses the specific organization.

- Organization–This paragraph furnishes specific information concerning the authority for establishing the unit, such as applicable MTOE or other staffing documentation.
- Succession of command–The guidance for determining the succession of command is discussed.
- Task organization–Task organization is contingent on the mission and will be approved by the headquarters ordering deployment.
- Joint, multinational, and interagency operations–This paragraph provides guidance on any issues concerning C2 and TF organization in joint, multinational, and interagency operations. It also provides guidance on liaison officer requirements.
- Organizational charts–Contained in Annex A.

C-9. The third section of the TSOP discusses the unit functions.

- Battalion headquarters–This paragraph discusses the C2 of the assigned, attached, and OPCON units.
- Headquarters detachment–This paragraph discusses the functions of the headquarters detachment element, such as supervising movements, internal arrangements, area layout, physical security, and operation of the headquarters and staff.
- Attached, assigned, and OPCON units–This paragraph discusses the missions and functions of these units (such as providing medical evacuation of patients, emergency movement of blood and blood products, biologicals, medical logistics, and transportation of medical personnel and equipment).

- Staff responsibilities–This paragraph lists the unit's key personnel and their duties as prescribed in FM 5-0 and any command-specific duties.

C-10. The fourth section of the TSOP pertains to staff operations and is subdivided into annexes.

Sample Tactical Standing Operating Procedures (Annexes)

C-11. Annexes are used to provide detailed information on a particular function or area of responsibility. The commander determines the level of specificity required for the TSOP. Depending upon the complexity of the material to be presented, the annex may be further subdivided into appendixes and tabs. If the annex contains broad guidance or does not provide formats for required reports, paragraphs may be used. The annex should not require further subdivision. However, as the material presented becomes more complex, prescribes formats, or contains graphic materials, the annex may require additional subdivision. This paragraph discusses the subdivision of the annex by appendixes. It does not contain examples of subdividing the information presented in the appendixes into tabs. Applicable references such as ARs, FMs, and technical manuals should be provided in each annex. The number of annexes and their subdivisions presented below are not to be considered as an all-inclusive listing. Different commands will have unique requirements; therefore, supplementation of the information presented is permitted.

C-12. Annex A. (Organizational Charts.)

C-13. Annex B. (Command Post.)

- General. The division may operate main and/or tactical command post CPs (TAC CP) depending on the mission and tactical situation. Characteristics of the CPs include—
  - Main CP–Normally located in the sustainment area. Personnel staffing is tailored to provide planning, coordinating, and C2 of assigned/attached/OPCON units. The area location for the main CP is selected by the battalion S3; the exact site is designated by the commander in coordination with the XO and battalion S3. The XO designates work areas within the main CP. The commander, HHD, serves as the headquarters commandant. His duties include coordinating for and obtaining construction, maintenance, and logistical services and support for the CP.
  - Tactical command post CP–Normally established in the objective area or at an intermediate staging base (ISB). The ISB is the forward designated staging base at which final preparation for assault operations are controlled and is generally a safe area for support operations. The TAC CP is tailored to extend the commander's span of control and to better assess the situation on the ground and coordinate the arrival and subsequent activities of AHS organizations within the AO.
- Battalion tactical operations center.
  - Definition–The TOC is the command element of the battalion containing communications and personnel required to command, control, and coordinate HSS operations. The TOC is located within a secure, controlled area whether at a main CP or TAC CP.
  - Purpose–The purpose of the TOC is to provide the commander, in a secure environment, current evaluated information and recommendations concerning HSS/FHP operations.
  - Responsibilities–The battalion commander has overall supervision and control over the TOC. The battalion XO operates the TOC and has primary staff responsibility in the absence of the commander.
  - Operations–The TOC will operate on a 24-hour basis. It is principally staffed by each primary staff section furnishing necessary manpower as required. Secure and nonsecure telephone communications connect the TOC to other staff sections within the CP area. Access to the TOC is strictly controlled by means of an access roster, and if available, security badges. Only essential personnel and authorized visitors are allowed to enter. Each staff section will maintain an SOP on the organization and operation of its element. All elements within the TOC will, when appropriate, maintain a current situational map of their specific operations. Discussion and portrayal of tactical plans outside of the security area is prohibited.

  - Composition of the TOC–This is a listing of those personnel comprising the TOC. It normally includes the commander, XO, command sergeant major, principal staff members, and other specific staff members such as the S3 (air) or the battalion aviation maintenance officer.
  - Tactical operations center configuration–This is a schematic representation of the physical layout of the TOC.
- Responsibilities for the main CP services–The HHD commander's main mission is to support the headquarters, medical evacuation battalion. He has the personnel and resources assigned to facilitate this mission. He plans and coordinates for providing shower, laundry, transportation, maintenance, and other required services. He is also responsible for establishing and maintaining security; supplying fuel and other POL products; establishing the support area for the main CP (orderly room, supply room, motor pool, billets, and dining facilities); providing and operating generators to meet electrical requirements.
- Camouflage–This appendix discusses what camouflage procedures are required, to include type and amount of required camouflage materials (such as nets and shrubs); display of the Geneva Conventions distinctive emblem on facilities, vehicles, and aircraft on the ground (STANAG 2931); and other pertinent policies, guidance, or procedures.
- Message center–This establishes procedures for the handling of classified messages; provides delivery and service of IMMEDIATE and FLASH messages to the appropriate staff section; establishes procedures for preparing outgoing messages; and establishes a delivery service to the servicing message center for transmission of outgoing messages.

C-14. Annex C. (Administration and Human Resources.) This annex outlines procedures relating to administrative and human resources matters and associated activities.

- Personnel accountability and strength reporting.
  - Personnel daily summary (PDS)–This provides the procedures for filling out and submitting a daily personnel status report. The instructions may include requirements for encrypting the report prior to transmission, and specific guidance on time of submission, corrections, or other administrative requirements.
- Personnel readiness management.
  - Maintain the personnel readiness of unit personnel to include projected gains and losses.
- Personnel information management.
  - Maintain 100 percent accountability of the unit's military personnel files.
  - Update records directly through the officer record brief/enlisted record brief.
  - Reception, replacement, RTD, R&R, and redeployment.
  - Track the flow and duty status of the five categories of Soldiers into and out of the unit.
  - Update deployed theater accountability software database as required.
  - Coordinate the movement of the five categories of Soldiers.
  - Replacements–Individual replacements will not be readily available during the initial phases of operations. The S1 will automatically initiate replacement requests for personnel who are reported on the PDS report as WIA, missing in action (MIA), or killed in action (KIA).
- Essential personnel services.
  - Issue identification cards and tags.
  - Personnel actions–All personnel actions will be channeled through the S1. Company XOs and 1SGs will be the company points of contact. Actions will be handled expeditiously and meet suspense dates (tactical situation permitting).
  - Efficiency reports–This paragraph provides pertinent information on the completion and submission of these reports.
  - Award recommendations–This paragraph delineates the responsibilities for and guidance concerning submitting recommendations for awards and for scheduling and conducting award ceremonies.
  - Promotions–This paragraph discusses the procedures for submitting recommendations for promotion and scheduling and conducting promotion ceremonies.

- Correspondence–All correspondence addressed to higher headquarters will be submitted through the S1. Requirements for submission, preparation, and approval are also provided.

- Casualty operations.
  - Casualty Feeder Card–This report is submitted on DA Form 1156 (Casualty Feeder Card). Instructions on the completion of the form and submission requirements are included.
  - Tracking of casualties will require coordination with CLT and CSH.
- Postal operations.
  - This includes hours of operation and services available.
- Morale, welfare, and recreation operations.
  - Sporting activities and morale and welfare activities.
  - American Red Cross.
  - This includes hours of operation and services available.
  - Post Exchange services–This includes hours of operation and availability.
- Management, and HR planning and staff operations.
  - Preparation of the personnel estimate and personnel portion of plans or OPORD.
  - Coordinate external HR support with HR operations cell within the sustainment brigade.
- Other administrative support.
  - Financial management–This service includes disbursements and currency control, payday activities, currency conversion, check cashing, and the appointment of Class A agents.
  - Legal services–Information and specific guidance on administrative boards, courts-martial authority and jurisdiction, legal assistance, and general services should be provided.
  - Religious activities–Religious activities include chaplain support, services available from different faiths, schedule of services, and hospital visitations.
  - Distribution–Pick up and delivery schedules and any command-specific issues/procedures are provided.
- Graves registration (GRREG)–Commanders at all levels are responsible for the recovery, identification, and evacuation of US dead. This section discusses the responsibilities and procedures for unit-level GRREG activities for assigned and attached personnel.

Note. This activity is for unit members only. Medical evacuation units do not accept nor transport the remains of nonunit members. Remains will not be transported in ambulances under any circumstances.

  - Responsibilities–This paragraph discusses both unit and battalion requirements.
  - Disposition–Specific guidance on procedures, GRREG collection points, transportation requirements, and the handling of remains are provided.
  - Hasty burials–Specific requirements for conducting hasty burials, marking, and reporting of grave sites are included.
  - Personal effects–Guidance on accounting for personal effects and requirements for burial should a hasty burial be required is contained in this paragraph.
  - Disposition of civilian and EPW remains–The local civilian government is responsible for the burial of remains of its citizens. The remains of EPW should be accomplished in separate cemeteries from US and allied personnel. If this is not possible, separate sections of the same cemetery should be used.
  - Contaminated remains–This paragraph discusses handling and disposition requirements (to include protective clothing), procedures, and marking and reporting of burial site.
- Public information–This annex contains procedures for obtaining approval on the public release of information to include the hometown news release program. In overseas locations, specific guidance on the interaction of unit members with the news media should be provided.

- Maintenance of law, order, and discipline–This appendix should provide applicable regulations, policy, and command guidance on topics such as serious incident reports, notifications, and submission formats, straggler control, confinement of military prisoners, and EPW and detainees (also discussed below).
- Enemy prisoners of war–This appendix discusses the unit responsibilities of EPW and detainees surrendered to the medical unit. These procedures do not pertain to EPW and detainee patients captured by other units. Medical personnel do not guard, search, or interrogate EPW and detainees while in the AHS; guards are provided by nonmedical personnel designated by the tactical commander for these duties. Until EPW and detainee personnel can be evacuated to an EPW collection point, medical personnel should remember and enforce the basic skills: segregate, safeguard, silence, secure, speed, and tagging. (The speed portion of evacuating EPW and detainees to designated collection points is of paramount importance to medical units.)

Note. The treatment of EPW and detainees is governed by international and US law and the provisions of the Geneva Conventions. Personnel should be aware of these requirements and have ready access to the applicable regulations and policy guidance.

- Allied, coalition, and interagency personnel–This paragraph should provide guidance on the eligibility of allied, coalition, and interagency personnel for evacuation by Army ambulance and destination locations.

C-15. Annex D. (Intelligence and Security.) This annex pertains to intelligence requirements and procedures and OPSEC considerations.

- Intelligence–The S2 has the responsibility of collecting information to assist the commander in reaching logical decisions as to the best COA to pursue. Priority intelligence requirement (PIR) may include the location, type, and strength of the air defense threat; location, type, and strength of enemy air defense radars; known or suspected CBRN activity; the health threat; and issues which the commander considers to be PIR. In addition to PIR, the commander's critical information requirements (CCIR) are also considered.
- Intelligence reports–The S2 is responsible for disseminating all applicable estimates, analyses, periodic intelligence reports, and intelligence summaries generated within the battalion or received from higher headquarters. Information on submission of reports and suspense on intelligence products and reports should also be addressed in this appendix.
- Weather data–Weather conditions can disrupt ground evacuation efforts and planners must recognize their impact when coordinating AA operations.
- Reports–These include information acquired during the routine performance of duty by pilots, ambulance drivers, and medics.
- Counterintelligence.
  - Camouflage–When ordered or directed by the tactical commander, all units will initiate and continually strive to improve camouflage operations of positions, vehicles, and equipment. Noise and light discipline is emphasized at all times.
  - Communications security–These measures will be enforced at all times. The specific requirements and considerations are included.
  - Signs and countersigns–This paragraph lists the signs and countersigns to be used during hours of darkness. It also includes reporting requirements and procedures if the sign/countersign is lost or compromised.
  - Signal operating instructions.
  - Document security–This paragraph discusses the procedures for marking and safeguarding classified material, both work documents and completed documents. (Reporting requirements in the event of compromise are also included.)
- Captured or surrendered personnel, equipment, supplies, and documents–This appendix provides specific guidance on the handling of captured personnel, equipment, supplies, and

documents. The disposition of captured medical equipment and supplies is governed by the Geneva Conventions and is protected against intentional destruction.

- Security–This appendix discusses weapons security and checks, safeguarding of patient weapons, aircraft security, signal operating instructions (communications) security, TOC security, Sensitive Item Status Report, and escape and evasion.

C-16. Annex E. (Operations.) This annex establishes procedures for S3 operations within the medical evacuation battalion, and provides a basis for standardization of medical evacuation operations in a tactical environment. It is essential that these procedures be standardized to ensure common understanding, facilitate control and responsiveness, and enhance mission accomplishment. Information on readiness levels, threat levels, warning levels, camouflage, security, area damage control, operations, and leader checklists is also included.

- Operational situation report–Requirements for preparation, format, and submission of this report are discussed in this appendix.
- Operations security–This appendix provides the guidance and procedures for secure planning and conduct of combat operations.
  - Priority intelligence requirements and security classification.
  - Responsibilities–The commander is ultimately responsible for denying information to the enemy. The S3 is responsible to the commander for the overall planning and execution of operations. The S2 has the principle staff interest in assuming the required degree of OPSEC and has the primary staff responsibility for coordinating the efforts of all other staff elements in this regards. The OPSEC officer is responsible for the preparation of the PIR and providing classification guidance. Additionally, the OPSEC officer identifies the priorities for OPSEC analysis and develops OPSEC countermeasures. The S2 coordinates with the S3 in planning OPSEC.
  - Hostile intelligence threat–The different sources of intelligence (human intelligence, signal intelligence, and so forth) are discussed.
  - Operational security program–This includes physical security, information security, and signal security.
  - Document downgrading/declassification and classification authority.
- Terrorist threat and countermeasures–This appendix provides guidance on security measures to counter the terrorist threat, both on an individual and unit level.
- Minefield threat–This appendix provides guidance on the potential of minefields being established in the AO and delineates reporting requirements. (Specific minefield extraction techniques are contained in Annex L of this TSOP.)
- Operations security and countermeasures–This appendix discusses camouflage, light discipline, noise discipline, physical security, information security, and signal security.
- Unit location update–his appendix provides timely information on the location of main and forward CPs, location of subordinate unit CPs, location of helipads, and location of POL points.
- Flight operations–This appendix provides information concerning Army aviation LZs throughout the corps area and division areas; required reports; airdrop information; and mission debriefing.
- Communications-electronics–This appendix establishes communications policies, procedures, and responsibilities for the installation, operation, and maintenance of CE equipment.
  - Responsibilities of the battalion CE noncommissioned officer in charge.
  - Concept of operations.
  - Command and control.
  - Radio communications.
  - Radio teletypewriter communications.
  - Message/communications center service.
  - Message handling procedures.
  - Wire communications.

- Switchboard operations.
- Communications security and operations.
- Intelligence security.
- Meaconing, intrusion, jamming, and interference (MIJI) reporting and electronic communications countermeasures.
- Security violations–This prescribes procedures for reporting any event of action which may have jeopardized the security of communications.
- Destruction of material.
- Daily shift inventory.
- Physical security.
- Transmission security.
- Security areas–This discusses access procedures and rosters, access approval requirements, and prohibited items.
- Inventory of classified materials.
- Communications security officers and custodians–The appointment procedures, orders requirements, and duties of personnel are described.
- Safety–This discusses requirements for grounding, handling, and storing COMSEC equipment.
- Power units.

- Rear battle responsibilities–This appendix discusses rear battle responsibilities, task organization, and support for reaction forces.

C-17. Annex F. (Chemical, Biological, Radiological, and Nuclear Operations.) This annex prescribes the policy, guidance, and procedures for CBRN defensive operations.

- Responsibilities.
- Chemical, biological, radiological, and nuclear reporting requirements and procedures.
  - Contamination avoidance.
  - Protection–Protection pertains to those measures each Soldier must take before, during, and after a CBRN attack to survive and continue the mission.
  - Decontamination–This discusses equipment requirements, procedures, and types of decontamination (such as hasty). This paragraph also provides guidance on deploying clean ambulances into a contaminated area.
  - Mission-oriented protective posture–This appendix provides guidance on the garments required for the different MOPP levels and identification procedures for personnel in MOPP.
  - Radiation exposure guidance–Establishes OEG. Discussion in this appendix includes determining what constitutes a radiological hazard, prescribing acceptable limits of potential casualty producing doses of radiation, minimizing exposure, and protecting against electromagnetic pulses.
  - Masking and unmasking procedures.
  - Radiological monitoring and survey operations.

C-18. Annex G. (Logistics.) This annex establishes logistics procedures for subordinate units when operating in a field environment.

- Nonmedical supply and services–A discussion of applicability, responsibilities, policy, classes of supply, requisition and delivery procedures, hours of operation, and other supply-relevant topics and available services (such as laundry and bath) can be addressed in this section.
- Medical logistics–This paragraph discusses supply and resupply procedures for Class VIII; policies and procedures for the backhaul of emergency Class VIII on medical vehicles; and hours of operation for the supporting medical logistics facility.
- Food service–This appendix discusses responsibilities, hours of operation, Class I supplies, sanitation requirements, layout of field kitchen, fuel storage, maintenance, safety precautions,

and administration, such as headcounts, required reports, shift schedules, meals, ready-to-eat (MRE), and inspections/visits of subordinate unit kitchens.

- Transportation/movement requirements–This appendix may cover the following areas: deployability data, applicability; responsibilities; policies on speed, vehicle markings, transporting flammable materials, transporting ammunition and weapons, and so on; convoy procedures; safety; and accident reporting.
- Fire prevention and protection–Guidance on the use of the tent stoves, flammable materials, use of cigarettes, matches, and lighters, electrical wiring and appliances, safety of tents and occupants, spacing of tents, stoves and ranges, installation of British thermal unit (BTU) heaters, and fire fighting equipment are presented in this appendix.
- Field hygiene and sanitation–This appendix provides uniform guidance and procedures for the performance of functions related to field hygiene and sanitation. It includes policies, communicable disease control, field water supply, water containers and cans, water purification bags, food sanitation, latrines, liquid waste disposal, and garbage and rubbish disposal.
- Conventional ammunition down/upload procedures–This appendix delineates responsibilities and provides guidance and procedures for the requisition, storage, and distribution of ammunition and weapons, reporting requirements, and safety.
- Petroleum, oils, and lubricants accounting.
- Medical logistics support. The medical logistics concept of operations, requisition and distribution procedures, accountability, and reports are provided in this appendix.
- Maintenance–This appendix includes information on the maintenance requirements of the battalion and the location and hours of operation of maintenance units and collection points. Maintenance for medical equipment, vehicles, aircraft, and communications and other categories of equipment are discussed.

C-19. Annex H. (Safety.) This annex establishes minimum essential safety guidance for commanders and units. It includes risk assessment, accident reporting, safety measures, emergency procedures, vehicle safety, ground guide procedures, fire prevention and protection, antennas, climate, survival training, animal and arthropods hazards, personal protective measures, hearing conservation, carbon monoxide poison, helicopter safety, and refueling operations.

C-20. Annex I. (Civil-Military Operations.) This annex discusses the participation in CMO. Medical elements are often involved in CMO, humanitarian and civic assistance, and disaster relief operations. The activities which may be covered include providing—

- Direct support medical evacuation for the operation.
- Guidance on developing a medical evacuation system in a HN.
- Training to a HN's medical infrastructure.

C-21. Annex J. (Mass Casualty Situations.) This annex discusses the procedures for providing medical evacuation support to MASCAL situations, to include coordination for nonmedical transportation assets and the augmentation of these assets with medical personnel to provide en route patient care. It discusses evacuation capabilities of US forces.

C-22. Annex K. (Minefield Operations.) This annex discusses notification procedures, training requirements, and techniques for clearing pathways to casualties, marking mines, and extracting personnel and equipment.

C-23. Annex L. (Religious Support.) This annex discusses procedures for processing unit requests for religious support and coordination of religious support to casualties, declared MASCALs, mortuary affairs support, unit memorial ceremonies/services, EPW, detainees, religious support logistics and supplies, and reports and administration.

Appendix D

Example of the Medical Evacuation Plan and of the Operations Order

SECTION I — EXAMPLE FORMAT OF THE MEDICAL EVACUATION PLAN

D-1. Figure D-1 provides an example of the medical evacuation plan.

(Classification)

Headquarters
Place
Date, time, and zone

Time Zone Used Throughout the OPLAN: *The time zone used throughout the OPLAN (including attachments) is the time zone applicable to the operation. Operations across several time zones use ZULU time. Place the classification and short title of the OPLAN at the top of the second and any subsequent pages of the base OPLAN.*

Task Organization: *List the number and coordinates of medical evacuation units here or in a trace or overlay. If you do not list units here, omit this heading.*

1. SITUATION. *(State the general medical evacuation factors affecting support of the operation. Include any information essential to understanding the current situation as it influences medical evacuation.)*

a. Enemy forces. *(Refer to the intelligence annex if it has been published or is to be published. List information about the composition, disposition, location, movements, estimated strengths, and identification of enemy forces. List enemy capabilities that could influence the medical evacuation mission.)*

(1) Strength and disposition. *(The size of the opposition force and its placement on the battlefield is important during the planning process for medical evacuation operations. When evacuation routes are selected, caution must be exercised to ensure medical evacuation assets are not compromised by going through enemy-held territory or by being ambushed by isolated pockets of resistance. Further, this information is vital in determining if medical evacuation assets [both ground and air] will require a security escort provided by movement and maneuvers forces before entering areas of the battlefield.)*

(2) Combat efficiency. *(Information on actual combat units or other forces, their training status, and their level of expertise and experience can be identified here. The level of AHS training and the development of a health care delivery system can also be included.)*

(3) Capabilities. *(This paragraph should discuss the potential capability to interfere with or disrupt medical evacuation operations.)*

(4) Logistics situation. *(This can include information on how well supplied the enemy/ opposition force is with food, clothing, or other vital logistics factors. It may also include the financial backing and availability of future support from outside individuals/groups/nations.)*

(5) Weapons. *(This should include a discussion of the enemy's weapons that present the greatest threat to air and ground evacuation personnel and vehicles. Ground assets are more likely to face a small arms threat, rocket propelled grenades (RPGs), and improvised explosive devices (IEDs) while performing their mission; air assets are vulnerable to surface-to-air weapons and when on the ground to the same threats as ground elements.)*

b. Friendly forces. *(List pertinent information concerning friendly forces that might influence the medical evacuation mission.)*

(1) Strength and disposition. *(This should include all forces [US, allied, coalition, and HN] and should be maintained on overlays. It may also include liaison officers, interpreter support, and coordination requirements.)*

(2) Combat efficiency. *(The health of the command has a significant impact on this factor. Additionally, when there are significant numbers of DNBI casualties, the medical evacuation workload will increase.)*

(3) Present and projected operations. *(Medical evacuation planners must be familiar with plans for current and projected operations. A risk management assessment of current and projected operations is conducted. Casualty collecting points and AXPs must be designated during the planning process. Medical evacuation planners must also be able to anticipate changing requirements to ensure continuous medical evacuation is available to the supported force.)*

(4) Logistics situation. *(Since medical evacuation vehicles conduct emergency resupply of forward deployed medical units, the medical evacuation planner must maintain visibility of the current medical logistics situation.)*

(5) Sustainment area protection plan. *(Medical units play an important role in the sustainment area protection plan and should be included in the planning process.)*

(6) Weapons. *(Medical units only have small arms for self-defense and defense of their patients. Medical vehicles and aircraft should not carry any automatic or crew-served weapons; to do so would result in the loss of the protections afforded under the Geneva Conventions for the unit, medical personnel, and patients under its care.)*

(7) Health of the command.

(a) Acclimation of troops. *(Medical evacuation crews require acclimation when introduced into an AO where the temperature range and elevation are different from their home station. Medical evacuation operations involve heavy lifting and may require extended periods of time to complete. To reduce the risk of injury to litters bearers and to facilitate the evacuation effort, medical and nonmedical personnel engaged in these operations should be acclimated to the AO and a work/rest schedule should be developed and implemented. Air ambulance crew rest requirements impact aircraft availability and must also be considered.)*

(b) Presence of disease. *(The presence of disease impacts medical evacuation operations in two areas. The presence of disease in the AO [usually at subclinical levels in the native population] contributes to the incidence of disease manifesting itself within the supported force. As DNBI rates increase, so do the requirements for medical evacuation. Medical evacuation personnel are also susceptible to the endemic and epidemic diseases within the AO and/or multinational force. High rates of DNBI for medical personnel will adversely impact the medical evacuation capability.)*

(c) Status of immunizations and/or chemoprophylaxis. *(Commanders must ensure appropriate measures are taken to protect their Soldiers from DNBI. The records of replacement personnel need to be screened to ensure all required immunizations have been received and appropriate chemoprophylaxis/barrier creams for the AO initiated/provided.)*

(d) Status of nutrition. *(The nutrition status of the troops involved impacts on the susceptibility to disease and environmental injuries, morale, and fatigue. The medical evacuation commander must also ensure that his evacuation crews carry sufficient supplies of MREs to sustain themselves and their patients should delays in evacuation be experienced [such as delays resulting from shifts in the battle or inclement weather].)*

(e) Clothing and equipment. *(Special requirements for clothing and equipment to support a particular operation should be obtained prior to the beginning of the operation [for example, cold weather clothing for extreme cold weather operations; additional blankets for patients being transported (either to add padding or warmth); or block and tackle equipment for medical evacuation operations conducted in mountainous terrain or in urban areas].)*

(f) Fatigue. *(Mandatory work/rest schedules and sleep plans for air crews and ground evacuation personnel must be developed and implemented.)*

(g) Morale.

(h) Status of training. *(This can include any specialized training required for the conduct of a specific operation [for example: helicopter crews require deck-landing qualifications to perform shore-to-ship evacuation with US Navy ships or orientation to the social, political, economic, religious, and ethnic issues of a HN or supported population, or training nonmedical personnel in the proper techniques for carrying litters].)*

c. Environment.

(1) Terrain. *(This paragraph should discuss any aspects of the terrain that will either hinder or enhance the execution of the evacuation mission. It should discuss both natural and man-made terrain, as medical evacuation urban areas can pose significant challenges not found on a natural battlefield. Medical personnel must be able to recover injured and wounded Soldiers from below ground and from upper levels of man-made structures. If UO are planned, armor ambulances provide added protection for medical personnel and their patients; augmentation should be requested if armor ambulances are not organic to the unit. If additional resources will be required to accomplish evacuation due to impassable or difficult terrain [such as in mountain operations where additional litter bearers and medical personnel are required], it should also be addressed here. Further, the type of terrain to be traversed, such as rugged mountain or jungle swamps may require a patient [who would otherwise be ambulatory] to be transported on a litter until easier terrain is encountered.)*

(2) Weather. *(This should include a discussion of current weather conditions and seasonal variants. Weather conditions impact both ground and air evacuation operations; however, the most significant impact may be on AAs as severely inclement weather can ground all aircraft. It should also discuss the impact that the weather has on the terrain [such as rivers being frozen in winter or tundra becoming impassable in spring]. The climate may pose problems with acclimation as well as place additional requirements to sustain personnel during evacuation on evacuation assets [litter evacuation in the mountains in extreme cold weather operations may require warming tents and the capability to sustain the patient during nighttime when evacuation is difficult].)*

d. Civil considerations. Dislocated civilian population, EPW, and detainees. *(Dislocated civilians fleeing an area of hostilities can clog roadways leading away from the conflict area. This congestion on road networks may make evacuation along these routes almost impossible. The establishment of camps to sustain these categories of personnel may interrupt the road network requiring detours and lengthened evacuation times. Injured, ill, or wounded EPW or detainees are evacuated using the same evacuation means but are segregated from US, allied, or coalition patients. Coordination for nonmedical guards for EPW/detainee* *patients being evacuated through medical channels must be accomplished with the echelon commander. A determination of eligible beneficiaries for evacuation should also be developed and disseminated.)*

2. **MISSION.** *(Statement of the medical evacuation mission.)*

3. **EXECUTION.**

a. **Concept of operations.** *(The concept of operations describes how the commander sees the actions of subordinate units fitting together to accomplish the mission. Commanders ensure that their concept of operations is consistent with their intent and that of the next two higher headquarters. The concept of operations describes any other details the commander considers appropriate to clarify the concept of operations and ensure unity of effort. If the integration and coordination are too lengthy for this paragraph, they are addressed in the appropriate annexes. When an operation involves two or more clearly distinct and separate phases, the concept of operations may be prepared in subparagraphs describing each phase.)*

b. **Coordinating instructions.** *(List only instructions applicable to two or more units and not covered in unit SOPs.)*

c. **Patient estimates.** *(Indicate rates and numbers by type unit/division/corps/EAC.)*

(1) **Number of patients anticipated.** *(The anticipated number of patients affects the number and type of medical evacuation resources required.)*

(2) **Distribution within the area of operations.** *(Distribution within the AO is an important consideration because varying types of terrain will generate different requirements for support. Operations conducted in mountainous terrain rely heavily on manual and litter evacuation techniques, are labor intensive, and require more time to complete. Differences in road composition, traffic density, and abundance/absence of paved roads will impact differently on the type of support required. Further, if the distance between Role 2 and Role 3 MTFs exceeds the capability of US Army AAs, the use of the USAF aircraft should be anticipated and preplanned. Coordination with the USAF is required.)*

(3) **Distribution in time during the operation (evacuation time).** *(As the battle progresses over time and space, medical evacuation assets may be reallocated to support the units in contact. The medical evacuation plan must be sufficiently flexible to enable the commander to shift resources as required.)*

(4) **Areas of patient density.** *(Areas with a high patient density may require augmentation of the medical evacuation assets supporting that location.)*

(5) **Possible mass casualties.** *(MASCAL situations should be anticipated when possible. Evaluation of the general threat and health threat in the AO can facilitate the medical evacuation planner in forecasting evacuation requirements. When dealing with MASCAL situations, the use of nonmedical transportation platforms should be included. Augmentation of these vehicles by medical personnel to provide en route care or CLS to provide first aid will assist in reducing the deterioration of the patient's medical condition until arrival at an MTF.)*

(6) **Lines of patient drift.** *(This indicates what routes injured Soldiers are most likely to take from the battlefield. It is usually the most direct route over the least demanding terrain, such as at the base of the hill rather than over the hill where climbing would be required.)*

(7) **Evacuation routes/corridors.** *(Evacuation routes should be preplanned, indicated on the medical evacuation overlay, and reconnaissance accomplished. Routes that provide lucrative targets, have significant obstacles to circumvent, or will be unduly congested due to fleeing refugees or other displaced persons should only be used if no other routes are available. Air corridors are provided by the supporting A2C2 element. Due the placement of hospitalization resources within the JOA, AA may be required to operate out of fixed sectors and coordination for air escorts must be coordinated.)*

(8) **Support requirements.** *(This paragraph discusses the type of direct support, general support, or operational control required for the operation and its command relationship. Specific guidance on trigger points and release points from a particular category of support should be included.)*

(9) **Medical evacuation procedures.**

(a) **Evacuation overlays.** *(Evacuation overlays must be developed to facilitate the evacuation effort. Both supporting and supported units must maintain and update [as required] overlays throughout the operation.)*

(b) **Communications.** *(Medical evacuation frequencies must be designated at the outset of the operation. Requirements for liaison officers, A2C2, and interpreters must be anticipated.)*

(c) **Casualty collecting points.** *(Location, staffing, and activation trigger [such as crossing a phase line] must be known to supported and supporting units.)*

(d) **Ambulance exchange points.** *(Location, staffing, and activation trigger [such as crossing a phase line] must be known to supported and supporting units).*

Note: If an AXP is staffed with a treatment element, it is no longer considered to be an AXP but rather it becomes the forward element of a BAS or Role 2 MTF. The AXP is only a location on the ground where the patient is transferred from one evacuation platform to another. If it is staffed, it is usually staffed by one member of the evacuation platoon to direct returning ambulances to the area where the last ambulance departed from. The element designating the AXP is responsible for staffing it. Ambulance exchange points may be rendezvous points to be used only once during the operation.

Note: Ambulance exchange points should not be used as resupply points for Class VIII. (Planning on pushing Class VIII to these points may result in the loss of the Class VIII materiel, as there may not be anyone at the point to receive the supplies.)

(e) Ambulance shuttle system. *(The ambulance [or litter] shuttle system is a management tool to facilitate the medical evacuation of forward areas.)*

(10) Casualty evacuation (transportation).

(a) Medical augmentation. *(Units and/or facilities capable of providing medical augmentation support should be identified and tasked as appropriate.)*

(b) Manual evacuation.

(c) Litter evacuation. *(A trauma specialist or CLS should be included as a member of the litter team when there is a significant distance to be covered.)*

(d) Pack animals. *(In some remote locations, the use of pack animals may be the most feasible form of transportation from the POI to a place where vehicles, aircraft, or watercraft become accessible. Whenever possible, medical personnel should accompany patients evacuated in this manner.)*

(e) Nonmedical vehicle. *(Nonmedical vehicles which could be preplanned for use for casualty transport should be identified. Further, guidance on the use of vehicles of opportunity should also be discussed. When possible nonmedical vehicles used for these purposes should be augmented with medical personnel or CLS.)*

(f) Nonmedical aircraft. *(Nonmedical aircraft which could be preplanned for use for casualty transport should be identified. Guidance on the use of aircraft of opportunity should also be discussed. When possible nonmedical vehicles used for these purposes should be augmented with medical personnel or CLS.)*

(g) Nonmedical watercraft. *(Nonmedical boats, ships, or other watercraft which could be preplanned for use for casualty transport should be identified. Further, guidance on the use of these platforms of opportunity should also be discussed. When possible nonmedical vehicles used for these purposes should be augmented with medical personnel or CLS.)*

(11) Resources available.

(a) Mode of transportation.

(b) Personnel. *(This includes a discussion on personnel strengths and the impact shortages have on the medical evacuation operations.)*

(c) Supplies and equipment. *(This includes both medical and nonmedical supplies and equipment. Litter and blanket exchanges should be covered here. If PMI are an issue, they should be discussed here.)*

(12) Use of smoke and obscurants. *(In most cases, medical units will not have a high priority for the use of smoke and obscurants; however, if the supported units have planned for its use, medical evacuation units can take advantage of the situation to clear the battlefield of casualties. Guidance on the use of colored smoke to identify the pickup location for patients can also be included.)*

d. Warfighting functions.

(1) Medical planners must take the following warfighting functions into consideration.

(a) Intelligence. (Known or suspected enemy situation. Terrain and weather. What kind of road network is in the AO? How will the predicted weather impact the traffic capability on and off the roads? Are there LZs along the axis of advance and in or around the urban area? How will the weather impact on the availability of aircraft?)

(b) Movement and maneuver. *(What is the combat scheme of maneuver and are there medical assets readily available to care for potential casualties? Particular attention should be paid to the employment of the armored ambulances.)*

(c) Fire support. *(Are there any buildings specifically targeted? If so, medical planners should ensure CCPs, evacuation routes, and treatment elements remain at a safe distance away from these buildings to prevent potential fratricide incidents.)*

(d) Protection. *(Preserving the force includes protecting personnel (combatant and noncombatant), physical assets, and information of the US and multinational partners.)*

(e) Sustainment. *(Typically, the BCT moves a forward logistics element toward the battle area to provide support. If this is the case, then how are medical assets integrated into the plan, and, more importantly, where are they positioned in the convoy and how are they secured?)*

(f) Command and control. *(The senior military on-site or the senior medical person at the scene, if available, assigns the evacuation precedence and determines if the patient should be evacuated by ground or air resources. What are the primary and secondary frequencies? This element is the central point for prioritizing assets and medical evacuation requests. It is imperative that this element is capable of continuous situational awareness through accurate battle tracking.) From information provided by the S2, medical planners must understand the tactics of the enemy and be prepared for casualties at all possible areas of contact. The medical planners should be able to gain insight into this matter during the war gaming process. The medical planners should provide input into the counterreaction portion of the action, reaction, counterreaction drill.)*

(2) Additional planning information.

(a) *Availability and numbers of ground and AA assets, evacuation routes and flight windows.*

(b) *Nonstandard ground and aircraft availability (for MASCAL situations) and how to trigger their involvement.*

(c) *Additional medical assets from higher levels such as treatment teams, ambulances, and combat configured loads of Class VIII.*

(d) *Existing and proposed locations of all available MTFs.*

(e) *Class VIII resupply plan.*

(f) *The need for litter bearer teams and special equipment for extracting casualties from buildings and rubble.*

e. **Casualty evacuation synchronization.**

(1) *Medical evacuation and CASEVAC assets must be incorporated into the evacuation plan. Ideally there should be a single dedicated controlling authority. This prevents duplication of effort and confusion and enables the battlefield to be cleared faster and more efficiently.*

(2) *In MASCAL situations CASEVAC assets once put into operation should be dedicated versus designated (lifts of opportunity). Potential problems that arise arising from relying on designated versus dedicated CASEVAC include excessive time spent acquiring the information needed to transport casualties. Medical regulating to higher roles of care facilities may have to be accomplished by blocks of beds rather than on a by name basis. If a facility is designated for certain specialty care [such as head trauma] the evacuator must know which facilities have been identified and where they are located.).*

(3) *Another consideration is aircraft configuration. Casualty evacuation may be a seats-out operation. By dedicating aircraft and crews, they have time to prepare for the mission. Casualty evacuation crews must be briefed on the details of their mission including LZs, PZs, air corridors, communications architecture, and supported units.*

(4) *Any CASEVAC plan should clearly articulate the following information.*

(a) *Who is responsible (such as the company 1SG) for coordinating the company CASEVAC mission?*

(b) *Combat lifesavers role.*

(c) *Class VIII and blood resupply procedures.*

(d) *Personnel to act as litter bearers.*

(e) *Locations of CCPs, forward treatment teams, BAS, and AXPs.*

(f) *Location and designation of MTFs.*

(g) *Medical regulating officer coordination.*

(h) *Medical treatment facility and MRO contact frequencies.*

4. **SERVICE SUPPORT.**

a. **Supply.** *(Refer to TSOP or another annex whenever practical.)*

(1) **General supply.** *(Provide special instructions applicable to medical evacuation units.)*

(2) *Consider supply levels for all classes of supply, in the event of mission requirements in an austere environment and at extended distances from the full compliment of logistics and sustainment resources.*

b. **Medical logistics.** *(Provide special procedures applicable to the operations.)*

(1) **Requirements.** *(For sustaining the ground and AA operations and personnel.)*

(2) **Distribution.** *(This should include the method of distribution and any limitations or restrictions that are applicable. Additionally, if special transportation requirements exist, they should also be noted.)*

(3) **Medical logistic activities.** *(This includes the location of the medical supply activity supporting the AO and the means of communicating requests for resupply.)*

c. **Supplies required for ground and AA operations and personnel.** *(Includes estimates of the number of patients anticipated to be evacuated.)*

d. **Health service support.** *(Include information and instructions for supported units. Prescribe the plan for ground and air medical evacuation.)*

(1) **Evacuation.** *(List CCPs and AXPs. Establish ambulances shuttles, routes, means, and schedules of evacuation and en route treatment policies for the use of nonmedical transportation assets; policies for evacuation by ground and air; and evacuation of CBRN-contaminated patients. Include information about medical evacuation request procedures and channels, and holding policies.)*

(2) **Treatment.**

(a) **Treatment of supported forces.**

(b) **Policies.** *(Should address treatment and hospitalization policies.)*

(c) **Units.** *(This includes information concerning the location, capabilities, and communications means of units providing support.)*

(d) **Other.** *(This can be information on other medical assets that may be used in emergencies.)*

e. **Personnel.**

(1) Current subordinate unit manning levels/critical MOS shortages (consider all manning influences—Task Organization, R&R flow, Boots On Ground data, and so forth).

(2) New personnel requirements resulting from the operation (language skills, additional skill identifier [ASI], and so forth).

(3) Casualty estimates (as developed).

(4) Forecasted replacement availability.

(5) Evacuation policy for the operation.

(6) Supporting HR organizations (location, command/support relationship, controlling element).

(7) Supporting medical elements

(8) Replacement fill priority (Coordinate w/S3/G3).

(9) Crew/key leader replacement.

(10) Projected postal flow/limitations.

(11) Change to established PASR reporting flow/times.

(12) Change to casualty reporting flow (CLT change, reporting changes).
(13) RSOI Reception HR impacts.
(14) R&R schedule/flow operational impact (during operations conducted during sustained operations).
(15) Civilian and JIM manning impacts on the operation focusing on strengths and personnel service requirements.
(16) Specific Army G1 personnel planning guidance impacts on the operation.
(17) Status of other personnel service or personnel support factors (postal, MWR, legal support, military pay support).

f. Coordinating instructions. *(List instructions applicable to two or more units and not routinely covered in unit SOPs. This is always the last subparagraph in paragraph 4. Place complex instruction in an annex.)*

(1) Time or condition when a plan becomes effective. *(Include the time or the conditions under which the plan is to be placed in effect.)*

(2) Commander's critical information requirements. *(List only in coordinating instructions and not in annexes.)*

(3) Risk mitigation control measures. *(These are measures unique to this operation and not included in unit SOPs. They must include MOPP, OEG, troop safety, vehicle recognition signals, and fratricide prevention measures.)*

(4) Rules of engagement. *(If required.)*

(5) Boundaries. *(Address measures established to protect units. Specify which tactical units are to provide protection.)*

g. Special reports. *(List reports requiring special emphasis that is not addressed elsewhere.)*

5. COMMAND AND SIGNAL.

a. Command. *(State the map coordinates for the CP locations and at least one future location for each CP. Identify the chain of command if not addressed in unit SOPs.)*

b. Signal. *(Include the headquarters location and movements, liaison arrangements, recognition and identification instructions, and general rules concerning the use of communications and other equipment, if necessary. Use an annex when appropriate.)*

(Classification)

Figure D-1. Example of a medical evacuation plan

SECTION II — EXAMPLE FORMAT FOR AN OPERATIONS ORDER

D-2. Figure D-2 provides an example of the medical operations order.

(Classification)

Place the classification at the top and bottom of every page of the OPLAN.

(Change from verbal orders, if any)

The phrases "No change from verbal orders" or "No change from verbal orders except paragraph #" are required. This statement is only applicable if the commander issues a verbal order.

Copy ## of ## copies
Issuing headquarters
Place of issue

Show the name of the town or place in capital letters, coordinates in parentheses, and the country in capital letters. You may encode both.

Date-time group of signature

(The effective time for implementing the plan or order is the same as the date-time group of signature unless the coordinating instructions state otherwise. Use time zone Z unless the order states otherwise. When orders apply to units in different time zones, use Z time. When an OPORD does not specify the actual time to begin an operation, state that time in terms of an event.)

OPERATION ORDER 00-001, Operation Liberation
References: OPORD 00-000
Time zone: Zulu
Task Organization

Ground Ambulance Units participating in Operation Liberation
123 MED CO (GA) assigned to 1st MMB and provides emergency support to the 1st Marine Expeditionary Force (MEF)
234 MED CO (GA) assigned to 1st MMB
Ambulance Platoons assigned to:
C CO BSMC
C CO BSMC
C CO BSMC
C CO BSMC

Air Ambulance Units participating in Operation Liberation
C 2-1 MED CO (AA) CAB assigned to 1st Infantry Division
C 2-2 MED CO (AA) CAB assigned to 2nd Infantry Division
C 2-3 MED CO (AA) DS to the 1st MEF
C 2-4 MED CO (AA) CAB assigned to 1st Corps support

Air Force aeromedical evacuation assets
2d AF Combat Search and Rescue (CSAR), 1st AF Squad

1. **SITUATION.**

   a. **Enemy forces.** *As addressed in OPORD 00-000, the ground-to-air threat is significant in the all enemy sectors. The air–to-air threat is minimal, in all sectors. All ground planned evacuation routes will be targets for ambush by enemy forces and sympathetic host nationals.*

   b. **Friendly forces.** *A six (6) aircraft (HH-60) Air Force CSAR unit will be staged from the APOD. They will respond to downed aircrew evacuation recovery requests from the Air Force Combat Control Center.*

   c. **Attachments and detachments.** *None.*

   d. **Assumptions.** *None.*

2. **MISSION.** *On order provide tactical and operational ground and air intratheater medical evacuation or patient movement in support of Operation Liberation and continue coverage of all combined joint task force (CJTF) assets until relieved.*

3. **EXECUTION. Intent: State the commander's intent.**

   a. **Concept of support operations.** *This is a three phase operation.*

   Phase 1—Will be prior to initiation of the attack. *All US forces will be in the staging areas and in preattack positions. There will be no or very limited direct contact with enemy forces. C 2-1, C 2-2, and C 2-3 will not be required to* *provide evacuation support until D-3 days. Units in the staging area will be directed to call C 2-4 for evacuation at all times prior to D-3 days. All other evacuation units will concentrate on training and preparation for hostilities as directed by their commanders.*

Phase 2—Will be the attack. *C 2-1, C 2-2, C 2-3 will provide DS to their assigned, GS to the corps sustainment area and the backhaul mission.*

Phase 3—Will be from the conclusion of the attack until stability operations commence. *Evacuation assets will begin transferring to an area support mission during this phase.*

Note. The concept of support operations takes into account both the tactical OPORD and the medical Annex to the OPORD. Information in both documents will shape the evacuation plan. Simultaneous planning should therefore occur.

---

b. **Tasks to maneuver units:** *Not applicable.*

c. **Tasks to movement and maneuvers units:**

(1) **Evacuation responsibilities.**

(a) **Phase 1.**

C 2-1, C 2-2, C 2-3 will provide direct support to their division assets from D-3 on. C 2-1, C 2-2, C 2-3 will respond to any request for evacuation that is geographically close to their positions even if not in their division area.

C 2-4 will provide support to the entire staging area (AO Buffalo). C 2-4 will perform the ship-to - shore mission as directed by the 11th MEDCOM TPMC

The 123 MED CO (GA) will operate in the 1st MEF sector in a direct support role. 1st MMB has will cover the 1st and 2nd Division footprint.

The 234th MED CO (GA) will provide support to Camps Jones, Smith, and Franks and perform patient transfers from camps to the APOD.

(b) **Phase 2.**

C 2-1, C 2-2, C 2-3 will provide direct support to their division assets. C 2-1, C 2-2, C 2-3 will be responsible for moving patients from within the division area back to the staging AO Buffalo until FLOT is beyond phase line alpha.

After FLOT is beyond phase line alpha (at which time one way flight time for the H-60 back to AO Buffalo will exceed 2 hours), C 2-1 will designate 1 FSMT to stage at XYZ Airfield to perform the backhaul and the area support mission until relieved. United States AF aeromedical transports will be coordinated for by 11th MEDCOM flying from XYZ airfield.

Beyond phase line Alpha C 2-2 will provide one FSMT to support the greater casualties expected in C 2-2 sector. Coordinate internally for link-up.

C 2-4 will provide support to the entire staging area (AO Buffalo). C 2-4 will perform the ship-to-shore mission as directed by the 11th MEDCOM MRO. C 2-4 will designate 4 aircraft to go forward as required to backhaul patients to AO Buffalo.

(c) **Phase 3.**

C 2-1, C 2-2, C 2-3 will provide direct support to their Divisions' assets.

The 1st MEF will identify areas of high casualty density conducive to ground evacuation (urban areas) and divide the resources of the 123rd MED CO (GA) to cover these areas efficiently, if required.

The 234th MED CO (GA) will conduct convoy support and provide coverage for AO Buffalo.

d. **Evacuation mission guidelines.**

(1) **Phase 1, 2, and 3.** *The senior military person advised by medical personnel on-scene will assign the evacuation precedence and determine the mode of evacuation (ground or air). Launch authority, a separate issue, based on risk will be according to the aviation SOP.*

(a) *Weather below acceptable minimums.*

(b) *No acceptable LZ in the immediate vicinity, except for hoist missions. Ground evacuation may be used for part of the evacuation.*

(c) *Chemical, biological, radiological and nuclear conditions at LZ would contaminate the aircraft. Ground evacuation may be used for part of the evacuation, followed by an air pickup after decontamination is complete.*

(d) *Patients are located at a LZ in direct enemy contact. Ground evacuation may be used for part of the evacuation, followed by an air pickup once clear of the enemy.*

(e) *Air evacuation assets are unavailable.*

(f) *Mission flight route requires an armed escort and one is not available.*

(2) **The following are additional guidelines for evacuation:**

(a) *Ground evacuation is the preferred method to clear casualties out of hostile urban areas. Air ambulance pick-up at a secured LZ outside of the urban area is preferred. For large urban areas, this may not be possible.*

(b) *If a mission requires the use of a hoist (such as thick forest or jungle) and the pick-up area has suspected enemy contact, ground evacuation is advised to a secure pick-up site.*

(c) *Prescheduled movement of ROUTINE or CONVENIENCE patients along secure routes of evacuation should be conducted using ground evacuation if the distance of evacuation is short.*

(3) **Coordinating instructions:**

(a) **Medical regulating.** *The 11th MEDCOM will oversee medical regulating during the operation.*

(b) **Tentative locations of MTFs.**

Phase 1

- *2nd CSH AB 1234 5678*

- *1st CSH AB 2345 6789*

Phase 2
- *2nd CSH AB 1234 5678*
- *1st CSH AB 2345 6789*

Phase 3
- *2nd CSH AB 1234 5678*
- *1st CSH AB 2345 6789*

(c) Reporting. *Evacuation units, through division G3 if assigned to a division, will submit a daily report of evacuation missions by 1200 the following day.*

(d) *If a MASCAL develops which overwhelms the resources of a single AA CO, contact GSAB for assistance.*

(e) Casualty collection. *Military treatment facilities can call 9 line missions to any medical evacuation asset for URGENT and PRIORITY missions however MTFs that need to clear beds will coordinate mass patient movements through CJTF G3 air for possible USAF support.*
- *1st CSH will call C 2-4 for evacuation requests during Phase 1.*
- *2nd CSH will call C 2-4 for evacuation requests during Phase 1.*
- *1st CSH will call C 2-1 for evacuation requests during Phase 2/3.*
- *2nd CSH will call C 2-2 for evacuation requests during Phase 2/3.*

(f) Ground evacuation. *Seasonal monsoons are expected to washout many of the main planned MSRs and cause flooding.*

(g) Air evacuation. *If air evacuation routes become so long as to seriously degrade the capabilities of an AA company, contact CJTF G3 Air for additional support and possible reallocation of evacuation assets. An AA CO is considered to be operating beyond its capability or inefficiently if:*
- *More than 50 per cent of missions require a one way trip time exceeding 2 hours.*
- *On more than 50 per cent of missions flown, routes of evacuation are flown through another AA company's area of direct support coverage.*

(h) Medical considerations for evacuation.
- US Army evacuation assets will be used to evacuate all injured personnel on the battlefield without prejudice.
- Human immunodeficiency virus is at epidemic proportions in Simula population. Take appropriate precautions during treatment and when dealing with biologicals.
- A significant number EPW and displaced civilians and detainees are predicted to be evacuated. Estimated to be a much as 75 per cent of total patient load.

(i) United States Air Force aeromedical evacuation.
- Combined joint task force G3 Air will coordinate for USAF intratheater aeromedical missions.

4. SERVICE SUPPORT.

a. Casualty estimates. Reference the Medical Annex for specific casualty estimates.

b. Aviation maintenance support. C 2-3 will receive AVUM/AVIM support from the 2-4 CAB.

c. Divisions G3 will report the following CCIR to CJTF G3 Air. Use the CJTF CCIR format.
- Any MASCAL.
- Any evacuation company with degraded capability for—
- Evacuation distance(defined above)
- Maintenance: below 70 percent operational readiness rate
- Personnel: number of crews less than available aircraft
- Any aircraft Class A accidents/events or shoot downs.

5. COMMAND AND SIGNAL.

a. Fragmentary orders (FRAGOs) to this order will be published to adjust the corps evacuation plan as necessary.

b. 11th MEDCOM MIRC Chat net name, Phone xxx-xxxx
- 1st Division Surgeon (G3), MIRC Chat net name
- 2-1 CAB MEDEVAC Cell, MIRC Chat net name
- 2nd Division Surgeon (G3), MIRC Chat net name
- 2-2 CAB MEDEVAC Cell, MIRC Chat net name
- 1st MEF, MEDEVAC Cell, MIRC Chat net name
- 1st Corps CAB MEDEVAC cell, MIRC Chat net name, phone xxx-xxxx

ACKNOWLEDGE:
PATRICK, LTG

Figure D-2. Example of a medical operations order

Appendix E

Use of Smoke and Obscurants in Medical Evacuation Operations

This appendix provides guidance for AMEDD personnel in the use of smoke and obscurants in medical evacuation operations.

TACTICAL COMMANDER

E-1. Throughout the battlefield, forces acquire and engage targets based on visual, laser, and microwave technologies. Friendly and enemy units use smoke and obscurants across the battlefield as a combat multiplier. The use of smoke and obscurants to mask combat operations is dictated by the tactical commander. He normally provides the operational guidance for units or elements operating in an area requiring obscuration. Permission to employ smoke and obscurants solely to mask medical evacuation operations may not be approved. However, if the tactical commander's plan indicates that smoke operations are to be employed in a given AO, the medical planner should consider both the advantages and disadvantages posed by their employment.

FACTORS TO CONSIDER

E-2. The medical planner should consider the factors which might impact the use of smoke and obscurants in medical evacuation operations. Factors to consider are the—

- Phase of the tactical operation in which smoke and obscurants will be employed.
- Effect on ground and air evacuation routes when operating in an obscured environment (such as limited hours of use, checkpoint or convoy requirements, or the elimination of NOE approaches).
- Potential for exploiting the use of the cover and concealment provided for clearing the battlefield of casualties, especially in defensive operations.
- Potential requirements for smoke generation to perform the medical evacuation mission which would not detract from the tactical capability and requirements.

E-3. Smoke can also be used to identify unit areas or LZs for which a medical evacuation request has been received. Further, smoke can indicate wind direction at a landing site for AA operations.

EMPLOYMENT OF SMOKE AND OBSCURANTS

E-4. Smoke and obscurants are employed to protect friendly forces from attack during the offense or defense. Smoke and obscurants disrupt enemy combat operations throughout the depth of the battlefield and across the operational continuum. They—

- Disrupt the ability to communicate.
- Conceal friendly forces.
- Deceive the enemy.
- Identify and signal.
- Degrade the effects of directed-energy weapons.

E-5. The benefit to AHS forces is derived through the tactical commander's use of smoke to obscure friendly tactical maneuvers. This obscuration—

- Prohibits the enemy from knowing how many casualties have been inflicted.
- Aids the movement of medical units and equipment.

- Enhances the ability to resupply forward deployed AHS elements.
- Aids in the tactical deception plan.

GENEVA CONVENTIONS AND THE USE OF SMOKE AND OBSCURANTS IN MEDICAL EVACUATION OPERATIONS

E-6. As discussed above, the 1949 Geneva Conventions for the Amelioration of the Condition of the Wounded and Sick in Armed Forces in the Field (GWS) provides protection of medical personnel and units from intentional attack so long as they carry out no duties harmful to the enemy (Article 21, GWS). In order to facilitate their identification so as to prevent their intentional attack, medical units, equipment, and personnel are authorized to display the distinctive emblem of the Red Cross (Article 41 and Article 42, GWS). Under tactical conditions, when requirements for concealment outweigh those for recognition, all distinctive emblems may be obscured or removed from medical equipment if ordered by competent military authority and authorized by Army regulations. Display of the distinctive emblem is not required to afford the right against intentional attack; attack of medical units, equipment, and personnel not displaying the distinctive emblem is prohibited if opposing forces realize that the forces about to be attacked are medical units performing humanitarian duties.

E-7. The use of smoke or obscurants in medical evacuation operations does not differ from the use of camouflage techniques and is not prohibited by the GWS. Its only effect will be to obscure the identity of units as they perform their humanitarian mission. Given the lethality of the modern battlefield, however, it would be difficult, if not impossible, to say that such obscuration of these units, equipment, and personnel would necessarily increase their risk from unintentional attack.

E-8. It is recognized that, with the advent of precision-guided munitions and electro-optical or laser target acquisition devices, there will be a substantial use of smoke and other obscurants on the modern battlefield as a result of normal combat operations. The legitimate use of obscurants by combatants to thwart the accuracy of precision guided munitions may increase the risk to units and equipment not employing obscurants. This may possibly place medical units and equipment at greater risk if they fail to employ them. Further, medical evacuation operations will have to be carried out on the battlefield as medical personnel find it, which will include obscurants employed for normal combat operations.

USE OF SMOKE IN AEROMEDICAL EVACUATION AND HOIST RESCUE OPERATIONS

EFFECTIVE USE

E-9. Smoke can be used effectively in aeromedical evacuation and overland hoist rescue operations to—

- Identify the landing site. Colored smoke is an excellent daytime marking method. The smoke generated from a smoke grenade is difficult to detect more than 2 to 3 miles away, but an aircraft in the area should have little difficulty in noting its location. As more than one unit may be operating in a given area, it is important that the unit requesting an aeromedical evacuation mission be able to signal the aircraft as to the correct landing site to use. Radio communications produce an electronic signature. The electronic signature created from a prolonged transmission to guide an AA to the landing site may not be an acceptable tactical risk.
- Ensure the LZ is controlled by friendly forces. When a unit employs colored smoke to mark a landing site, the aircrew should identify the color and confirm it with the ground personnel. The transmission time required for this procedure is limited, thereby reducing the electronic signature.
- Determine surface wind direction. The employment of smoke at the landing site also enables the aircrew to determine the wind direction.
- Provide cover and concealment. In some environmental conditions (such as desert operations), the phenomenon of inversion occurs. When this occurs, the smoke and obscurants used in

normal combat operations may provide an upper layer of smoke under which the AA can operate.

DISADVANTAGE

E-10. The use of smoke on aeromedical evacuation operations can be a disadvantage if incorrectly employed or if the smoke generated in the tactical operation interferes with the medical evacuation mission. Smoke can obscure the landing site and make NOE approaches unusable. Further, smoke on the battlefield can force aircraft to fly at higher than planned heights. This increases the risk of being acquired by the enemy.

OVERWATER HOIST OPERATIONS

E-11. In overwater hoist rescue operations, the employment of smoke for marking the patient pickup area, for determining surface wind conditions, and for spatial orientation is essential. The smoke employed by the aircrew must not interfere with the conduct of the operation or mask the location of the individual to be rescued.

EMPLOYMENT OF SMOKE IN GROUND MEDICAL EVACUATION OPERATIONS

E-12. The employment of smoke during ground evacuation operations must be in consonance with the tactical commander's plan. Smoke can mask medical evacuation operations on the battlefield, but must not interfere with the tactical mission. In all combat operations, but especially in UO, smoke can be employed to cover and conceal—

- Movement across open areas.
- Extraction of casualties from vehicles and/or buildings.
- Entry and exit into/out of structures.

Appendix F

Evacuation Capabilities of United States Forces

This appendix provides the evacuation capabilities of US forces and aircraft.

EVACUATION CAPABILITIES OF UNITED STATES AIR FORCE AIRCRAFT

F-1. Evacuation capabilities of USAF aircraft are provided in Table F-1. As the majority of USAF aircraft are not dedicated medical evacuation platforms, when used for medical evacuation the crew must be augmented with medical personnel to provide in-flight care.

F-2. The evacuation capability of Civil Reserve Air Fleet aircraft is provided in Table F-2.

Table F-1. Evacuation capabilities of United States Air Force aircraft

Transport Aircraft	*Litter*	*Ambulatory*	*Combination*
C-130 Hercules	74	92	Variety
C-141 Starlifter	48	170	Variety
C-5 Galaxy	70		
C-17A	*36	*54	
KC-135 and KC-10	8	24	
*Normal configuration			

Table F-2. Civil Reserve Air Fleet capabilities

Transport Aircraft	*Litter*	*Ambulatory*	*Combination*
Boeing B-767	*111	22	87 Litter 22 Ambulatory
* When activated, these aircraft are preconfigured for medical evacuation. The crew must be augmented with medical personnel			

EVACUATION CAPABILITIES OF UNITED STATES ARMY VEHICLES AND AIRCRAFT

F-3. The evacuation capabilities of US Army vehicles and aircraft are provided in Table F-3.

Table F-3. Evacuation capabilities of United States Army vehicles and aircraft

Vehicle/Aircraft	*Litter*	*Ambulatory*	*Combination*
Ground Vehicles			
M996 Truck, Ambulance	2	6	1 Litter 3 Ambulatory
M997 Truck, Ambulance	4	8	2 Litter 4 Ambulatory

M1010 Truck, Ambulance	4	8	
M792 Truck, Ambulance	3	6	2 Litter 3 Ambulatory
M1133, Stryker MEV	4	6	
Bus, Motor, 44 Passenger	18	37	
M113 Carrier, Personnel	4**	10	
M998 Truck, Cargo/Troop Carrier (Four Man)	3	4	
M998 Truck, Cargo/Troop Carrier (Two Man)	5		
Truck Cargo, 2 1/2 Ton, 5 Ton	12	16	
M997 Heavy Expanded Mobility Tactical Truck	9		
M871 Semitrailer, Cargo	16		
M1085 Truck, Cargo, Medium Tactical Vehicle, Long Wheel Base, 5 Ton	12	22	
M1093 Truck, Cargo, Medium Tactical Vehicle Air Drop/Air Delivery, 5 Ton	8	14	
M1081 Truck, Cargo, Medium Tactical Vehicle Light Air Drop/Air Delivery, 2 1/2 Ton	7	12	
Rotary-Wing Aircraft			
***UH-60 Blackhawk	6	7	*4 Litter 1 Ambulatory
****UH-60A Blackhawk	3	4	4 Litter 1 Ambulatory
UH-1H/V Iroquois	6	9	*3 Litter 4 Ambulatory
CH-47 Chinook	24	31	#
Fixed-Wing Aircraft			
C-12 Huron		8	

LEGEND

* Normal configuration
** Spall liner must be removed
*** without internal hoist installed, requires 6 man litter configuration kit
**** With internal hoist installed
# See Table F-5

Railway Car Capabilities

F-4. Although railway cars are not available within the US Army inventory, it is important to know the approximate capacities of rail transport in the event they become available and/or required in civil support operations or through wartime HN support agreements. The approximate capabilities of railway cars are provided in Table F-4.

Table F-4. Capabilities of railway cars

Rail Transport	*Litter*	*Ambulatory*	*Combination*
Pullman Car (US)	32	48	
Sleeping Car (NATO/HN)	32	32	
Ambulance, Railway Car (NATO/HN)	24	30	
Ambulance, Railway Car, Personnel	21	21	
Railbus			40 Litter 16 Ambulatory

EVACUATION CAPABILITIES OF UNITED STATES NAVY SHIPS, WATERCRAFT, AND ROTARY-WING AIRCRAFT

F-5. The evacuation capabilities of USN ships, watercraft, and aircraft are provided in Table F-5. The entries for ambulatory and litter patients on ships are the same because all patients require a bunk.

Table F-5. Evacuation capabilities of United States Navy ships, watercraft, and aircraft

Ships/Watercraft/Aircraft	*Litter*	*Ambulatory*
Hospital Ships		
T-AH 19, US Naval Ship Mercy	1000	1000
T-AH 20, US Naval Ship Comfort	1000	1000
Casualty Receiving And Treatment Ships		
Amphibious Assault Ship (LHD)(Multipurpose)	604	604
Amphibious Assault Ship (LHA)(General Purpose)	367	367
Amphibious Assault Ship (LPH)	222	222
Amphibious Transport Dock (LPD)	14	14
Dock Landing Ship (LSD)	108	108
Limited Medical Capability Watercraft		
Amphibious Cargo Ship (LKA)	12	12
Amphibious Command Ship (LCC)	24	24
Rotary-Wing Aircraft		
CH-46 Sea Knight	15	25
CH-53 Sea Stallion	24	*55
V22 Osprey	12	24
* With centerline seating installed		

Appendix G

Selection of Patients for Aeromedical Evacuation and Patient Classification Codes and Precedence

Patient classification codes provide information to evacuators and treatment personnel in an abbreviated form. They can be used to complete administrative reporting requirements pertaining to the evacuation of patients.

Paragraphs G-2 through G-5 implement STANAG 3204.

BRIEFING OF PATIENTS PRIOR TO AEROMEDICAL EVACUATION

G-1. When patients are evacuated by aircraft (routinely in Roles 2, 3, 4, and 5), they should be briefed on the following points—

- A number of ambulatory patients will be detailed to assist with the evacuation of litter patients in any emergency.
- Safety belts and litter straps are to be properly fastened according to orders given by the pilot.
- Patients are instructed on the proper position to assume in preparing for an emergency. Flight crews and CCATT ensure that seat safety harnesses have been tightened.
- Ambulatory patients, with the exception of those designated to assist litter patients, are the first to leave a downed aircraft.
- Immobilized litter patients are freed from litters and assisted in leaving the aircraft. Litters will not normally be removed from their fastenings in view of the limited time available to evacuate the aircraft.
- Mentally disturbed patients should be quieted so that the orderly removal of other patients will not be jeopardized.

INTERNATIONAL STANDARDIZATION EVACUATION AGREEMENT

G-2. Table G-1 provides the patient classification codes defined in international standardization agreements.

Table G-1. Patient classification codes

Code	Classification
Class 1. Psychiatric Category	
1A	Stretcher patients who are frankly disturbed and inaccessible, requiring close supervision and sedation, or the use of restraint equipment.
1B	Patients who are not at the time of request grossly disturbed, but who may react badly to air travel, or who may commit acts likely to endanger themselves or the safety of the aircraft and its occupants.
1C	Sitting patients who are cooperative and have proved themselves to be reliable under preflight observation.
Class 2. Stretcher Patients Other Than Psychiatric	
2A	Patients, who in an emergency, would be unable to leave the aircraft without assistance.
2B	Patients, who in an emergency, would be able to leave the aircraft unaided.
2C	Patients who can walk onto an aircraft but on a long flight would benefit from a litter.
Class 3. Sitting Patients Other Than Psychiatric	
3A	Sitting patients, including handicapped persons, who may need medical or nursing attention en route and who, in an emergency, would require assistance to escape.
3B	Sitting patients who may need medical or nursing attention en route and who would be able to escape unassisted in an emergency.
Class 4. Sitting Patients Other Then Psychiatric	
4	Patients who will not need medical or nursing attention en route and are capable of traveling unescorted. Class 4 patients are normally required to make their own way from the deplaning airfield to their destination.

INTERNATIONAL STANDARDIZATION EVACUATION PRECEDENCE

G-3. Patients for AE will be given appropriate degrees of priority so that, if aircraft space is limited, the more urgent patients may be evacuated before those whose conditions are less serious. The degrees of priority are depicted in Table G-2. (The evacuation precedence used by the USAF is essentially the same as this listing. It contains a few word changes and introduces specific time limits. It does not contain Priority 4.)

Table G-2. Patient priorities as designated in STANAG 3204

PRIORITY 1/URGENT	Emergency patients for whom speedy evacuation is necessary to save life or limb, to prevent complication of serious illness, or to avoid serious permanent disability.
PRIORITY 2/PRIORITY	Patients who require specialized treatment not available locally and who are liable to suffer unnecessary pain or disability unless evacuated with the least possible delay.
PRIORITY 3/ROUTINE	Patients whose immediate treatment requirements are available locally but whose prognosis would definitely benefit by air evacuation on routine scheduled flights.
PRIORITY 4	Patients for whom air evacuation is a matter of medical convenience rather than necessity.

PATIENT CLASSIFICATION

G-4. Table G-3 provides the patient classification codes used aboard USAF aircraft and that can be used in completing DD Form 601 (Patient Evacuation Manifest). These codes are expanded to include categories of patients and other personnel which may or may not apply on the battlefield (such as infants, relatives, or friends).

United States Air Force Evacuation Precedence

G-5. The evacuation precedence used by the USAF is dramatically different than that employed by the US Army medical evacuation system. These precedence should not be confused. Table G-4 (page G-5) provides the evacuation precedence and time frames used by the USAF.

Table G-3. Patient classification codes

Code	*Classification*
Class 1.	***Neuropsychiatric Patients***
1A	Severe psychiatric litter patients requiring the use of restraining apparatus, sedation, and close supervision at all times.
1B	Psychiatric litter patients of intermediate severity requiring tranquilizing medication or sedation, not normally requiring the use of restraining apparatus, but who react badly to air travel or who may commit acts likely to endanger themselves or the safety of the aircraft. Restraining apparatus should be available for use.
1C	Psychiatric walking patients of moderate severity who are cooperative and who have proved reliable under observation.
Class 2.	***Litter Patients (Other Than Psychiatric)***
2A	Immobile litter patients unable to move about of their own volition under any circumstances.
2B	Mobile litter patients able to move about of their own volition in an emergency.
Class 3.	***Walking Patients (Other Than Psychiatric)***
3A	Nonpsychiatric and nonsubstance abuse patients who require medical treatment, assistance, or observation en route.
3B	Recovered patients who are returning to their units and require no medical attention en route.
3C	Ambulatory drug or alcohol substance abuse patients.
Class 4.	***Infant Category***
4A	Infants under three years, occupying a seat or in a bassinet or a car seat secured in an ambulatory seat.
4B	Recovered infants under three years, occupying a seat or in a bassinet or car seat secured in an ambulatory seat.
4C	Infants in an incubator.
4D	Infants under three years on a litter.
4E	Outpatients under three years on a litter for comfort.
Class 5.	***Outpatient Category***
5A	Ambulatory, nonpsychiatric, and nonsubstance abuse outpatients who are traveling for an outpatient visit and do not require a litter or medical assistance in flight.
5B	Ambulatory drug or substance abuse outpatients going for treatment.
5C	Psychiatric outpatients going for treatment.
5D	Outpatients on a litter for comfort or safety.
5E	Returning outpatients on a litter for comfort or safety.
5F	Other returning outpatients.
Class 6.	***Attendant Category***
6A	Medical attendants, either physician, nurse, or technician, who are assigned to give specialized medical treatment or nursing care to a particular patient.
6B	Nonmedical attendants, either relatives or friends, who may assist with the patient's care and who may also require support.

Table G-4. Evacuation precedence used by the United States Air Force

Evacuation Precedence	*Definition And Time Frames*
URGENT PRECEDENCE	Applies only to the need for immediate life, limb or eyesight saving. These patients should be picked up and delivered to the designation facility as soon as possible.
PRIORITY PRECEDENCE	Applies to the need for prompt medical care not available locally. These patients should be picked up within 24 hours and delivered to the destination facility with the least possible delay.
ROUTINE PRECEDENCE	Applies to all other patients. Routine patients will be picked up and delivered on regularly scheduled flights.

Appendix H

Troop Leading Procedures, Precombat Checks/Precombat Inspections, and Leader Checklists

Troop leading procedures (TLPs) are universal in their application. They are used by leaders at all levels regardless of branch or MOS. For purposes of this appendix the leader titles of ambulance platoon leader and ambulance platoon sergeant are used for familiarity.

SECTION I — TROOP LEADING PROCEDURES

H-1. Leaders must conduct detailed planning and preparation before each mission in order to be successful. That is why the Army developed an eight-step process of TLP for leaders to follow.

H-2. Each step of the TLP process is designed to give leaders a framework on which to thoroughly develop their plans and orders. If these steps are not followed the resulting plans and orders may be missing information which may be critical to successfully execute the mission.

THE EIGHT STEPS OF TROOP LEADING PROCEDURES

STEP ONE: RECEIVE THE MISSION

H-3. Once staff planning is initiated the ambulance platoon leader gathers information concerning the operation from the company commander and disseminates to his subordinate leaders. Examples of this information include but are not limited to intelligence preparation of the battlefield (IPB) and METT-TC. If the company commander has not completed a full plan, he may still issue a warning order (WARNO) to the ambulance platoon leader.

- In keeping with the rules of time management, the ambulance platoon leader uses one-third of his overall time for planning the operation. The reverse planning sequence is recommended.
- The ambulance platoon leader's OPORD must also be understandable to subordinate personnel.
- Throughout the planning and development process the ambulance platoon leader may find it necessary to change or refine the plan as the tactical situation changes, as long as it follows the commander's intent.

STEP TWO: ISSUE A WARNING ORDER

H-4. The ambulance platoon leader issues a WARNO based on information received from the company commander's WARNO or OPORD.

- The ambulance platoon leader can present the order in any format.
- As a minimum, the platoon should use the company or battalion tactical standing operating procedures (TACSOP). The best method is to use already developed and implemented platoon level TACSOP that cover preparation for offensive or defensive missions. These TACSOPs should be thoroughly trained and rehearsed.
- Whatever the mission, the platoon can begin work on its preparation by assessing equipment needs through the various military supply classes (such as Class I, III, IV, V, or VIII).
- Every Soldier in the platoon should have a specific task and purpose (identified in the platoon SOP) once the ambulance platoon leader has issued a WARNO.

STEP THREE: MAKE A TENTATIVE PLAN

H-5. The ambulance platoon leader considers the following while preparing the tentative plan:

- Mission analysis, IPB, METT-TC considerations, higher mission and intent (two levels up).
- The ambulance platoon leader must know what systems are available for his use and how to employ them. He must fully understand his own equipment and capabilities as well as those of his enemy.
- Consider **Observation** and Fields of Fire, **Avenues** of Approach, **Key** Terrain, **Obstacles** and **Cover** and Concealment, (man-made and natural), (OAKOC), weather, morale, and flexibility to react to last-minute changes. As the commander updates the ambulance platoon leader, the ambulance platoon leader, in turn, continually updates the platoon with FRAGOs.
- Experience is a valuable asset for mental preparation. Knowing how Soldiers are going to react to situations, based on past experiences, can help any officer or NCO develop his plan.
- During the planning process, the leader should have a *vision* of the enemy and understand the threat using available intelligence. The better the ambulance platoon leader understands the enemy, the greater his opportunity for success.

STEP FOUR: START NECESSARY MOVEMENT

H-6. Even though listed as Step Four, necessary movement can begin as soon as the mission is received.

- If movement is required, the ambulance platoon sergeant, with the assistance of squad leaders, can move the platoon to a new tactical assembly area (TAA). This allows the ambulance platoon leader more time to prepare his order and graphics.
- At the new TAA, preparations continue with precombat checks (PCCs). This is a good opportunity to conduct PCCs on squad equipment by using the platoon SOP checklists and to certify squad rehearsals. If the ambulance platoon sergeant is unable to conduct squad inspections, the platoon SOP should designate someone else, such as the senior squad leader, for the task.
- If platoon linkups are time driven, movements and time management become critical. Failure to plan or allow for ample movement time creates late linkups and results in late rehearsals, late PCCs, and various other events that will not be accomplished on time or to standard.
- If movements were conducted late or linkups were late, the mission timeline will suffer.

Note. Time management is mission essential!

STEP FIVE: RECONNOITER (CONDUCT RECONNAISSANCE)

H-7. Although listed fifth, this step may be implemented at any time and as often as necessary during TLP. When the risk of enemy contact is too great, or establishing additional orders for reconnaissance parties is time consuming leaders may have to rely on map reconnaissance and or photographic imagery to reconnoiter an AO. Ambulance platoon leaders must ensure that ground evacuation routes are reconnoitered and strip maps developed for security escort vehicle drivers and ambulance drivers.

- A leader's reconnaissance must be performed! Leaders who see more of the AO are better able to develop more detailed and useful plans.
- Information gained by the leader's reconnaissance allows more accurate terrain models to be built and greater detail can be provided when the ambulance platoon leader is briefing the OPORD.

STEP SIX: COMPLETE THE PLAN

H-8. Complete the plan—

- With reconnaissance completed, movement accomplished, and the initial plan established, the ambulance platoon leader should have most of the information that is needed to finish writing the OPORD.
- Any additional resources needed should be en route or already on-hand so the ambulance platoon leader, when briefing the OPORD, has positive control of additional equipment or manpower.
- Graphics must be finished before the OPORD is briefed to the platoon and quickly handed down to squad leader level. Logistics and sustainment graphics must include CBRN decontamination points, MTFs locations, and supported unit CCP and AXP.

STEP SEVEN: ISSUE THE ORDER

H-9. Avoid rushing through the OPORD briefing as this may result in pertinent information being left out of the brief.

- When completed, distribute the OPORD with graphics to each squad and team leader. Also include a terrain model or sand table to enhance visualization of the terrain.
- Minimize distractions. Having subordinate leaders in the right mindset to receive the order is critical to allow them to collect and retain all of the information presented at the OPORD brief.
- If possible, present the OPORD with a representation of the terrain in the background to enhance visualization of the mission.
- After the OPORD has been presented, the ambulance platoon leader ensures that the information has been disseminated to every Soldier. Backbriefs from Soldiers and NCOs to the ambulance platoon leader should reveal any portion or portions of the OPORD that were missed or not fully understood. Everyone leaving the OPORD brief should have a clear understanding of the mission.
- Contingencies need special consideration. If, during the OPORD brief, a potential activity is discussed that has no clear plan, subordinates can help develop a contingency plan, unless the activity is already specified in the platoon SOP. Defer making a decision on topics discussed during the OPORD brief which require more thought and possibly guidance from higher headquarters. If at all possible, allow a follow-up period to develop clarity on the situation and the techniques needed to accomplish the mission.

STEP EIGHT: REHEARSALS

H-10. Supervise rehearsals and plan to run as many rehearsals as time permits. After each rehearsal, conduct an after-action review (AAR). When conducting platoon rehearsals with attachments, ensure that the ambulance platoon leader/sergeant supervises the AAR with attachments. When conducting the AAR, ensure Soldier involvement. By this time all Soldiers understand the concept of the operation.

- When rehearsing the mission, ensure that the rehearsal lane coincides with what was briefed during the OPORD brief. Based on the latest intelligence, the rehearsal lane should match (as much as possible) what will be encountered during the actual mission. Do not allow for complacency. If there is information about the terrain and the obstacle that can be placed in the rehearsal lane to add realism, ensure that obstacle information is built into the lane. A company-level rehearsal is of limited value if the enemy obstacle does not resemble what was reconnoitered or templated.
- If time does not allow for a full rehearsal, backbriefs must be conducted. A successful understanding of the mission can be accomplished with key personnel by walking troops through the mission on the terrain model and asking questions. The backbrief is never a substitute for a rehearsal, but is an acceptable recourse when time is limited.
- Conduct final precombat inspections (PCI).
- If there are any last-minute concerns, they should be satisfied during this time.
- Conclude the brief.

H-11. Some steps of the TLPs require more development than others. Regardless of the mission, the ambulance platoon leader must not, for any reason, omit any step from the TLP.

SECTION II — PRECOMBAT CHECKS AND PRECOMBAT INSPECTIONS

H-12. Precombat checks and PCIs are tools which leaders have to ensure that their Soldiers—

- Fully understand the mission and their role.
- Have the equipment that is necessary to accomplish the mission on hand and that it is serviceable.

H-13. Leaders must conduct PCCs and PCIs for every mission. Failure to perform these vital checks could result in unnecessary casualties, loss of equipment, and failure to accomplish the mission.

H-14. When conducting PCCs/PCIs, as deficiencies are identified corrections must be made immediately.

H-15. Who performs PCCs and PCIs are both intended to accomplish the same thing, which is to have a leader ensure that personnel and equipment are ready. The only difference is who is doing the inspecting.

- Commanders conduct PCIs. Ambulance platoon leaders, ambulance platoon sergeants, and squad leaders conduct PCCs.
- Why do we identify a commander's inspection separately? Equipment readiness is essential for mission success. Squad leaders and ambulance platoon leaders are certainly capable of ensuring equipment readiness, but the commander's first-hand knowledge of mission requirements cannot be substituted.

H-16. Units need to establish standard PCC/PCI checklists based on their units mission profile. They should be included in the TACSOP and issued to all leaders. They should include routine daily checks and checks that are mission specific. They do not need to be extremely detailed; a list of items to check and the leader's knowledge of the equipment and the appropriate manuals should suffice. The PCCs checklists provided below are examples that units may use to develop their own or they can be modified as appropriate to meet their own requirements.

SECTION III — LEADER CHECKLISTS

H-17. This section provides sample checklists which may be used as is, or modified, to meet unit specific requirements. As with any checklist they are only effective if they are incorporated into a unit's TACSOPs and then used routinely.

GENERAL CONSIDERATIONS

H-18. Leader receives the mission from the next higher headquarters.

___ Clarifies any questions.

___ Coordinates with next higher headquarters as required.

___ Conducts mission analysis.

___ Produces the estimate.

___ Revises mission statement.

___ Produces tentative time schedule.

___ Issues a WARNO.

H-19. Unit members perform readiness, maintenance, and functional checks under the supervision of unit leaders which includes:

___ Medical equipment sets are present and all perishable and dated items are current and on-hand.

___ Vehicles/aircraft/generators.

___ Night vision devices.

___ Communications equipment.

___ Weapons and ammunition.

___ Field sanitation equipment and supplies.

___ Special equipment (such as hoist and forest penetrator).

___ Common table of allowances equipment.

H-20. Leader makes a tentative plan.

___ Uses estimate of the situation to depict plan of support.

___ Develops courses of action (COA).

___ War games COA.

___ Determines best COA.

___ Completes his plan.

H-21. Leader issues OPORD.

H-22. Leader uses sand table or sketches to depict plan of support.

H-23. Leader affects coordination for the mission.

___ Support requirements.

___ Current intelligence (to include health threat) update.

___ Control measures.

___ Communications and signal information.

___ Time schedule.

H-24. Leader receives attachments/augmentation (personnel), if appropriate.

___ Attachments are oriented to unit.

___ Attachments are briefed on the mission.

H-25. Leader supervises AHS mission preparation.

___ Key leaders brief back unit leader.

___ Key personnel rehearsals are conducted.

H-26. Unit leaders:

___ Supervise.

___ Inspect.

___ Ensure adequate security.

___ Conduct brief backs.

___ Rehearse.

___ Continue coordination.

H-27. Unit plans for support of combat operations.

___ Analyze mission.

___ Patient acquisition and medical evacuation requirements.

___ Area support requirements.

___ Requirements for water, Class VIII, and other supply classes.

___ Requirements for CBRN defense.

___ Transportation requirements.

H-28. Unit leaders:

___ Execute a work/rest plan based on work priorities and statutory (crew rest) requirements.

___ Monitor current situation.

___ Issues appropriate FRAGOs based on intelligence or operational updates.

___ React to messages or orders from higher headquarters.

___ Execute any actions and coordination resulting from change.

H-29. Unit headquarters remains current on positions and missions of higher, adjacent, and subordinate units.

Ground Ambulances Inspected For

___ Authorized MES are on hand.

___ Medical equipment is complete and serviceable.

___ Oxygen is on hand and serviceable.

___ Authorized medications are on hand and current.

___ Packing list is available.

___ Strip maps and/or road maps are available (with overlays).

___ On-vehicle equipment is on hand.

___ Log book is present and current.

___ All drivers are licensed.

___ Situational awareness equipment (position locator) is on hand and serviceable.

___ Communications equipment is on hand and serviceable and set to correct frequency.

___ Medical unit identification markers (according to the Geneva Conventions) are displayed.

Note. Markers are red on a white background only; camouflaged or subdued markers are not authorized.

Chemical, Biological, Radiological, And Nuclear Equipment

___ Individual protective equipment is on hand and serviceable. (One set is issued; the other is maintained in support.)

___ Protective masks are issued and serviceable.

___ Nerve Agent Antidote Kit is available and distributed, if required.

___ Convulsant antidote for nerve agent is available and distributed, if required.

___ Decontamination apparatus is complete and serviceable.

___ Basic load of decontamination supplies is on hand.

___ Chemical agent alarms are on hand and serviceable.

___ M256A1 detector kits are issued.

___ Chemical, biological, radiological, and nuclear contamination marking kits are distributed.

___ Chemical agent monitors are on hand, if authorized.

___ Replacement filters for protective masks are on hand.

___ Biological and chemical warfare agents prophylaxis, immunizations, and barrier creams have been accomplished, if appropriate.

___ Nerve agent pretreatment packets are available.

___ Radiac sets are on hand.

MISCELLANEOUS EQUIPMENT

___ Inspect binoculars.

___ Inspect camouflage nets and support systems, if appropriate.

___ Inspect night vision devices.

___ Ensure batteries are on hand and serviceable.

___ Inspect tentage.

___ Inspect global positioning systems, if available.

PERSONNEL

___ Ensure Soldiers are in the correct uniform.

___ Ask questions to ensure that Soldiers have been briefed on mission and situation.

___ Implement appropriate MOPP level.

___ Check for drivers licenses.

___ Brief Soldiers on operations safety and environmental injuries.

___ Individual equipment is on hand and stowed properly.

___ Soldier has eaten and is briefed on future field feeding.

___ Identification cards and tags are on hand and serviceable.

___ Camouflage self and equipment, when directed. (If medical vehicles are camouflaged, the Geneva Convention emblem must be removed.) (Refer to STANAG 2931.)

___ Work/rest plan (to include crew rest) is implemented.

___ Water discipline plan is implemented, if appropriate.

COMMUNICATIONS EQUIPMENT

___ Radios are operational (communications check completed).

___ Telemedicine equipment is available and operational, if available.

___ Speech security equipment functions, if available.

___ Frequencies are set.

___ Matching units are operational.

___ Antennas are tied down properly.

___ Connectors are clean and serviceable.

___ Field telephones (TA-312) are on hand and serviceable, if appropriate.

___ Batteries are on hand and charged.

___ Man-pack radio sets are complete.

___ Switchboard is on hand and serviceable.

___ Communications wire (WD-1) is on hand and serviceable.

___ Antennas and remotes are present and serviceable.

___ Signal operating instructions are available and secured.

___ Call signs, frequencies, and challenge passwords have been disseminated.

___ Perform communications check again.

VEHICLES

___ Loads are according to load plan; load plan is posted in the vehicle.

___ Hazardous cargo is properly identified and stored toward the rear of the vehicle for easy access and inspection.

___ Ammunition is issued and properly stored.

___ Vehicle fuel tank is topped off.

___ Package POL products and small arms lubricant are present.

___ Water cans are full.

___ Meals, read-to-eat are issued and stowed.

___ First-aid kits are present and complete.

___ Operator's manuals and lubrication orders are present for the vehicle, radios, and associated equipment.

___ Critical on-vehicle equipment and basic issue items are present.

___ Vehicle dispatch is complete.

___ Department of the Army (DA) Form 2404 (Equipment Inspection and Maintenance Worksheet) is complete.

___ No deadline deficiencies exist.

___ Before operation PMCS has been completed.

INDIVIDUAL WEAPONS

___ Clean and functional.

___ Cleaning tools/kits, bolts, and ruptured cartridge extractors are present.

___ Range cards are on hand.

___ Ammunition is issued, accounted for, and secured.

___ Magazines are issued.

___ Blank adapter installed, if appropriate.

___ Function check performed.

MEDICAL EVACUATION SUPPORT PLAN

___ Mission statement/commander's intent/task organization.

___ Intelligence preparation of the battlefield considerations.

___ Patient estimates by phase.

___ Identify expected areas of patient density.

___ Anticipate and plan for likely patient conditions.

___ Formulate/evaluate/implement/reevaluate COA.

___ Location of supported units.

___ Location of supporting MTFs/FSTs/hospitals.

___ Location of supporting evacuation units.

___ Augmentation of personnel and vehicles coordinated supporting MTFs.

___ Class VIII supply/resupply coordinated and available.

___ Location of supported unit CCPs and AXPs.

___ Evacuation routes primary and alternate.

___ Trafficability (enemy activity, hazards, condition, obstacles).

___ Security (gunship escort, MP escort, convoy support).

___ Strip maps/overlays/maps.

___ Landing zones designated for rotary-wing evacuation aircraft (primary and alternate)

___ Ensure that platoon/company/battalion/treatment facility litter teams are available to assist ambulance crews in loading patients.

___ Available litters and types (NATO standard, folding nonrigid, SKED®, and so forth).

___ Ground ambulances available (type/number/carrying capacity).

___ Air ambulances available (type/number/carrying capacity).

___ Supporting evacuation assets available (requested/ prepositioned).

___ Nonmedical vehicles for CASEVAC (type/number/carrying capacity).

___ Available aircraft (UH-60, CH-47).

___ Medical evacuation support plan for units without evacuation resources.

___ Treatment and evacuation of EPW and detainees.

___ Chemical, biological, radiological, nuclear considerations (contaminated evacuation vehicle plan, patient decontamination site, decontamination sites, contaminated aircraft).

___ Other.

CASUALTY CARE CONSIDERATIONS

___ Self-aid/buddy aid/CLS/trauma specialist.

___ Combat lifesavers trained and CLS bags stocked and issued.

___ Emergency care specialists and flight medics are rehearsed, aid bags fully stocked.

___ Trauma specialist Class VIII resupply sets fully stocked and pushed forward to company command post.

___ Locations of platoon/company/battalion/treatment facility CCPs are known.

___ Other.

Mass Casualty Plan (Resourced/Rehearsed/Coordinated)

___ Litter teams.

___ Litters (all).

___ Medical evacuation/CASEVAC vehicles (all).

___ Class VIII supplies (on hand).

___ Class VIII resupply (Prepackaged and prepositioned).

___ Company/battalion command net.

___ Brigade support medical company command net.

___ Personal effects, weapons, equipment, and property exchange.

___ Chemical, biological, radiological, and nuclear casualty/evacuation plan.

___ Location of decontamination site.

LANDING ZONE OPERATIONS

Safety

___ At all times take all the necessary safety precautions.

___ Platoon that is responsible for running the LZ provides all required safety gear. Included are—

  ___ Safety goggles.

  ___ Visual Signal, VS-17 panels/chemical lights.

  ___ Smoke (Appendix E).

___ Is the LZ secure.

Marking the LZ

___ Use colored smoke to identify the LZ and wind direction.

___ Use an inverted "Y" to mark the LZ.

___ During daylight, camouflage the LZ so the VS-17 panels cannot be seen until evacuation aircraft are confirmed and inbound.

___ At night, emplace chemical lights only when an AA is confirmed and is inbound.

Landing Requirements for Light Helicopters

___ Minimum real estate required is a cleared area 30 meters in diameter.

___ Area must have an approach and departure zone clear of obstructions and not over the established company area.

Appendix I

Safety on the Battlefield

The tactical environment provides ever-changing demands and unpredictable problems, often under stressful conditions. The interface of man, machine, and environment is constantly shifting. In this environment, mission accomplishment requires continuous leader involvement and flexible decision making. Not surprisingly, accidents and injuries increase during tactical operations. Safety in the tactical environment depends upon compliance with established standards. However, due to fluid conditions in the tactical environment, safe mission accomplishment relies heavily on complete integration of risk management into both planning and execution phases. Risk management assists commanders in anticipating and controlling hazards in the planning phase and in dealing with unexpected hazards as they arise in the execution phase.

SECTION I — GENERAL

COMMON ACCIDENTS

I-1. During all operations, it is critical commanders and leaders minimize risks through appropriate risk management procedures to protect the force and preserve the unit's war-fighting capability. Over half of all Army accidents during tactical operations are accredited to the following operations.

Vehicle Operations

I-2. Most accidents in this category are caused by excessive speed for conditions (weather, traffic). Other causes are recklessness, fatigue, unfamiliarity with roads, and untrained and inexperienced drivers. A lack of equipment/vehicle handling characteristics knowledge also contributes to accidents. Strict enforcement of standards is needed for all vehicle operations. The senior occupant must be responsible for this enforcement. Only trained, licensed personnel should be assigned to operate vehicles or equipment. Ground guides are mandatory during movement in bivouac and assembly areas, when backing, and during periods of reduced visibility. Operators must be familiar with proper operation and maintenance of commercial equipment.

Materiel Handling

I-3. These accidents occur when an object is too large or too heavy to handle for those attempting to move it. As a result, someone may sustain a muscle or back injury, or may be crushed. Overconfidence in one's ability, a lack of planning, and fatigue are common factors in such accidents.

Maintenance

I-4. These accidents are often caused by failure to follow procedures, using the wrong tools, and/or fatigue.

SEVERE ACCIDENTS

I-5. The following activities produce fewer accidents than those listed above; however, when they do occur, they often result in catastrophic damage or death. Commanders and leaders must pay special attention to unit missions which involve these activities or put Soldiers in situations or locations where they will come in contact with associated hazards and ensure they properly brief their personnel and mitigate risk to the fullest extent.

AMMUNITION AND EXPLOSIVES HANDLING

I-6. Horseplay, mishandling, disassembly, unauthorized use, and improper storage of ammunition and explosives account for many personnel injury accidents. It is essential to enforce accountability and security procedures for unexpended ammunition and explosives and to comply with explosive storage safety standards in AR 385–64 and DA Pam 385–64.

EXPLOSIVE SOUVENIRS

I-7. Educate Soldiers to dangers involved and the serious consequences of collecting unexploded ordnance on the battlefield or ranges. Post-tactical training shakedown inspections for this type of materiel are a must. Amnesty boxes are also useful. Platoon sergeants and squad leaders policing their Soldiers can prevent most of these accidents through proper training.

FIELD EXPEDIENTS

I-8. Be suspicious of shortcuts. Although tactical operations frequently involve employment of field expedients, risks and benefits must be carefully weighed. In many cases, field expedients are the result of a weak supply system or inadequate planning.

FIELD HEATERS AND STOVES

I-9. Operators of all types of heaters and stoves should be trained and licensed in advance. Equipment should be maintained and operated in accordance with operating instructions including use of proper fuel. Ensure combustible material is kept well away from heaters and stoves and fire fighting equipment is available for each heater and stove. Heaters or stoves with self-contained fuel supplies (small space heater) should not be refilled while the heater is on or still warm. Do not use heaters or stoves in tents or other confined spaces without use of proper ventilation, such as tent vent flaps, doors, or windows.

PETROLEUM, OILS, LUBRICANTS STORAGE AND HANDLING

I-10. The POL handlers must know and practice safety rules and procedures. Frequent inspections ensure safe storage and transfer of POL products. Proper grounding procedures must be followed. Field manual 10–67–1 describes use of protective equipment to prevent personnel exposure.

SOLDIER FATIGUE

I-11. When a Soldier's sleep time is dependent upon the tactical situation, debilitating fatigue can occur. Soldiers suffering from sleep loss experience various symptoms of fatigue including decreased coordination, narrowed attention span, and reduced performance to standard. Anticipate fatigue-related errors and take action to prevent them.

TACTICAL SLEEP PLAN

I-12. Commanders need to develop and enforce sleep plans. Control sleeping areas to prevent Soldiers from being crushed by moving vehicles in and out of the area.

WATER OPERATIONS

I-13. Plan water operations carefully, as risks of drowning and equipment loss are high during water operations. Pair strong swimmers with weak ones to protect personnel. Secure equipment and float it across rather than requiring individuals to carry their equipment. Use safety lines and personal floatation devices.

WEAPONS

I-14. Most accidents occur when cleaning or clearing individual weapons, entering or exiting vehicles, or running with loaded weapons. Provide guidance for weapons handling, loading, and clearing, and ensure it is strictly enforced. Do not load weapons that are not essential for the current mission.

WEATHER-RELATED CASUALTIES

I-15. Consider the effects of weather during planning. Unit effectiveness is quickly lost through weather-related casualties such as frostbite, heat stroke, lightning strikes, and falls. Instruct Soldiers in awareness, prevention, and first aid for weather-related injuries and when these conditions can be expected.

VEHICLE CONVOY OPERATIONS

I-16. Convoy operations can be very dangerous if not properly planned. Control of convoy speed and proper separation between vehicles is critical to reducing risk of an accident. See FM 4-01.45 for in-depth information on tactical convoy operations.

CONVOY COMMANDER RESPONSIBILITIES

I-17. Convoy accidents are most commonly caused by a leader failing to perform his or her duties as the commander of a vehicle movement as required in FM 55-30. This includes failure to:

- Control the group's movement.
- Ensure vehicles maintain proper march speed.
- Properly mark unit vehicles.

I-18. Convoy commanders are responsible for ensuring the safe movement of the convoy. To do this, they must positively control the convoy's movement by using communications equipment among the vehicles. This includes ensuring each vehicle has a properly trained, equipped, and supervised crew leading from the front in the absence of radios and/or other means of managing the march. Control of the movement includes enforcing speed limits, march intervals, and seat belt usage. It also requires the ability to stop the march if an unexpected hazard is encountered along the route.

I-19. Preparing vehicles and Soldiers for movement is a leader responsibility. Inexperienced Soldiers, personnel turbulence, and ever increasing training requirements have caused some units to become complacent in managing risks associated with vehicle movements. According to the risk management process, leaders must identify hazards associated with the mission and develop, implement, and supervise control measures to mitigate those risks. These control measures include marking the vehicles in accordance with local SOPs, briefing crews on hazardous conditions expected along the route, precombat checks of personnel and equipment, and developing preaccident emergency contingency plans.

I-20. Unit SOPs should address leader responsibilities during movements of any number of vehicles and identify means to implement common controls.

DRIVER SKILLS

I-21. Operators should be taught the specific skills needed for tactical vehicle operations, which, in addition to requirements outlined in AR 600-55, include—

- Pulling and backing trailers.
- Vehicle recovery operations.

- Loading and lashing of cargo.
- Methods of negotiating difficult terrain such as sand dunes, rice paddies, mountainous terrain, icy roads.
- Ground-guide procedures and signals.
- Methods and procedures for retrieving vehicles stuck in snow, mud, sand, or other restrictive terrain.
- Proper parking and use of the proper-sized chock blocks.

Use of Safety Equipment

I-22. When the tactical situation allows, flashers should be turned on immediately if a vehicle is disabled or impedes traffic, and every effort made to move the disabled vehicle from the roadway. Highway warning kits should be provided for each vehicle in a convoy. When a vehicle is disabled, place the warning triangle a minimum of 100 meters to the rear of the vehicle and ensure personnel remain clear of the road and rear of the vehicle.

Night Operations

I-23. Personnel who are required to operate motor vehicles while wearing NVG must be trained and tested on use and operation of such devices, and the training recorded on the individual's driver training records. Other factors include:

- Ground guides should be used when moving vehicles at night in areas where troops are present.
- Blackout driving should be prohibited on roads open to the public.

FIRE PREVENTION

I-24. The risk of fire is high in areas where a large number of Soldiers are in tents. The following guidance will reduce the risk of fires:

- Establish a fire prevention and protection plan including procedures for inspecting and recharging fire extinguishers during tactical operations.
- Appoint a fire marshal for each bivouac area and train them in their duties. Train Soldiers in fire prevention techniques as well as emergency procedures in the event of a fire.
- Establish safe distances between tents to reduce risk of multiple losses from one fire.
- Provide available fire fighting equipment (portable extinguishers, sand, water buckets, and shovels) to contain small fires. Ensure personnel are trained on their use.
- Establish procedures for sounding fire alarms.
- Ensure no-smoking areas are established and enforced.
- Establish an inspection system ensuring compliance with fire prevention standards.
- Ensure flammable materials are stored according to appropriate directives and checklists.
- Ensure vehicle fire extinguishing/suppression systems are operational and crews are proficient in using the systems.
- Provide a designated fire plan, equipment, and trained personnel for POL storage, ammunition supply points, motor pools, hospitals, and hangars.

SECTION II — ASSEMBLY AREA OPERATIONS

SAFETY OFFICER/NONCOMMISSIONED OFFICER

I-25. The safety officer and NCO must establish, in coordination with the 1SG, platoon leaders, and platoon sergeants, procedures for ground guides and incorporation of these procedures into the unit SOP. Company commanders must weigh mission requirements and personnel available while in combat operations and use risk management analysis to validate procedures for implementation into the SOP.

Operating within the modular force structure may require assistance from support units to meet necessary personnel and vehicle requirements to safely ground guide and move aircraft.

GROUND VEHICLES

I-26. Commanders and leaders must be aware of hazards from ground vehicles in AAs. Vehicles moving in the AA should always have a ground guide leading the vehicle until out of the AA. This prevents injury to personnel or damage to equipment. Use of a one-way path through the unit area will assist greatly in preventing accidents. Additionally units may be operating in the vicinity of armored vehicles and Soldiers must take additional precautions due to the inability of armored vehicle crews to see personnel close to their vehicles.

I-27. During initial entry into the AA, platoon ground guides lead convoy parties to their platoon locations with each vehicle dismounting one person to ground guide their vehicle. During this period of movement into the AA, there is a higher probability of accident or injury due to the amount of traffic and unfamiliarity of the area. Commanders must ensure movement into the AA is thoroughly briefed.

AIRCRAFT

I-28. Aircraft in the AA present movement hazards and contribute to an additional hazard of blowing debris. Blowing debris is a hazard to personnel, aircraft, and equipment. The variation in environments in which a unit may operate determines the type and amount of debris. Desert environments, for instance, produce brownout conditions which are a hazard to personnel, equipment, aircraft, and crews. Preventive measures should be incorporated into the unit SOP mitigating this risk. An example of risk mitigation for aircraft operations, in and around the AA parking areas, is brownout mats. These mats reduce the brownout effect of operating in a desert environment. Commanders must ensure SOPs are implemented for each of the environments the unit may operate within. This will allow the unit to establish ground guide procedures for aircraft, whether while hovering into and out of the AA or during ground movement of the aircraft within the AA.

I-29. Ground guide procedures for flight into or out of the AA will be based on the distance between aircraft and proximity to hazards (natural or manmade). In environments where the AA allows for wide scatter of aircraft, the need for ground guides while hovering into and out of the parking area is negligible. In contrast, when the AA parking areas require close interval spacing of aircraft, the need for guides is critical to safety. Additionally, if the AA requires aircraft to park in close proximity to natural or manmade hazards, ground guides are required. The need for ground guides is more critical with operations under night or with NVG devices.

I-30. Upon initial entry into the AA, ground guides should meet incoming aircraft. The situation may dictate crews entering the AA know limited information on where they will park their aircraft. The ground guides guide the aircraft into their assigned parking spot. In desert environments, the ability of personnel to guide aircraft into the parking location is decreased. Brownout upon landing requires a thorough prebrief to the flight crew on the AA parking plan and as exact a location as possible in relation to a ground marking, such as an inverted Y, of each aircraft parking location. Certain situations may require aircraft to land directly on a parking spot. The company safety officer and NCO must be involved in the design and procedures planned for aircraft operations into and out of the AA.

WING WALKING

I-31. Ground guides will also be utilized while moving aircraft with ground handling equipment. Ground guiding aircraft while moving with a vehicle requires additional personnel to ensure rotor blades and aircraft do not come in contact with another aircraft or object. Commanders must ensure procedures designed by the safety officer/NCO are integrated into the SOP. The plan should address guides necessary for various settings (such as define how many are necessary for close proximity parking and scattered parking not in close proximity to another aircraft or hazard). Ground movement of aircraft is normally associated with aviation maintenance operations. The maintenance officer should be involved in the design of ground guide/wing walking procedures for the unit SOP. A critical component of the aircraft ground movement plan is communication. Personnel conducting these operations must conduct a team prebrief and ensure that the team leader has a device that will signal all members to immediately halt the operation (survival vest whistle).

GROUND GUIDES

I-32. Commanders and leaders must ensure ground guide operations, vehicle and aircraft, are incorporated into their company SOPs. During AA operations, ground guides perform a key role in preventing personnel injury, and vehicle and equipment damage. Ground guide operations are critical to maintaining combat readiness for the unit.

PYROTECHNICS

I-33. Aviation units normally use pyrotechnics while in an AA. Units utilize pyrotechnics along likely enemy avenues of advance within their security sector as a warning of activity along these avenues. Additionally, aircrews utilize pyrotechnics as signaling devices in downed aircrew rescue operations. Commanders, unit leaders, and the ALSE officer must train personnel in proper use of pyrotechnics and its associated hazards. These items will be addressed in the unit SOP.

I-34. Pyrotechnics require protection against moisture, dampness, and high temperature. Pyrotechnic items must be given high priority for the best available protection because of their sensitivity. Wet pyrotechnic material is hazardous to store; consequently, any boxes showing signs of dampness will be removed from a storage site and inspected. If the pyrotechnic materiel is wet, it will be destroyed. Certain kinds of materiel deteriorate with age and have expiration dates on the containers. Loose pyrotechnic tracer composition, flare composition, and similar mixtures spilled from broken containers should be carefully taken up and covered completely with SAE 10 (EO–10) engine oil and removed for appropriate disposal.

I-35. Pyrotechnics are hazard division 1.3 which includes items that burn vigorously and cannot usually be extinguished in storage situations. Explosions are confined to pressure ruptures of containers and do not produce propagating shock waves or damaging blast overpressure beyond the magazine distance specified in DA Pam 385-64, Table 5–16. Tossing about of burning container materials, propellant, or other flaming debris may cause a severe hazard of spreading fire.

I-36. The fire fighting extinguishing agent for pyrotechnic materials, Class D (combustible metals [magnesium, potassium, and so forth]) hazard, should be a dry powder agent. Pyrotechnics are a mass fire hazard due to their chemical agents. The mass fire hazard can be more hazardous based on other likely types of ammunition stored with pyrotechnics. These fires may be fought if:

I-37. Explosives are not directly involved.

- White phosphorous munitions are involved, there may be an explosion (immerse/spray with water continuously).
- Pyrotechnics and magnesium incendiaries do not use carbon dioxide, Halon extinguishers, or water on or near munitions. Allow magnesium to cool unless upon flammable materiel. In this case, use a 2-inch layer of dry sand or powder on the floor and rake the burning material onto this layer and re-smother.

WARNING

DO NOT DISCHARGE PYROTECHNICS IN THE VICINITY OF AIRCRAFT FLYING IN THE AREA.

SECTION III — PROTECTING THE SOLDIER

ENVIRONMENTAL CONSIDERATIONS

I-38. As the military looks into the next century, and even today, military units should try to avoid unnecessary environmental damage, not only in training, but also across the spectrum of operational missions. A mission's success may be determined by political or socio-economic stability, both of which are affected by environmental factors and resources. The Army must be able to identify ways to protect the natural environment while executing full range of their missions:

- Considering the environment in planning and decision making in conjunction with other essential considerations of national policy.
- Protecting the environment of home stations and training areas as a means of retaining resources for mission purposes.
- Using environmental risk assessment and management principles to integrate environmental considerations into mission performance.
- Instilling an environmental ethic in Soldiers.
- Understanding linkages between environmental protection issues and their associated impact on safety, force protection, and AHS.

I-39. Soldiers must be prepared to respond across full spectrum operations in any component (offense, defense, stability or civil support operations), sometimes within the same operation. Deployed forces must be able to conform to environmental protection requirements of the theater commander without impairing combat effectiveness. For in-depth information on environmental considerations and Soldier protection, see FM 3-100.4.

I-40. When deployed, company commanders will often deal with base camps. Base camps, while not installations, are comparable to small towns and require many of the considerations applied to installations. A mayor (often the HHC commander) assists base camp commander with control of base operations. A base camp coordination agency will provide expertise and support to the commander, largely through its subordinate base camp assistance/assessment team. Environmental expertise is resident or aligned with this team and available to support the base camp commander and designated mayor of the base camp. They provide technical recommendations and maintain appropriate standards. More information about this phenomenon is provided in the Center for Army Lessons Learned Newsletter 99-9, Integrating Military Environmental Protection.

I-41. Army Regulation 200-1 and the Army's Commander's Guide to Environmental Management specify commanders' environmental responsibilities. To carry out these responsibilities, commanders:

- Comply with an installation's environmental policies and legally applicable and appropriate federal, state, and local laws and regulations or country-specific final governing standards if outside the continental United States.
- Demonstrate a positive and proactive commitment to environmental stewardship and protection.
- Provide environmental training required by law, regulation, or command policy.
- Ensure all personnel perform their duties in compliance with environmental laws and regulations, and respond properly to emergencies.
- Promote proactive environmental measures and pollution prevention.
- Supervise compliance with environmental laws and regulations during operational, training, and administrative activities.
- Include environmental considerations in mission planning, briefings, meetings, execution, and AARs.
- Understand the requirements of Army environmental programs.
- Identify and assess environmental risks of proposed programs and activities.
- Coordinate unit activities with higher headquarters environmental elements.
- Appoint and train an environmental compliance officer (ECO) coordinator for the unit.

- Ensure SOPs contain all environmental considerations and regulatory requirements appropriate for the level of command.
- Conduct environmental self-assessment or internal environmental compliance assessments.
- Understand linkages between environmental considerations and their associated impact on safety, force protection, and AHS.

ENVIRONMENTAL DUTIES

I-42. The key to fulfilling environmental requirements successfully at unit level is the ECO. Army Regulation 200-1 directs all Army unit commanders "appoint and train ECOs at appropriate levels to ensure compliance actions take place." In company-sized units, this duty will generally translate into an extra duty. The ECO manages environmental issues at unit level and ensures environmental compliance. He also coordinates, through the respective chain of command, with the supporting installation environmental staff to clarify requirements and obtain assistance.

I-43. The ECO accomplishes environmental compliance requirements on behalf of the commander. He also coordinates with supporting installation environmental staff to clarify requirements and obtain assistance. While this position of responsibility is not a formal staff position, the ECO is critical to the commander's environmental program. The ECO:

- Advises the unit on environmental compliance during training, operations, and logistics functions.
- Serves as commander's eyes and ears for environmental matters.
- Coordinates between unit and higher/installation headquarters environmental staffs.
- Manages information concerning unit's environmental training and certification requirements.
- Performs unit environmental self-assessment inspections.
- Performs environmental risk assessments.

UNIT PLANNING

I-44. Staffs integrate environmental protection into planning for larger units. Unit leaders integrate environmental protection into unit planning for battalion- and company-level units. Soldier safety should be addressed in SOPs, OPORDS, and training plans.

STANDING OPERATING PROCEDURES

I-45. Unit leaders develop SOPs reflecting environmental protection considerations for routine tasks and activities. The SOPs provide information to Soldiers on how to accomplish routine tasks in an environmentally sound manner. The SOPs incorporate local requirements. As local requirements change, unit leaders update their SOPs. The SOPs also help define environmental protection requirements for all unit activities—facility operations, field operations, deployment, and combat. Unit leaders ensure SOPs comply with local requirements by coordinating with higher headquarters staff—usually the environmental office, the surgeon and his staff, preventive medicine personnel, and the SJA or engineer coordinator.

I-46. Unit leaders conduct environmental risk assessments when planning operations or activities. Risk assessment is a standard element of TLP. Unit leaders perform environmental risk assessments for activities not addressed in the SOP or when conditions differ significantly from those described in the SOP. A maintenance unit does not perform a risk assessment every time it performs a lubrication or service. Rather, the SOP describes the correct manner to perform these actions. Risk assessments apply to garrison operations as well as field operations.

OPERATION ORDERS

I-47. Unit leaders address environmental protection in their plans and orders including: WARNOs, OPORDs, OPLANs, concept of operation plans (CONPLANs), and FRAGOs. The higher headquarters staff develops an environmental appendix/annex to its OPORD/OPLAN/CONPLAN. Subordinate unit leaders draw environmental information from the environmental appendix to the OPORD/OPLAN/

CONPLAN, or from Annex L in a joint operations planning and execution system document. Field Manual 5-0 directs the inclusion of Appendix 2 (Environmental Considerations) to Annex F (Engineer) of the OPLAN/OPORD/CONPLAN (battalion level and above) and specifies that lower-level unit leaders/staffs include environmental information in coordinating instructions and service and support paragraphs.

Risk Management Plan

I-48. Risk management is an integral part of the military decision-making process. The risk management process provides the framework for making risk management a routine part of planning, preparing, and executing operational missions and everyday tasks. Assessing environmentally-related risks is part of the total risk management process.

I-49. Knowledge of environmental factors is a key to planning and decision making. With this knowledge, leaders quantify risks, detect problem areas, reduce risk of injury or death, reduce property damage, and ensure compliance with environmental laws and regulations. Unit leaders should conduct risk assessments before conducting any training, operations, or logistical activities.

Tactical Risk/Accident Risk

I-50. When assessing the risk of hazards in operations, the commander and leaders must look at two types of risk.

- Tactical risk is risk concerned with hazards existing because of the presence of either the enemy or an adversary, thus involving the considerations of force protection. It applies to all levels of war and across the spectrum of operations. For example, during the Gulf War, the enemy's demolition of oil fields created a significant health and environmental hazard to the surrounding countryside and those units maneuvering through the area.
- Accident risk includes all operational risk considerations other than tactical risk. It includes risk to friendly forces and risk posed to civilians by an operation, as well as the impact of operations on the environment. It can include activities associated with hazards concerning friendly personnel, civilians, equipment readiness, and environmental conditions. Examples of environmentally related accident risk are improper disposal of hazardous waste, personnel not properly trained to clean up a spill, and units maneuvering in ecologically sensitive terrain. Preventive medicine considerations also fall into this area of risk.

Note: Tactical risk and accident risk may be diametrically opposed. The commander may accept a high level of environmentally-related accident risk on reducing the overall tactical risk. For example, a commander may decide to destroy an enemy's petroleum storage area to reduce its overall tactical risk.

Training Plans

I-51. Formal planning for training culminates with publication of the training schedule. Informal planning and coordination (preexecution checks) continue until training is performed. During rehearsals, leaders ensure all safety and environmental considerations are met.

I-52. To conduct effective, meaningful training for Soldiers, leaders, and units, thorough preparation is essential. This means well prepared trainers, Soldiers, and support personnel are ready to participate, and their facilities, equipment, and materials are ready to use.

I-53. A unit executes training the same way it executes a combat mission. The chain of command is present, in charge, and responsible. During operations, leaders ensure environmental practices and preventive measures are being employed. Once Soldiers understand what is expected of them, these practices become merely another measure of unit proficiency and level of unit discipline.

FIGHTER MANAGEMENT

I-54. Fighter management describes a process essential to effective and efficient battle staff operations. At company level, commanders do not have staffs. The battle staff becomes those personnel in the unit who operate the CP. Successful continuous operations at the company level are more demanding than higher level organizations. The unit requires a TACSOP allowing rest, especially for critical personnel. At company level, the commander and 1SG will find it very demanding to try to rest. Their rest plan must be a priority for the organization to be lead effectively and their units to be successful on the battlefield.

I-55. The cycle of recurring events within a CP focuses the CP officer in charge/noncommissioned officer in charge (OIC/NCOIC) on meeting information and action requirements. Company personnel are normally required to attend battalion/company recurring meetings. This has an impact on company CP operations and schedules. These recurring events include—

- Shift changes.
- Battle update briefings to the commander.
- Battalion targeting meetings.
- Battalion reports.
- Battalion battle updates without the commander.
- Battalion commanders' collaborative sessions.
- Company CP OIC/NCOIC shift change briefing/collaborative sessions.

I-56. The company CP OIC/NCOIC must achieve battle rhythm for updating and viewing information and understand how to use it to affect operations. A well-established battle rhythm aids the commander and leaders with CP organization, information management and display, decision making, and fighting the battle. Battle rhythm demands careful planning and design. Many competing demands must be deconflicted. Even subordinate platoons and sections affect a company battle rhythm based on their needs and unit procedures.

I-57. Command discipline is required to enforce sleep cycles and create an environment where sound sleep can be achieved. Battle rhythm and associated planning is a proactive, not reactive, exercise.

SLEEP PLANS

I-58. Units must develop detailed rest plans and enforce them. Leaders must rest to maintain their effectiveness; however, some leaders attempt to get involved in every aspect of planning and execution. An integral part of the planning process is determining when senior leader presence is required. It is just as important to identify when a leader's presence is not required. The planning process should include the following supporting techniques:

- Include a sleep plan in the METT-TC analysis.
- Ensure leaders have confidence in second and third echelons of leadership and their ability to make routine decisions.
- Instill trust and confidence in the officers, junior NCOs, and specialists by effective training and SOPs.
- Consider contingencies and establish criteria for waking leaders.
- Post sleep plans in CPs, platoon, and section areas; everyone needs to know the plan.
- Synchronize sleep plans with higher headquarters and subordinate platoons/sections.

CONTINUOUS OPERATIONS

I-59. Continuous operations are combat operations continuing around the clock at a high pace, requiring Soldiers to fight without relief for extended periods. Opportunities for sleep are scattered throughout the day and night.

SUSTAINED OPERATIONS

I-60. Sustained operations are operations conducted 24 hours a day with little or no opportunity for sleep. Sustained operations are when the same Soldiers or small units engage in combat operations with no opportunity for the unit to stand down and little time for Soldiers to sleep. Aviation units routinely conduct sustained operations to an extent even greater than the infantry due to aviation's maneuver, maneuver support and sustainment roles. To combat the negative effects of sustained operations, commanders and leaders must enforce fighter management principles on their organizations. A fighter management policy should be addressed in the unit SOP, and higher headquarters policy should correlate with company policy.

DEGRADATION OF COMBAT CAPABILITY

I-61. As sustained operations continue, all Soldiers begin to show effects of general fatigue and lack of sleep. Unless counteracted, unit performance of combat tasks decline. Recent studies indicate performance is degraded by 25 percent for each 24-hour period without sleep. After 96 hours, performance can be expected to be near zero. Determination to endure must be supplemented by countering the adverse effects to slow the rate of decline. It becomes more difficult to perform assigned tasks to the required standard. Leaders need to recognize signs of serious sleep deprivation in their subordinates. Studies show performance in all duty positions does not degrade the same. Performance in a duty position where there is a heavy load of mental tasks, such as aviation duties (determining, calculating, thinking, decision-making), degrades faster than the performance in a position whose tasks are mainly physical (firing, running, lifting, digging).

I-62. In addition to the degradation caused by fear, fatigue, and loss of sleep, there is a significant loss of effectiveness caused by operation in MOPP 4. When units are using full CBRN protective equipment, judgment is degraded, communications are less effective, and information flow between units is reduced.

FATIGUE IN FLYING OPERATIONS

I-63. Aviation operations are inherently dangerous. Commanders and leaders must be aware of increased dangers present in combat aviation continuous and sustained operations. Additionally, leaders must be able to recognize the symptoms of fatigue and how to deal with fatigued aircrews. Field Manual 3-04.301 provides an in-depth review of aeromedical factors associated with aviation.

OPERATIONAL TEMPO AND BATTLE RHYTHM

I-64. The aviation company should be staffed for 24-hour operations; however, it also conducts cyclical missions. Standard operating procedures establish methods of ensuring the right personnel are available for either cyclical or 24-hour operations. Regardless of methods used, practice during exercises must determine strengths and weaknesses of headquarters personnel for CP operations and training for additional personnel who may be utilized to staff the CP during continuous or sustained operations. Such knowledge allows leaders to focus on critical areas and personnel requiring additional training.

I-65. In planning, the battalion staff must consider battle rhythm requirements of subordinate companies and platoons. Depending on the situation, the battalion may schedule missions allowing company or platoon rotations to maintain their battle rhythm.

ABSENCE OF BATTLE RHYTHM

I-66. Without procedures establishing battle rhythm, leaders and units reach a point of diminished returns. This typically occurs between 72 and 96 hours of operations. As leader fatigue sets in, information flow, planning process, execution, and sustainment suffer—often greatly. Symptoms of diminished battle rhythm include the following:

- Leader fatigue.
- Leaders not fully aware of critical decision points (DPs).
- Leaders not available at critical DPs.
- Disjointed time lines between various levels of command.

PRESENCE OF BATTLE RHYTHM

I-67. Battle rhythm allows units and leaders to function at a sustained level of efficiency for extended periods. Effective battle rhythm permits an acceptable level of leadership at all times. It can focus leadership at critical points in the fight or during particular events. Procedures and processes facilitating efficient decision making and parallel planning are critical to achieving battle rhythm. Every component of battle rhythm contributes uniquely to sustained operations.

TRAINING

I-68. It is difficult, if not impossible, to establish battle rhythm while simultaneously conducting operations. Preplanning makes it happen. Planning, preparing, and training before deployment lays a solid foundation for viable battle rhythm during operations. Commanders must ensure all company personnel are trained on TOC set-up and operations, and they understand the team concept required to conduct CP operations during combat operations for extended periods. Additionally, company commanders must coordinate with battalion staffs during training to deconflict staff operations and requirements which are counterproductive to the company battle rhythm.

BATTLE RHYTHM ELEMENTS

I-69. Battle rhythm is a multifaceted concept including the following elements:

- Sleep/rest plans.
- Trained second- and third-tier leadership in CPs.
- Synchronized upward multiechelon timelines.
- Parallel planning.
- Established processes and SOPs.

COMMAND POST PERSONNEL DEPTH

I-70. Established processes and SOPs relieve many antagonistic effects of extended operations. Standard operation procedures establishing and maintaining battle rhythm by facilitating routine decisions and operations are a step in the right direction. Soldiers trained to act appropriately in the absence of leaders or orders can relieve commanders and leaders of many of the time-consuming tasks robbing them of essential rest. Examples of areas NCOs and junior officers can accomplish for the commander include—

- Battle summaries and updates during a fight.
- Intelligence updates before, during, and after a battle.
- Sustainment updates before, during, and after a battle.
- Updates to the next higher commander.
- Shift change briefings.

CHALLENGES TO BATTLE RHYTHM

I-71. Challenges to battle rhythm include enlisted, NCO, and junior officer duties; and synchronization of planning, execution, rehearsal timelines, equipment maintenance, and sleep plans.

NONCOMMISSIONED OFFICER AND JUNIOR OFFICER RESPONSIBILITIES

I-72. At the company level, all personnel are critical to the success of the operation and provide valuable contributions. It is imperative commanders make certain each Soldier understands the importance of even the most menial tasks, such as CP security and TOC setup and teardown. Given the amount of personnel in a company, 24-hour operations and force protection requirements, commanders must utilize all personnel available to successfully accomplish CP operations and continuous missions. This becomes even more critical when a company has platoons or sections staged forward in support of BCTs. The improper use of personnel produces the following results:

- Key leaders become exhausted.

- CP operations trained personnel become exhausted.
- The initiative of trained subordinates is stifled, and the incentive to train is diminished.

I-73. The following techniques ensure proper use of personnel:

- Appropriate tasks are assigned to junior NCOs and specialists.
- Effective training and SOPs instill trust in officers and confidence in junior NCOs and specialists.
- Effective command guidance to the company that it is a team effort and all personnel available contribute to the effort.

CONTINUOUS OPERATIONS AND TIMELINES SYNCHRONIZATION

I-74. Timelines for the operation at hand must allow for not only the next operation, but also extended continuous operations. Synchronized, multiechelon timelines assist units in achieving battle rhythm. If units do not address critical events at least one level up and down, disruption results. An example of an unsynchronized timeline is a battalion rehearsal that conflicts with platoon precombat inspections or other events in their internal timeline. Lower echelon units seldom recover from a poor timeline directed by a higher headquarters, as well as platoons and sections at company level. Company commanders must coordinate with their battalion staffs on development of SOPs that include planning, rehearsal, and execution timelines one level below battalion to prevent these conflicts.

STANDING OPERATING PROCEDURES UTILIZATION

I-75. SOPs must be practiced and reviewed during professional development and sergeants' time. The existence of an SOP will not resolve troop-leading challenges unless the SOP is practiced often and internalized by unit members. Checklists are critical, as many leaders will often find themselves rushed, physically fatigued, distracted, and deprived of sleep. Checklists ensure each step is considered even when leaders are exhausted.

SECTION IV — ACCIDENT/INCIDENT REPORTING

I-76. Persons involved in, or aware of, an accident must immediately report it to the commander or supervisor directly responsible for the operation, materiel, or person(s) involved. Table I-1 and Table I-2 identify reporting requirements for aviation and ground accidents/incidents (peacetime and combat). See AR 385-40, DA Pam 385-40, and AR 385-95 for in-depth instructions on Army accidents. Company commanders ensure unit SOPs cover accident/incident reporting, and it is fully addressed by the company aviation safety officer.

Table I-1. Aviation accident notification and reporting requirements and suspenses

Peacetime				***Combat ****	
Accident Class	***Notification Telephonic Worksheet***	***DA Form 2397***	***Reporting AAA report***	***Notification Telephonic Worksheet***	***Reporting AAA report***
A	Immediate–to USASC (telephonic notification–no hardcopy notification required (DSN 558-2660/2539/3410 or Commercial (334) 255-2660/2539/3410	(CAI/IAI) 90 calendar days	Aircraft ground ACDTS only-30 calendar days	Same as peacetime to USASC or Safety Rep, forward	(Only when CDR determines DA Form 2397 investigation/ report not feasible) submit as soon as conditions, situation permit-Do not exceed 30 calendar days.
B	Immediate–to	(CAI/IAI) 90	Aircraft ground	Same as	(Only when CDR

Peacetime				*Combat **	
Accident Class	***Notification Telephonic Worksheet***	***DA Form 2397***	***Reporting AAA report***	***Notification Telephonic Worksheet***	***Reporting AAA report***
	USASC (telephonic notification–no hardcopy notification required (DSN 558-2660/2539/3410 or Commercial (334) 255-2660/2539/3410	calendar days	ACDTS only 30 calendar days	peacetime to USASC or Safety Rep, forward	determines DA Form 2397 investigation/ report not feasible) submit as soon as conditions, situation permit- Do not exceed 30 calendar days.
C	Immediate–to USASC (telephonic notification–no hardcopy notification required (DSN 558-2660/2539/ 3410 or Commercial (334) 255-2660/2539/3410	N/A	30 calendar days	Same as peacetime to USASC or Safety Rep, forward	Same as peacetime
D	N/A (Unless SOF issue involved/ suspected)	N/A	10 calendar days	Same as peacetime	Same as peacetime
E	N/A (Unless SOF issue involved/ suspected)	N/A	10 calendar days	Same as peacetime	Same as peacetime
F	N/A (Unless SOF issue involved/ suspected)	N/A	10 calendar days	Same as peacetime	Same as peacetime
Submis-sion Methods	Class A-C telephonic (immediate) Class D, E, F-IF SOF	Mail	Typed or hand printed AAA reports by mail/FAX or courier/ message/ format/elec-tronic submission, Include attachments as required	Same as peacetime	Same as peacetime
Notes * Only when the senior tactical commander determines that the situation, conditions, and/or time does not permit normal peacetime investigating and reporting.					

Table I-2. Ground accidents notification and reporting requirements and suspenses (3)

Peacetime				***Combat (2)***	
Accident	***Telephonic Notification***	***AGAR***	***DA Form 285***	***Telephonic Notification***	***AGAR only*** ***By any means possible***
Class	***Worksheet***			***Worksheet***	***(Message, Electronic, FAX, Phone, Hand Carry, Mail)***
On-Duty					
A	Immediately (1)	Not required	IAI/CAI-90 days	Immediately (1)	As time permits (Not to exceed 30 days)
B	Immediately (1)	Not required	IAI/CAI-90 days	Immediately (1)	As time permits (Not to exceed 30 days)
C	Not required	Within 30 days	Not required	Not required	As time permits (Not to exceed 30 days)
D	Not required	Within 30 days	Not required	Not required	As time permits (Not to exceed 30 days)
Off-Duty					
A	Immediately (1)	Within 30 days	Not required	Immediately (1)	As time permits (Not to exceed 30 days)
B	Immediately (1)	Within 30 days	Not required	Immediately (1)	As time permits (Not to exceed 30 days)
C	Not required	Within 30 days	Not required	Not required	As time permits (Not to exceed 30 days)
D	Not required	Within 30 days	Not required	Not required	As time permits (Not to exceed 30 days)

Notes:
(1) USASC must be notified IMMEDIATELY by phone at DSN 558-2660/2539-3410 or Commercial (334) 255-2660/2539/3410 or notify USASC Safety Rep forward (during combat).
(2) ONLY when the senior tactical commander determines that the situation, conditions, and/or time does not permit normal peacetime investigating and reporting.
(3) Army civilian injury only accidents should be reported on appropriate Department of Labor Forms according to this regulation.

Glossary

SECTION I — ACRONYMS AND ABBREVIATIONS

A2C2	Army airspace command and control
AA	air ambulance/active Army
AAA	abbreviated aviation accidents
AAR	after-action review
AATF	air assault task force
ABCA	American, British, Canadian, and Australian
ACDTS	accidents
ACR	armor cavalry regiment
AD	air defense
AE	aeromedical evacuation
AECT	aeromedical evacuation control team
AELT	aeromedical evacuation liaison team
AF	Air Force
AGAR	abbreviated ground accident report
AGPU	aviation ground power units
AHS	Army Health System
ALCT	airlift control team
ALSE	aviation life support equipment
amb	ambulance
AMEDD	Army Medical Department
AO	area of operations
AOC	air operations center
AOE	Army of Excellence
APOD	aerial port(s) of debarkation
AR	Army regulation
ASI	additional skill identifier
ARSOF	Army special operations forces
ARNG	Army National Guard
ASCC	Army service component command
ASF	aeromedical staging facility
ASMC	area support medical company
ASOS	Army support to other Services
ASTS	aeromedical staging squadron
ATM	advanced trauma management
ATS	air traffic service

attn	attention
aug	augmentation
avail	available
AVIM	aviation intermediate maintenance
AVN	aviation
AVUM	aviation unit maintenance
AXP	ambulance exchange point
BAE	brigade aviation element
BAS	battalion aid station
BCT	brigade combat team
bde	brigade
BH	behavioral health
BIFV	Bradley infantry fighting vehicle
bn	battalion
BOA	basis of allocation
BSA	brigade support area
BSB	brigade support battalion
BSMC	brigade support medical company
BSS	brigade surgeon section
BTU	British thermal unit
C2	command and control
CA	civil affairs
CAB	combat aviation brigade
CAI	centralized accident investigation
CASEVAC	casualty evacuation
CASF	contingency aeromedical staging facility
CBRN	chemical, biological, radiological, and nuclear
CBRNE	chemical, biological, radiological, nuclear, and high yield explosives
CCATT	critical care air transport team
CCIR	commander's critical information requirements
CCP	casualty collecting point
CE	communications-electronic
CJA	Command Judge Advocate
CJTF	combined joint task force
CLS	combat lifesaver
CLT	casualty liaision team
CM	campaign module
CMCC	corps movement control center
CMD	command
CMO	civil-military operations

CO	company
COA	course(s) of action
COCOM	combatant command
COMSEC	communications security
CONPLAN	concept of operation plan
CONUS	continental United States
COSC	combat and operational stress control
COSR	combat and operational stress reaction
CP	command post
CRTS	casualty receiving and treatment ships
CSH	combat support hospital
CSAR	combat search and rescue
CTC	Combat Training Center
CZ	combat zone
DA	Department of the Army
DCO	defense coordination officer
DCSCOMPT	Deputy Chief of Staff, Comptroller, US Army
DCSIM	Deputy Chief of Staff for Information Managment, US Army
DCSLOG	Deputy Chief of Staff for Logistics, US Army
DCSPER	Deputy Chief of Staff for Personnel, US Army
DCSSPO	Deputy Chief of Staff for Support Operation, US Army
DD	Department of Defense
DHS	Department of Homeland Security
div	division
DMOC	division medical operations center
DNBI	disease and nonbattle injuries
DOD	Department of Defense
DODD	Department of Defense Directive
DP	decision point
DS	direct support
DSN	defense switched network
DSS	division surgeon section
EAB	echelons above brigade
EAC	echelons above corps
EAD	echelons above division
ECO	environmental compliance officer
EEE	early entry element(s)
EEH	early entry hospital
EEM	early entry module
EMEDS	expeditionary medical support

EMT	emergency medical treatment
EPS	essential personnel services
EPW	enemy prisoner(s) of war
1SG	first sergeant
F	Fahrenheit
FARP	forward arming and refueling point
FAX	facsimile
FEMA	Federal Emergency Management Agency
FHP	force health protection
FM	field manual/frequency modulated
FMC	field medical card
FMI	field manual interim
FRAGO	fragmentary order
FRSS	forward resuscitation
FSB	forward support battalion
FSMC	forward support medical company
FSMT	forward support MEDEVAC team
FST	forward surgical team
G3	Assistant Chief of Staff (Operations and Plans)
GA	ground ambubance
GAMC	ground ambulance medical company
GCC	geographic combatant commander
GRREG	graves registration
GS	general support
GSAB	general support aviation battalion
GWS	Geneva Convention for the Amelioration of the Condition of the Wounded and Sick in Armed Forces in Field
HBCT	heavy brigade combat team
HD	homeland defense
HHD	headquarters and headquarters detachment
HMMWV	high mobility multipurpose wheeled vehicle
HN	host nation
hosp	hospital
HQ	headquarters
HR	human resources
HS	homeland security
HSS	health service support
IAI	installation-level accident investigation
IED	improvised explosive device
IG	Inspector General
IM	information management

IPB	intelligence preparation of the battlefield
IPMC	intratheater patient movement center
ISB	intermediate staging base
IT	information technology
JFC	joint force commander
JHSS	joint health service support
JMD	joint manning document
JMTB	Joint Military Transportation Board
JOA	joint operations area
JP	joint publications
JPMRC	joint patient movement requirements center
JTF	joint task force
LHA	amphibious helicopter assault carrier
LHD	amphibious helicopter assault carrier dock
LIN	line item number
LOC	lines of communication
LOTS	logistics over-the-shore
LPD	amphibious transport dock
LPH	amphibious assault helicopter carrier–aviation
LSD	dock landing ship
LST	tank landing ship
LZ	landing zone
MACA	military assistance to civil authorities
MASCAL	mass casualty
MASF	mobile aeromedical staging facility
MBA	main battle area
MC3T	medical command, control, and communication telemedicine
MC4	medical communications for combat casualty care
MCB	movement control battalion
MCO	major combat operation
MCP	main command post
med	medical
MEDBDE	medical brigade
MEDCOM	medical command
MEDEVAC	medical evacuation
MEDLOG	medical logistics
MEF	Marine expeditionary force
MES	medical equipment sets
METT-TC	mission, enemy, terrain and weather, troops and support available, time available, and civil considerations

MEV	medical evacuation vehicle
MH	mental health
MHS	military health system
MIA	missing in action
MIJI	meaconing, intrusion, jamming, and interference
MMB	multifunctional medical battalion
MOC	medical operations cell
MOPP	mission-oriented protective posture
MOS	military occupational specialty
MP	military police
MRE	meals, ready-to-eat
MRI	medical reengineering initiative
MRO	medical regulating office
MSB	main support battalion
MSC	Military Sealift Command
MTF	medical treatment facility
MTOE	modification table of organization and equipment
MWD	military working dogs
MWR	morale, welfare, and recreation
N/A	not applicable
NATO	North Atlantic Treaty Organization
NAVAIDS	navigational aids
NBC	nuclear, biological, and chemical
NCO	noncommisioned officer
NOE	nap-of-the-earth
NP	neuropsychiatric
NSHS	National Strategy for Homeland Security
NVG	night vision goggles
OCP	operational command post
OEG	operational exposure guidance
OEH	occupational and environmental health
OMF	originating medical facility
OPCON	operational control
OPLAN	operation plan
OPORD	operation order
OPS	operations
OPSEC	operations security
OPTEMPO	operational tempo
OR	operating room
OVE	on vehicle equipment

PASR	personnel accountability and strength reporting
PAD	patient administrator
PCC	precombat checks
PCI	precombat inspections
PDS	personnel daily summary
PERS	personnel
PIM	personnel information management
PIR	priority intelligence requipment
PMCS	preventive maintenance checks and services
PMI	patient movement items
PMM	preventive medicine measures
PMRC	patient movement requestment center
POI	point of injury
POL	petroleum, oils, and lubricants
PR	personnel recovery
prep	preparation
PRM	personnel readiness management
PT	platoon
PSYOP	psychological operations
PZ	pickup zone
PVNTMED	preventive medicine
QSTAG	Quadripartite Standardization Agreement
R5	reception, replacement, return to duty, rest and recuperation, redeployment
R&R	rest and recuperation
RC	reserve component
rep	representative
RPG	rocket propelled grenade
RSTA	reconnaissance and target acquisition
RSOI	reception, staging, onward movement, and integration
RTD	return to duty
S&RO	stability and reconstruction operations
S1	Adjutant (US Army)
S2	Intelligence Officer (US Army)
S3	Operations and Training Officer (US Army)
S4	Supply Officer (US Army)
S6	Information Management (US Army)
SBCT	Stryker Bridage Combat Team
SEC	section
SF	special forces
SIMLM	single integrated medical logistics manager

SINCGARS	Single Channel Ground/Airborne Radio System
SJA	Staff Judge Advocate
SOA	special operations aviation
SOF	special operations forces
SOP	standing operating procedures
SPT	support
SRC	standard requirements code
STANAG	standardization agreement
SURG	surgical
SUST	sustainment
TAA	tactical assembly area
TACC	tanker airlift control center
TAC CP	tactical command post
TAC SOP	tactical standing operating order
TAES	Theater Aeromedical Evacuation System
TC	training circular
tm	team
TMIP	Theater Medical Information Program
tmt	treatment
tng	training
TO	theater operations
TOC	tactical operations center
TOE	tables of organization and equipment
TPFDL	time-phased force deployment list
TPMC	theater patient movement center
TPMRC	theater patient movement requirements center
TSOC	Theater Special Operations Command
TSOP	tactical standing operating procedures
UO	urban operations
US	United States
USAF	United States Air Force
USAMEDDC&S	United States Army Medical Department Center and School
USASC	United States Army Safety Center
USN	United States Navy
USNS	United States Navy Ship
USNORTHCOM	United States Northern Command
USTRANSCOM	United States Transportation Command
vet	veterinary
WARNO	warning order
WIA	wounded in action

WMD	weapons of mass destruction
XO	executive officer
ZULU	universal time

References

SOURCES USED

These are the sources quoted or paraphrased in this publication.

NATO STANAGs

2040. *Stretchers, Bearing Brackets, and Attachment Supports*, 23 September 1982 (Latest Amendment, 6 October 2004)

2087. *Medical Employment of Air Transport in the Forward Area*, 29 June 1983 (Latest Amendment, 10 September 1999)

2128. *Medical and Dental Supply Procedures*, 15 November 1982 (Latest Amendment, 21 November 1991)

2132. *Documentation Relative to Medical Evacuation, Treatment, and Cause of Death of Patients*, 7 August 1974 (Latest Amendment, 15 September 1986)

2454. *Road Movements and Movement Control*, 27 January 2005

2931. *Orders for the Camouflage of the Red Cross and Red Crescent on Land in Tactical Operations*, 18 October 1984 (Latest Amendment, 3 April 1998)

3204. *Aeromedical Evacuation*, 26 September 1973 (Latest Amendment, 10 September 2001)

ABCA QSTAGs

436. *Minimum Labeling Requirements for Medical Materiel* (Amendment 1), 8 October 1980

470. *Documentation Relative to Medical Evacuation, Treatment, and Cause of Death of Patients*, 23 February 1979 (Latest Amendment, 14 August 1989)

PRESIDENTIAL DIRECTIVE

Homeland Security Presidential Directive-5, *Management of Domestic Incidents*, February 28, 2003

DODD

DoD Directive 3025.15, *Military Assistance to Civil Authorities*, February 18, 1997

DoD Directive 5200.31E, *DoD Military Working Dog (MWD) Program*, March 29, 2006

JOINT AND MULTISERVICE PUBLICATIONS

Joint Pub 1-02, *DOD Dictionary of Military and Associated Terms*, 23 March 1994

Joint Pub 4-02, *Doctrine for Health Service Support in Joint Operations*, 26 April 1995

Joint Pub 5-0, *Joint Operation Planning*, 26 December 2006

Joint Pub 5-00.1, *Joint Doctrine for Campaign Planning*, 25 January 2002

Joint Pub 5-00.2, *Joint Task Force (JTF) Planning Guidance and Procedures*, 13 January 1999

ARMY PUBLICATIONS

AR 40-5, *Preventive Medicine*, 22 July 2005

AR 71-32, Force Development and Documentation-Consolidated Policies, 3 March 1997

AR 200-1, *Environmental Protection and Enhancement*, 21 February 1997

AR 385-40, *Accident Reporting and Records*, 1 November 1994

AR 385-64, *U.S. Army Explosives Safety Program*, 1 February 2000

AR 385-95, *Army Aviation Accident Prevention*, 12 December 1999

AR 600-55, *The Army Driver and Operator Standardization Program (Selection, Training, Testing, and Licensing)*, 31 December 1993

DA Pam 385-40, *Army Accident Investigating Reporting*, 1 November 1994

DA Pam 385-40, *Ammunition and Explosives Safety Standards*, 15 December 1999

FM 1-202, *Environmental Flight*, 23 February 1983

FM 1-564, *Shipboard Operations*, 29 June 1007

FM 3-0, *Operations*, 14 June 2001

FM 3-04.301, *Aeromedical Training for Flight Personnel*, 29 September 2000

FM 3-05, *Army Special Operations Forces*, 20 September 2006

FM 3-06.11, *Combined Arms Operations in Urban Terrain*, 28 February 2002

FM 3-07, *Stability Operations and Support Operations*, 20 February 2003

FM 3-11.5, *Multiservice Tactics, Techniques, and Procedures for Chemical, Biological, Radiological, and Nuclear Decontamination*, 4 April 2006

FM 3-34.2, *Combined-Arms Breaching Operations (Including Changes 1—3)*, 31 August 2000

FM 3-97.6, *Mountain Operations*, 28 November 2000

FM 3-100.4, *Environmental Considerations in Military Operations (Including Change 1)*, 15 June 2000

FM 4-01.45, *Multi-Service Tactics, Techniques, and Procedures for Convoy Operations*, 24 March 2005

FM 4-02, *Force Health Protection in a Global Environment*, 13 February 2003

FM 4-02.1, *Combat Health Logistics*, 28 September 2001

FM 4-02.4, *Medical Platoon Leader's Handbook—Tactics, Techniques, and Procedures (Including Change 1)*, 24 August 2001

FM 4-02.6, *The Medical Company—Tactics, Techniques, and Procedures (Including Change 1)*, 4 September 2004

FM 4-02.7, *Health Service Support in a Nuclear, Biological, and Chemical Environment—Tactics, Techniques, and Procedures*, 10 January 2002

FM 4-02.12, *Health Service Support in Corps and Echelons above Corps*, 2 February 2004

FM 4-02.17, *Preventive Medicine Services*, 28 August 2000

FM 4-02.18, *Veterinary Services—Tactics, Techniques, and Procedures*, 30 December 2004

FM 4-02.21, *Division and Brigade Surgeon's Handbook (Digitized)—Tactics, Techniques, and Procedures*, 15 November 2000

FM 4-02.24, *Area Support Medical Battalion—Tactics, Techniques, and Procedures*, 28 August 2000

FM 4-02.25, *Employment of Forward Surgical Teams—Tactics, Techniques, and Procedures*, 28 March 2003

FM 4-25.12, *Unit Field Sanitation Team*, 25 January 2002

FM 4-02.43, *Force Health Protection Support for Army Special Operations Forces*, 28 November 2006

FM 5-0, *Army Planning and Orders Production*, 20 January 2005

FM 8-42, *Combat Health Support in Stability Operations and Support Operations*, 27 October 1997

FM 8-55, *Planning for Health Service Support*, 9 September 1994

FM 8-250, *Preventive Medicine Specialist (Including Change 1)*, 27 January 1986
FM 10-67.1, *Concepts and Equipment of Petroleum Operations*, 2 April 1998
FM 21-10, *Field Hygiene and Sanitation*, 21 June 2000
FM 31-70, *Basic Cold Weather Manual*, 12 April 1968
FM 55-30, *Army Motor Transport Units and Operations (Including Change 1)*, 15 September 1999
FM 90-3, *Desert Operations*, 24 August 1993
FM 90-13, *River Crossing Operations*, 26 January 1998
ST 4-02.121, *Multifunctional Medical Battalion*, October 2005
TC 1-400, *Brigade Aviation Element Handbook*, 27 April 2006
TB MED 507, *Heat Stress Control and Heat Casualty Management*, 7 March 2003
CALL Newsletter 99-9, *Integrating Military Environmental Protection*, 1999
FMI 4-90.1, *Heavy Brigade Combat Team Logistics*, 15 March 2005
FMI 5-0.1, *The Operations Process*, 31 March 2006

DOCUMENTS NEEDED

These documents must be available to the intended users of the publication.

DA Form 285 Series, *U.S. Army Accident Report*
DA Form 1156, *Casualty Feeder Card*
DA Form 2397 Series, *Technical Report of U.S. Army Aircraft Accidents*
DA Form 2404, *Equipment Inspection and Maintenance Worksheet*
DD Form 601, *Patient Evacuation Manifest*
DD Form 1380, *U.S. Field Medical Card*

Index

References are to paragraph numbers unless otherwise stated.

S

T

FIELD MANUAL
No. 7-85

HEADQUARTERS
DEPARTMENT OF THE ARMY
Washington, DC, 9 June 1987

FM 7-85
RANGER UNIT OPERATIONS

IMPORTANT

U.S. Army Infantry School Statement on
U.S. NATIONAL POLICY CONCERNING ANTIPERSONNEL LAND MINES

TABLE OF CONTENTS

 DISTRIBUTION RESTRICTION: Approved for public release; distribution is unlimited.

PREFACE

This publication contains doctrinal guidelines for the employment and training of ranger units. The intended audience is intermediate and higher level commanders, staffs of those headquarters, and newly assigned officers and NCOs of the ranger regiment. It is to be used with all other published US doctrine concerning NBC operations and the climatic or geographic influences on combat. The fundamentals apply to the employment of ranger units on worldwide operations and across the spectrum of conflict from low to high intensity warfare. This includes all aspects of the integrated battlefield: nuclear, biological, and chemical. It is not intended to apply to the employment of long-range reconnaissance patrol (LRRP) companies, even though such units carry the designation "ranger." These LRRP companies, as well as the long-range surveillance company (LRSC) (corps level) and the long-range surveillance detachment (LRSD) (division level), are organized, trained, and equipped for long-range reconnaissance in enemy territory.

The scope of this publication addresses the mission, organization, equipment, capabilities, limitations, planning, training, operations, and logistical support of ranger units. It describes how the ranger regiment and its subordinate battalions train, receive taskings, plan missions, task-organize, deploy, and integrate supporting elements to perform the mission. Once in combat, the ranger regiment maneuvers to accomplish the mission in the manner of other light infantry units. Ranger units make full use of published light infantry doctrinal guidance. This manual will repeat procedures or techniques contained in other publications only if necessary for understanding or continuity.

This publication does not contain everything the ranger needs to know to successfully execute a unit mission. Rather, it supplements unit training and the ranger unit commander's guidance; it does not supersede it.

The proponent of this publication is HQ, TRADOC. Submit changes for improving this publication on DA Form 2028 (Recommended Changes to Publications and Blank Forms) and forward it to Commandant, US Army Infantry School, ATTN: ATSH-B-ID, Fort Benning, GA 31905.

Unless otherwise stated, whenever the masculine gender is used, both men and women are included.

CHAPTER 1

Introduction

The ranger regiment is a major component of the US Army's special operations forces. It is a unique light infantry unit tasked to conduct special military operations in support of national policies and objectives. These operations require highly trained, well-disciplined units capable of employment in any environment, either alone or in concert with other military forces. Within this publication, the term *ranger force* describes any size force consisting mainly of members of the ranger regiment and led by a member of the ranger regiment's chain of command. A ranger force may be a TOE unit or it may be a specially organized task force for a specific mission.

1-1. Mission.

a. The mission of the ranger regiment is to plan and conduct special military operations. These operations are conducted by specially trained, equipped, and organized forces against strategic or tactical targets in pursuit of national military, political, economic, or psychological objectives. They may support conventional military operations or they may be performed independently when conventional forces cannot be used. (See Figure 1-1.)

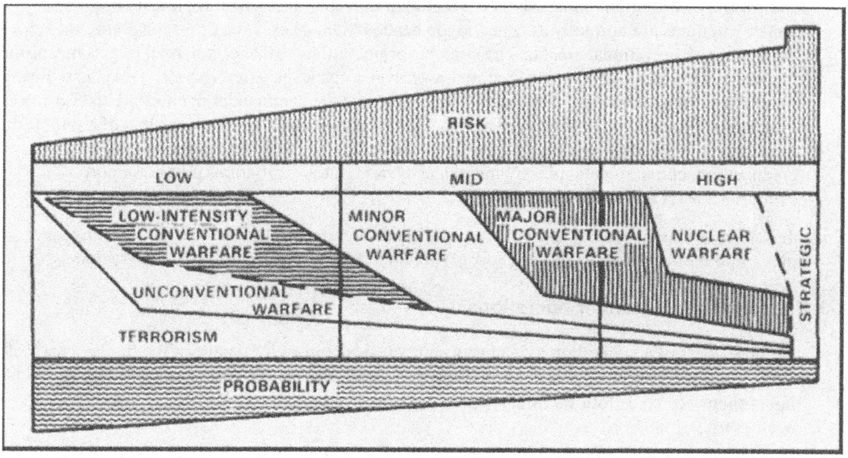

Figure 1-1. Spectrum of conflict.

b. Special military operations conducted by the ranger regiment include strike operations, usually deep penetration, and special light infantry operations. Strike operations include raids, interdiction, and recovery operations. Special light infantry operations include many of the light infantry missions assigned to airborne, air assault, or light infantry battalions and brigades. These operations are conducted in support of the AirLand Battle at all levels of intensity.

1-2. Purpose and function.

The ranger regiment provides the national command authority (NCA) the ability to move a credible military force quickly to any region in the world. The regiment uses the entire spectrum of intelligence support, from national systems to organic assets. Ranger units maintain a readiness posture that supports their immediate commitment to battle once deployed. They are often tailored for specific missions and may require augmentation from external sources. Tactical mobility may be augmented by USAF or Army special operations aviation (SOA) aircraft.

a. The ranger regiment conducts light infantry operations on the integrated battlefield as well as in low intensity conflicts. Special light infantry operations are conducted to accomplish certain contingency missions during peacetime or as part of the echelons above corps (EAC) battle.

b. The regiment conducts both strategic and operational missions. Its efforts are combined into an overall plan to destroy, delay, and disorganize the enemy, or to cause him to divert his attention and combat forces to rear area security. The limited number of ranger units and the diverse targets dictate a careful assignment of missions.

c. The mission, enemy, terrain, troops and time available (METT-T) factors govern the command and control arrangement under which the regiment operates. The ranger regiment or separate ranger battalions are normally assigned to the headquarters whose area of responsibility includes the regiment's operational area(s). The strategic or operational value of potential targets means that ranger units are normally employed at no lower than corps level. Employment at EAC is routine. The EAC headquarters employing ranger units could be the commander in chief's (CINC's) special operations command or field army. The controlling headquarters could also be that of a joint task force (JTF) or the Army component command (ARFOR) of a JTF. The ranger regiment's organization, communications equipment density, and training programs do not support operational level reconnaissance missions.

d. Ranger battalions are not oriented to a specific theater. Current force structure, contingency plans, and training needs (see Appendix G) preclude committing battalions to one region.

1-3. Fundamentals of ranger operations.

a. The success of an operation by a ranger unit depends on the observance of the fundamentals of the US Army's AirLand Battle doctrine (see FM 100-5). The ranger regiment fights the enemy as a light infantry force. It follows infantry doctrine and observes all the basic rules of infantry operations.

b. Along with initiative, depth, agility, and synchronization, ranger operations require--

(1) Detailed planning and coordination that allow the ranger unit to discern and exploit the enemy's weaknesses while avoiding his strength. Both permit the evaluation and use of information gathered from all sources. Detailed coordination integrates all supporting units and services. During planning, the ranger unit conducts pre-mission training, briefings, and rehearsals for all personnel. The use of special mission equipment is practiced and perfected.

(2) Decentralized execution IAW mission orders and the commander's intent. Special operations forces must use individual and unit initiative. Mission-type orders give the ranger force commander the flexibility to take advantage of opportunities on the battlefield.

(3) Surprise, achieved through the ranger unit's ability to move by uncommon means, along unexpected routes, and over rough terrain. Ranger units normally conduct operations during poor weather and reduced visibility, aided by night vision devices. This adds to the attainment of surprise.

(4) Survivability, achieved by using the classic infantry combat techniques of stealth and concealment. The ranger unit engages the enemy at the time and place of its own choosing. It takes full advantage of terrain, and destroys or suppresses enemy weapons. The ranger unit seeks to destroy or neutralize the enemy's command and control systems, his surveillance assets, and his mobility assets. Survivability is enhanced by rapid mission accomplishment and a prompt departure from the objective area.

(5) Mobility, speed, and violence of execution that allow the ranger unit it to close quickly on the objective area and complete the mission before the enemy can react. The speed at which events take place confuses and deceives the enemy as to the intent of the ranger unit. This forces the enemy to react rather than to take the initiative. Tactical mobility lets the ranger unit break contact and withdraw from the objective area.

(6) Shock effect, which is a psychological advantage achieved by the combining of speed and violence with the precision of the ranger attack. The ranger unit strives to apply its full combat power at the decisive time and place, and at the point of the greatest enemy weakness. The ranger unit strives to achieve maximum physical and psychological effect on the enemy by exhibiting aggressiveness and reasoned audacity.

(7) Multiple methods of insertion and attack, trying not to repeat operations thus decreasing the chance the enemy will detect a pattern. This is achieved through imaginative training and planning.

(8) Deception, practiced by the ranger unit during all phases of its operations, from deployment through the insertion phase, to the actions in the objective area and extraction. The enemy is kept unaware of the ranger unit's intentions, is fed conflicting and wrong information, and is kept from knowing the size or mission of the ranger force. The ranger force commander makes full use of ruses, feints, false insertions, electronic countermeasures, and dummy transmissions. Concurrent military operations by US or allied forces may be used as cover for ranger operations. Ranger units are careful to hide every possible aspect of an operation and to disguise those that cannot be hidden. While in the operational area, or when departing any location, ranger units ensure that no material is left that could provide the enemy with information concerning the ranger force.

(9) Audacity, achieved by a willingness to accept risk. The ranger force commander considers what the enemy expects the unit to do and the actual ranger abilities, He then chooses a course of action that may confuse the enemy, while remaining within the capability of the ranger unit. This fundamental is combined with deception and surprise to disrupt the enemy command and control.

c. Detailed planning and coordination finds the enemy. Surprise, mobility and speed, variety, deception, and audacity combine to shock and disorient the enemy, fixing him in place. The violence and precision of the ranger attack finishes the enemy while ensuring the ranger force survives.

1-4. Capabilities and limitations.

a. The ranger regiment has the following capabilities:

(1) Deploying quickly to conduct operations on all types of terrain and in all kinds of weather.

(2) Establishing a credible American presence in any part of the world to show US interest or resolve.

(3) Infiltrating and exfiltrating an area of operations, and assaulting an objective by land, sea, or air.

(4) Conducting strike operations to include raids, personnel and equipment recovery operations, and interdiction of key areas.

(5) Conducting special light infantry operations to include seizing and securing airfields, communications centers, command and control facilities, and key bridges; and other special light infantry operations.

(6) Performing short-duration reconnaissance of assigned ranger objectives for the ranger force commander.

(7) Operating for up to three days without resupply, and for longer periods when provided with accompanying or airdropped supplies.

(8) Providing liaison, communication, and coordination personnel and equipment to integrate the deployed ranger force into the logistical, intelligence, and operational system of the theater or joint task force (JTF) commander.

(9) Assuming operational control, for a limited time, of other US military forces such as engineers or infantry, airborne, or air assault battalions.

(10) Providing the focal point for all-source intelligence support to attached and assigned units of the regimental task force.

(11) Conducting limited combat operations under conditions of chemical, nuclear, or biological contamination.

b. The ranger regiment has the following limitations:

(1) Limited capability against armored or motorized units in open terrain.

(2) No organic transportation.

(3) Limited sustained combat capability due to the shortage of organic combat support and combat service support elements.

(4) Limited organic air defense weapons.

(5) Limited organic indirect fire support.

(6) No casualty evacuation capability.

(7) Reconstitution and retraining needed to replace combat losses.

1-5. Employment considerations.

a. Ranger units are characterized by the quality, motivation, training, and individual skill of their members. This produces units with superb collective abilities, able to adapt well to changing, complex situations.

b. Ranger units can conduct either deliberate or quick-response operations.

(1) Deliberate operations rely on careful planning, reconnaissance and surveillance of the target area, deception, secrecy, thorough preparation and rehearsals, and violent execution. A deliberate operation aims to complete the mission even though the enemy may have heavy forces on or near the objective area. Deliberate operations allow for detailed planning, evaluation, rehearsal, and coordination before insertion. A deliberate operation is likely to succeed against targets that the enemy has protected in depth, that have strong natural defenses, or that need a detailed and long insertion process.

(2) Quick-response operations rely on the high level of training and readiness of the ranger regiment to execute a mission before the enemy can react. These operations are conducted when there is little time for long, detailed planning. They rely on set procedures set forth by the ranger regiment and its supporting elements. A quick-response operation aims to complete the mission before an enemy can react. This type of operation may be chosen due to the time-sensitive nature of the target, political or military goals, the time frame of other operations, or the increased chance of enemy detection.

(3) Whether an operation is to be deliberate or quick response is often a difficult and time-sensitive decision by a high-level command authority. Decision-makers must consider the enemy's strength in the area, his intentions, his ability either to reinforce or to alter the target area, and the consequences of success or failure of diplomatic or military initiatives in related areas. The ranger unit commander tries to use the existing time, manpower, and resources to complete a detailed and coordinated plan. He refines that plan up to the insertion into the objective area.

c. Ranger units train to operate in any environment or weather condition. They regularly perform operations during periods of limited visibility. Ranger units maintain a high state of physical fitness and often train in close quarters combatives. In addition to completing advanced marksmanship training with standard US weapons, each member of a ranger unit trains with many foreign weapons. Ranger units are trained to operate on urbanized terrain, becoming specialists in entry and clearing techniques and quick-fire methods, especially during periods of limited visibility.

d. Because ranger units have limited vehicles, logistics operations capability, indirect fire support, and heavy weapons systems, they are not designed for continuous operations. During all phases of operations and training, ranger units need responsive external support.

e. Ranger units are normally employed against targets and under conditions that need their unique skills. Although targeting priority is set by the overall commander, ranger units are not normally assigned missions that can be done by conventional aerial bombardment or by other units.

f. Ranger units are oriented toward offensive operations. They are not normally employed as a rear area protection force. Although the ranger regiment has a small reconnaissance unit, ranger units do not normally conduct long-range reconnaissance missions. The structure, communications, and training of the ranger unit do not prepare it for LRRP missions. Ranger units engaged in strike and special light infantry operations have a secondary mission to collect and report combat information.

g. Ranger units can be deployed worldwide when US military presence or participation with a host national military activity would serve US interests. This deployment shows a readiness to commit forces into a threatened area or proves US national resolve. After the deployment, other activities include staging operations, rehearsing combat operations, securing base areas for use and deployment of other forces, and so on that provide a clear signal of US intent. Ranger units are not trained or organized to provide mobile training teams (MTT) to train indigenous forces. The US Special Forces or other special operation forces are trained to conduct such security assistance operations.

h. Ranger units can serve as an example to a host country and provide limited military advice and training. They will normally be augmented with linguists and technicians to increase their abilities. The regiment would normally still function as a unit. Most of its assistance would be through short-term, high-impact, unit-oriented operations. They would not be long-term individual efforts associated with advisory-type activities.

i. Ranger units may be deployed to engage in combined training exercises with allies. This enhances US national image by demonstrating the outstanding abilities of the American ranger. These activities may include ranger, light infantry, airborne, air assault, or amphibious operations.

j. Ranger units use standard US nuclear, biological, chemical (NBC) warning; detection; protection; and decontamination equipment and doctrine when operating on the integrated battlefield.

1-6. Command and control.

a. **External.**

(1) The ranger regiment is a key component of the US Army's special operations forces. The other elements of special operations forces are special forces (SF), psychological operations (PSYOP), civil affairs (CA), and special operations aviation.

(2) The 1st Special Operations Command (1st SOCOM) is responsible for command of all active component special operations forces elements, to include the ranger regiment, in peacetime. The 1st SOCOM has command and control of all assigned special operations forces, less operational command (OPCOM) of special operations forces units forward deployed. (See Figure 1-2.) The ranger regiment, while based in the continental United States (CONUS), is under the control of 1st SOCOM. During peacetime, the OPCOM of deployed ranger units is through channels chosen by the NCA through the Joint Chiefs of Staff (JCS) to the unified command.

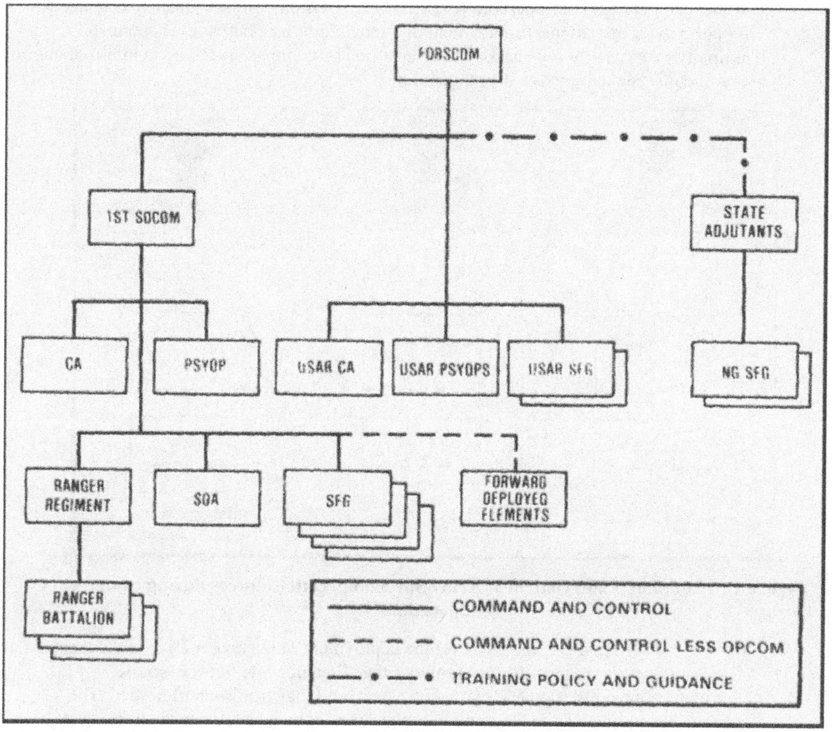

Figure 1-2. Command and control of Army Special operations forces during peacetime.

(3) In support of national policy, the NCA may direct contingency operations involving US

forces. These are politically sensitive military operations normally characterized by the short-term rapid projection or employment of forces in conditions short of conventional war. One aim of the NCA is to conclude them without going to war. The flexibility and effectiveness of the ranger regiment make it possible to use military force quickly to complete the mission and limit the spread of conflict. Command and control of special operations forces in a contingency operation is usually by a special operations task force (SOTF). (See Figure 1-3.) The SOTF and designated special operations forces are under the OPCOM of the highest level of command responsible for the operation. This command may be the unified command responsible for the contingency area (Option A), or a JTF designated by the NCA through the JCS (Option B). If no SOTF is formed, the ranger regiment could be OPCOM to the Army component commander (Option C). The ranger regiment or its battalions are not normally placed OPCOM to a division. A division's area of influence rarely contains appropriate targets. The division also lacks the assets to adequately support ranger operations. Employment of a ranger unit by a corps-level command is normally on a case-by-case basis. The theater or JTF commander retains centralized control over the mission assigned to the ranger unit.

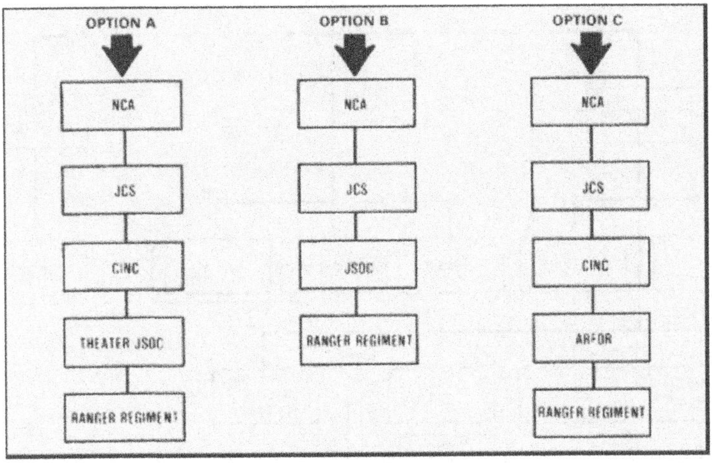

Figure 1-3. Command and control of Army Special operations forces during contingency operations.

(4) The 1st SOCOM is responsible for the deployment of the ranger force to the theater area. The headquarters having OPCOM provides the all-source intelligence, secure communications, insertion and extraction assets, and logistical support needed. The responsible commander in chief, through the Army component commander, provides support during the contingency operation.

(5) During wartime, ranger units are deployed as strategic assets to the theater of operations.

Strategic mission support is provided to the unified commander (CINC) from the NCA through the Joint Chiefs of Staff. It may involve the execution of sensitive operations. The employment of ranger units is controlled at the highest level of command directly responsible for the mission. Ranger units are assigned to a REMAB based on mission, force protection, and OPSEC considerations. The employing command and other elements provide timely and accurate intelligence, communications, logistical, and administrative support. Command and control of these units conforms to the provisions of JCS Pub 2. (See Figure 1-4.)

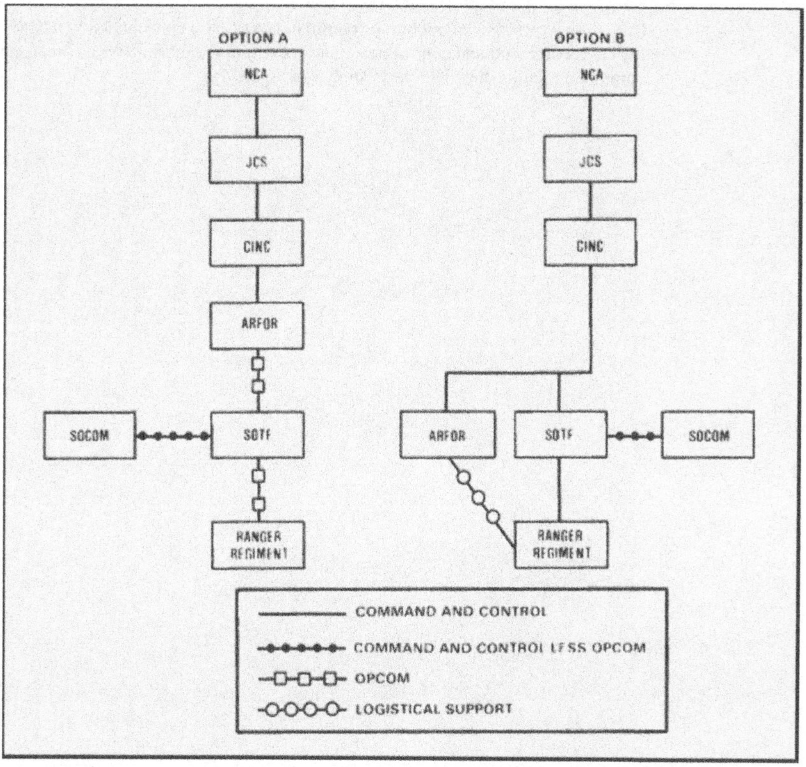

Figure 1-4. Command and control of Army Special operations forces during wartime.

(a) A theater joint special operations command (JSOC) is subordinate to the unified command to supervise the command and control, employment, and support of all US and allied special operations forces. The JSOC is a key element in the command and control of ranger units performing strike missions or special light infantry operations (Option A).

(b) The JSOC may be placed on a lateral line with other in-theater service components (Option B). This can occur when ranger units are -employed far from conventional forces and against strategic targets. It can also occur when the JCS or a CINC forms a JTF to conduct a limited-duration operation in an area where ranger units and conventional forces may operate near each other, or where operations must be coordinated on a regional level.

(c) Under the Army special operations command (ARSOC) concept, 1st SOCOM deploys as a command and control headquarters within the theater. When this occurs, the command relationships will be as shown in Figure 1-5.

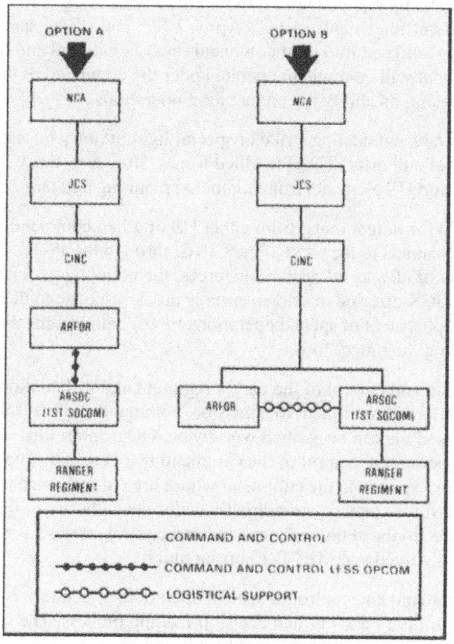

Figure 1-5. Command and control of Army Special operations forces under the ARSOC concept.

Figure –1-5

(d) Recently passed legislation will modify existing high level SOF command and control arrangements. All US Army, USN, and USAF special operations forces will be consolidated into joint commands at both national and theater level. The ranger regiment will continue to operate under the command of the 1st SOCOM, while increasing its ability to conduct joint operations.

(6) Ranger units conducting strike or special light infantry missions are not normally offered for transfer of authority (TOA) to allied forces. However, this will not preclude allied commands and US elements from combined planning and targeting for ranger missions.

(7) Requests for ranger units, from either US or allied commands, go through normal command channels to the CINC. The CINC, through the JSOC, is responsible for the employment of all special operations forces, including ranger units. This provides flexible response to JCS-directed missions, ensures quick response to the CINC's needs, enhances the joint employment of special operations forces, and lessens the chance of overcommitting and degrading the ranger force.

(8) Command and control of the ranger regiment and its battalions is IAW AirLand Battle doctrine. Before commitment to a mission, command and control is kept at a level where the unit's unique skills can be applied worldwide. Once committed, the ranger unit is placed under the operational control of the command that is responsible for the mission. This normally corresponds to the command whose area of influence includes the target area. Ranger task forces may operate directly under the authority of the senior American representative in the country. Command and control arrangements may vary according to the theater involved and METT-T considerations.

(9) The command and control of special operations is centralized and begins at the highest level of the military and civilian decision-making process. The specific command relationships are normally based on the mission.

b. **Internal.**

(1) The ranger regimental headquarters is organized to operate like that of a brigade headquarters. In addition to commanding and controlling all three ranger battalions, the regimental headquarters can assume operational control (OPCON) of conventional combat and combat support units, and other special operations forces for limited periods. It provides the regimental commander with a well-balanced responsive staff. It is structured as an operational headquarters that can deploy to the mission area and act as the ground tactical headquarters.

(a) The regimental headquarters prepares for combat by forming two small command and control groups. Each of these can deploy to an objective area and control combat operations. Other elements of the staff operate the tactical operations center (TOC) and the logistical operations center (LOC). These may be collocated depending on the situation. The TOC and LOC are normally located well behind the FEBA, at the REMAB or at the intermediate staging base (ISB). The ranger regiment also has two small liaison teams consisting of LNOS, staff representatives, and communications elements. (See Appendix B.)

(b) When a single ranger battalion is committed to an operation, the regimental commander normally exercises control from the TOC. A liaison and communications team is detached from the regimental headquarters to augment the ranger battalion. This team is attached to the deployed battalion until the end of the mission.

(c) If two or more ranger battalions are used in a single operation, the regimental commander normally deploys both command and control groups. He commands from the objective area, The use of two small, mobile control elements makes up for the possible loss of the regimental commander and the primary control element.

(d) Whenever the regimental commander is controlling an operation from the REMAB or the objective area, he normally provides a liaison team with a communications element to the higher headquarters exercising operational control. This cell can provide ranger representation to the controlling staff, operations interface, logistics interface, fire support planning and coordination, intelligence analysis and dissemination, and communications between the higher headquarters and the ranger battalion.

(2) The ranger regimental headquarters can serve as a ranger or special operations force task force headquarters. With augmentation, it can function as the Army special operations force command element. The regimental headquarters can also function as the Army component command of a JTF.

1-7. Communications.

a. Ranger missions need unique and dedicated communication support assets. Along with the ranger regiment's organic communications means, communications support assets are organized within the 1st SOCOM that-can deploy in support of the ranger regiment. This ensures that there are reliable communications to all elements of the SOTF. A special operations communications battalion (SOCB), deployed by the 1st SOCOM, provides more communications systems to support the special operations plan. The SOCB does not replace or duplicate existing in-theater communications. It augments the CINC's communications with the US Army's portion of the critical special operations force command, control, and communications link unique to the SOTF.

b. The ranger regiment's communication platoon can provide secure, long-range communication links from the CINC to a deployed ranger force. It can also enter the SOCB and the joint communications support element (JCSE) nets to provide a direct link with the NCA. Its equipment consists of four light vehicles equipped with various multiple communications systems. The two regimental liaison teams use one of these vehicles each. The TOC and LOC each use one.

c. Secure satellite communications (SATCOM) and radios with AM or FM capability are the main means of communication within the ranger regiment. These are also used for communications within the ranger battalions. The regimental commander has two light vehicles equipped with multiple communications systems. One of these is used to support each of the two mobile command and control elements of the regimental headquarters. Some of the radio systems can be dismounted and man-packed.

d. The ranger regimental headquarters has both man-packed and vehicular tactical satellite communication systems. Each of the ranger battalions has one base station and three man-packed systems.

(1) The man-packed radio is battery-operated. It provides two-way communications in both line-of-sight (mobile) and satellite (at-halt) modes. The digital message device group (DMDG) is the input and output device for data transmission. Secure voice communications is achieved using various encryption devices.

(2) The vehicular net control station uses the basic receiver/ transmitter unit installed on a vehicle. Each can serve up to 15 terminals in a communications net.

e. The reconnaissance platoon is equipped with multiple, man-packed communications systems. These let its teams operate in most of the communications nets within the regiment. Each five-man reconnaissance team has this communications ability, as does the platoon headquarters.

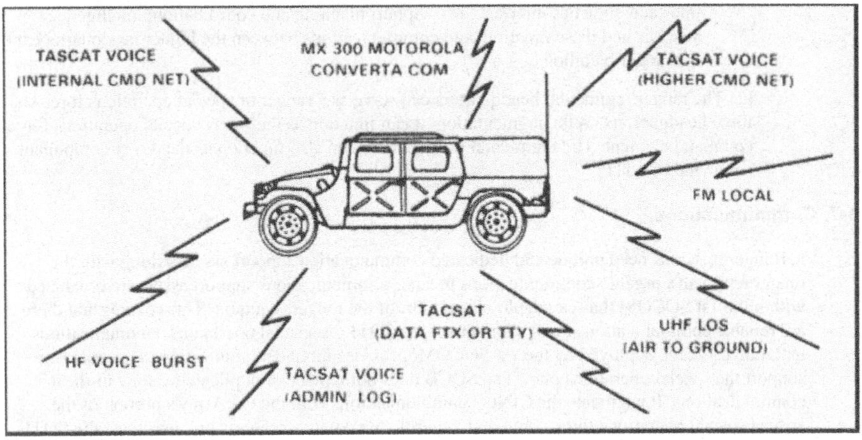

Figure 1-6. Communications capabilities within the regimental headquarters.

CHAPTER 2

Organization and Equipment

The ranger regiment is a specially organized and trained light infantry force consisting of about 2,000 personnel. It includes a regimental headquarters and headquarters company and three ranger battalions. (See Appendix A.) The regiment can operate as a single unit, separate battalions, or separate companies on specific missions. Independent missions are not normally assigned to ranger platoons.

2-1. Headquarters and headquarters company.

The HHC of the ranger regiment consists of 130 personnel, the majority being senior noncommissioned officers, commissioned officers, or soldiers with special skills to support the operations of the ranger battalions. (See Appendix A.) The HHC consists of two elements: the regimental headquarters and the headquarters company (minus).

2-2. Regimental headquarters.

a. The regimental headquarters consists of the regimental commander and his staff. The primary staff is composed of five sections: personnel (S1), intelligence (S2), operations and training (S3), supply (S4), and civil affairs (S5). These staff sections perform the standard functions and are supervised by the executive officer (XO). The regimental commander has a deputy commander and a special staff consisting of a communications officer, fire support officer, surgeon, and staff judge advocate. The S2 officer has a large intelligence analysis and counterintelligence section. The S3 section has extra liaison officers and assistant operations officers to coordinate the widespread and complex ranger missions. A United States Air Force (USAF) staff weather officer and a tactical air control party (TACP) are permanently attached.

b. The regimental headquarters:

- Analyzes assigned missions.
- Plans, coordinates, and controls training and operations.
- Coordinates the infiltration, resupply, exfiltration, and debriefing of the reconnaissance teams.
- Provides liaison support.
- Provides administrative, medical, NBC, and logistical planning and coordination support to the committed ranger battalions.
- Coordinates special support requirements (rigger, transportation, communications security [COMSEC]) for the ranger battalions.

2-3. Regimental headquarters company.

The regimental headquarters company (minus) consists of the company commander and his administrative personnel, plus the limited combat support (CS) and combat service support (CSS) elements assigned to the regiment. These are the fire support element, the communications platoon, the reconnaissance platoon, and the medical treatment team.

a. The company headquarters:

(1) Provides command and control of assigned elements.

(2) Provides administrative, personnel, logistical, and maintenance support to the assigned elements.

b. The fire support element:

(1) Plans fires to support the commander's ground tactical plan.

(2) Coordinates the integration of all available fire support means (mortars, artillery, armed helicopters, close air support, naval gunfire) to support mission execution.

c. The communications platoon (see Appendix A):

(1) Provides secure SATCOM teletype (TTY) and facsimile transmissions (FAX) abilities at the remote marshalling base or intermediate staging base (REMAB or ISB).

(2) Provides secure SATCOM, TTY, and FAX to the regimental headquarters and ranger battalions in the objective area.

(3) Provides cryptographic support to the regimental headquarters.

(4) Provides limited repair of radio and cryptographic equipment.

d. The reconnaissance platoon (see Appendix A):

(1) Trains, equips, and controls the employment of three reconnaissance teams.

(2) Infiltrates the objective area by parachute (high-altitude, low-opening [HALO]; high-altitude, high-opening [HAHO]; static line), helicopter, fixed-wing aircraft, self-contained underwater breathing apparatus (SCUBA), small boat, or other means.

(3) Remains undetected in the objective area for up to five days.

(4) Performs reconnaissance.

(5) Performs demolition target analysis.

(6) Uses small boats and inflatable rafts.

(7) Emplaces and monitors unattended ground sensors.

(8) Uses night image enhancement devices.

(9) Uses photographic equipment.

(10) Emplaces electronic target acquisition and designation devices.

(11) Collects specific combat information to satisfy priority intelligence requirements (PIR).

(12) Reports information using secure, long-range communication equipment.

Figure 2-1. Reconnaissance platoon communications.

(13) Performs drop zone (DZ) and landing zone (LZ) selection, marking, and reception duties.

(14) Escapes and evades enemy areas.

(15) Performs limited attacks or ambushes when tasked by the regimental commander.

(16) Reports weather in the objective area.

e. The medical treatment team:

(1) Provides medical support to the regimental headquarters.

(2) Coordinates required medical support and evacuation.

2-4. Ranger battalion.

a. The ranger battalion is organized and equipped to conduct both special and conventional combat operations. It contains organic combat, CS, and limited CSS elements. It consists of a battalion headquarters, a headquarters company, and three ranger rifle companies. (See Appendix A.)

b. The ranger battalion is the combat element of the ranger regiment. It can conduct independent combat operations or subdivide into its separate ranger rifle companies and their subordinate elements who can conduct semi-independent operations.

c. The ranger battalion headquarters and headquarters company (see Appendix A) includes:

(1) **Battalion headquarters.** The battalion headquarters consists of the commander; the executive officer; the command sergeant major; the S1, S2, S3, S4, and S5 officers; the chaplain; the communications officer; the fire support officer; and two radio operators.

(2) **Battalion headquarters section.** The battalion headquarters section consists of the S3 air and assistant S8, an assistant S2, an intelligence sergeant and a senior intelligence analyst, two operations sergeants and an operations assistant, a personnel staff noncommissioned officer (PSNCO), the chaplain's assistant, two clerk typists, two radio operators, a chemical officer and NCO, a personnel administration center (PAC) NCO, three PAC clerks, and a legal specialist. There is a USAF tactical air control party permanently attached.

(3) **Communications section.** The communications section consists of one telecommunications technician, one tactical communications chief, two combat signal radio team chiefs, one tactical communications systems supervisor, three radio operators, one switchboard operator, one single channel radio operator, and two equipment mechanics. The section provides communications support and associated maintenance.

(4) **Battalion medical section.** The battalion medical section consists of one general medical officer, one physician's assistant, two medical specialists, and twelve aidmen. The section provides unit-level medical support and trains selected members of the battalion in advanced medical procedures.

(5) **Battalion support platoon.** The battalion support platoon consists of a battalion support platoon leader, an ammunition NCO, and a food service section. The platoon provides limited CSS for assigned and attached units. This platoon does not have organic transportation. The battalion food service section consists of a food service sergeant (E7), a first cook (E6), and eight cooks.

(6) **Fire support team headquarters.** The fire support team (FIST) headquarters is supervised by the battalion fire support officer. It consists of the fire support sergeant and two fire support specialists. The headquarters controls three 9-man fire support teams. These teams are normally with the ranger rifle companies in a habitual support relationship. When operations are conducted near water, the fire support team headquarters may be augmented by spotter teams from the naval shore fire control party.

(7) **Company headquarters.** The HHC company headquarters consists of a commanding officer, an executive officer, a first sergeant, a supply sergeant, and an armorer.

d. The ranger rifle company (see Appendix A) includes a company headquarters, three rifle platoons, and a weapons platoon. Each company can conduct semi-independent operations for a short time when augmented by assets from the ranger battalion headquarters company.

(1) **Ranger rifle company headquarters.** The ranger rifle company headquarters consists of a commanding officer, an executive officer, a first sergeant, a supply sergeant, a tactical communications chief, two radiotelephone operators, an armorer, and an NBC NCO. It also has a two-man sniper team equipped with a sniper weapon. It normally has one of the 9-man FISTs attached from the ranger battalion headquarters company. Elements of this FIST may be further attached to the ranger rifle platoons.

(2) **Ranger rifle platoon.** The ranger rifle platoon consists of a headquarters section, three rifle squads, and a machine gun squad. The headquarters section personnel include a platoon leader, a platoon sergeant, and a radiotelephone operator. Each rifle squad consists of a squad leader and two balanced fire teams of four men each. The machine gun squad consists of a squad leader and three M60 machine gun crews of three men each. The machine gun squad is normally employed as a single element to provide quick, accurate, long-range supporting fire. One or more crews may be attached to other squads within the platoon.

(3) **Ranger weapons platoon.** The ranger weapons platoon consists of a headquarters section, a mortar section, and an antiarmor section. The headquarters section consists of a platoon leader, a platoon sergeant, and a radiotelephone operator. The mortar section has two 3-man 60-mm mortar squads. The mortar squads are normally employed as a section. They may be attached by section or squad to other platoons. The antitank section normally consists of a section sergeant and three 3-man 90-mm recoilless rifle or Dragon missile teams. They are normally employed as a section. The antiarmor section may also be attached by section or team to other platoons.

2-5. Special teams and elements.

The ranger regiment forms special teams to do critical tasks. These teams are made up of members of the unit and are used when needed. Examples of these special teams and elements are--

a. SCUBA teams whose members are graduates of the Special Forces Underwater Operations School. They conduct special insertion and reconnaissance duties.

b. HALO teams whose members are graduates of the Special Forces Freefall School and who conduct high-altitude, low-opening parachute operations.

c. HAHO teams whose members are also graduates of the Special Forces Freefall School and who conduct high-altitude, high-opening parachute operations, allowing offset parachute insertions.

d. Scout swimmer teams whose members are specially trained by the Landing Force Training Command (United States Marine Corps [USMC]) to conduct operations along coastlines, waterways, and inland waters. These teams conduct special insertion, reconnaissance, and security tasks often in support of amphibious operations.

e. Sniper teams, whose members are authorized by the table of organization and equipment (TOE), are used for both conventional and special operations. Along with the two-man sniper teams in the rifle company headquarters, each rifle platoon has two trained snipers.

f Demolition teams whose members are trained in advanced demolition and incendiary techniques and the use of special explosives,

g. Air defense teams trained in the use of man-portable air defense system (MANPADS) weapons. Such teams give the ranger battalion limited self-defense against air attack.

h. Liaison teams that are used by the ranger regimental commander to augment the existing command and control links from the employing unit headquarters to the ranger regiment. These teams are placed where they are most needed and may assist the higher headquarters staff in planning and target selection, as well as performing other liaison duties. See Appendix B for a detailed listing of the personnel and equipment included in the ranger liaison teams.

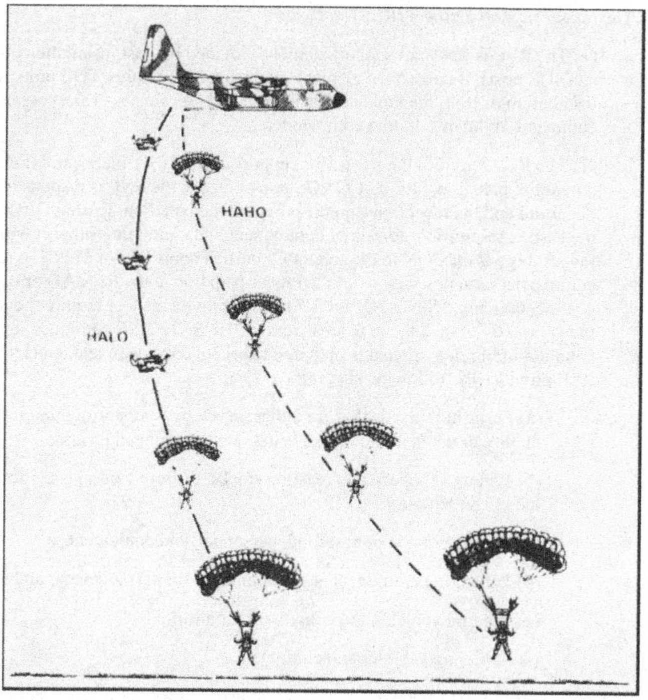

Figure 2-2. HALO and HALO parachute operations.

Figure 2-2

i. The ranger support element (RSE).

(1) The RSE is a special support element provided by host installation commanders at CONUS posts. It consists of elements and teams from either TOE units stationed at the installation or from the table of distribution and allowances (TDA) organizations under the command of the installation commander.

(2) The RSE provides the necessary support to ensure a timely and efficient deployment of the ranger battalion. During CONUS deployments, the RSE is responsible for both the departure and the reception support provided the battalion. During OCONUS deployments, the RSE is responsible for initial deployment and continuing support until the deploying ranger force is OPCON to the supported unified command or JTF. The responsible theater commander assumes support of the ranger battalion at the REMAB or the ISB if outside continental United States (OCONUS). On deployments that require the ranger battalion to move directly to the objective area from CONUS, the RSE continues support of the ranger battalion until released, and is prepared to accept considerable augmentation assets. The RSE provides the following support:

(a) Furnishing meals that are either served by a supporting unit in an established facility or taken to the ranger battalion at an isolated location.

(b) Issuing supplies from existing stocks, or from contingency items stored for use by the ranger battalion.

(c) Transporting supplies, food, personnel, and equipment.

(d) Palletizing equipment and ammunition for air movement and airdrop.

(e) Rigging vehicles and equipment for airdrop.

(f) Refueling vehicles and equipment.

(g) Issuing barrier and construction material for building of rehearsal sites.

(h) Helping in the construction of rehearsal sites and targets.

(i) Securing, storing, moving, and issuing ammunition and other Class V items.

(j) Maintaining ranger weapons and equipment, to include communications equipment. This is normally done by maintenance contact teams.

(k) Setting up and operating the RSE emergency operations center.

(l) Securing the REMAB if it is in CONUS.

CHAPTER 3
Deployment

Ranger units can quickly deploy by any means of transportation to any location in the world.

3-1. Readiness.

One ranger battalion is always in an advanced readiness condition as the Ranger Ready Force (RRF) available for immediate worldwide deployment. A second battalion is prepared to deploy later and a third battalion to follow. The regimental headquarters maintains command and control, liaison, communications, and reconnaissance elements immediately available for deployment. Higher status of readiness in response to specific world situations can be achieved.

3-2. Ranger ready force.

In order to maintain advanced readiness conditions, ranger battalions are rotated as the RRF.

a. Before taking over the duties of the RRF, battalions must do the following:

(1) Complete preparation for oversea movement (POM) qualification, to include an individual records check and required inoculations.

(2) Test fire all weapons, confirm battlesight settings, and repair or replace faulty weapons.

(3) Assemble and load on pallets the unit basic load (Classes I and V) and other supplies and equipment.

(4) Check unit-derived packing list components for accountability and serviceability.

(5) Prepare special operations troop listings.

b. During an RRF period, ranger units:

(1) Have a positive recall system for all personnel.

(2) Store enough air items for a battalion-size parachute assault.

(3) Conduct no off-post exercises, unless approved by the regimental commander.

c. Each unit keeps an alert roster and a telephone notification system for recalling personnel. The rosters ensure that the duty officer and the unit charge of quarters make a minimum number of calls. Each commander briefs personnel assigned to his unit on the recall system and tests it often.

d. Ranger units can maintain an increased readiness condition for long periods, but they should not be kept in that condition longer than needed. The effort to maintain this level of readiness will degrade the unit's training and effectiveness. If the needs of the operations dictate long periods at the highest readiness condition, provisions must be made for sustainment training. Also, added support will be required from the RSE.

e. The ranger regiment maintains a strong family support system when the unit is training or deployed on combat operations. Each member of the chain of command is responsible for family support.

3-3. N-hour sequence.

The N-hour sequence is a departure time sequence designed to help the ranger force complete all required deployment actions within a certain time. The length of the N-hour sequence depends on the readiness condition of the battalion before deployment. The N-hour sequence begins when the battalion is alerted and ends when the first elements depart. The N-hour sequence does not include travel time to the objective area.

3-4. Intelligence and security.

a. **Essential elements of friendly information.** Items of sensitive information and subject areas, if not secured, can be collectively analyzed by a hostile force. This analysis can reveal the intentions of a unit or compromise a mission. These items of information are designated as essential elements of friendly information (EEFI). They are never transmitted without encryption. The EEFI associated with a ranger unit alert normally are the mission, the use of staging bases, nature of the target, special terrain or weather conditions, and time limits. The ranger force commander must ensure that EEFI are not discussed outside of secure areas or made available to personnel without a security clearance and a need to know.

b. **Security plans.** Security plans for billets, marshalling areas, and out-loading areas must remain flexible to adapt to any situation. Certain items are always addressed in security plans. These are access control and physical security.

(1) **Access control.** Ranger units control access to their respective areas by allowing only authorized personnel to enter or leave. All entrances are secured and an armed guard with the access roster is posted. Access to telephones is restricted.

(2) **Physical security.**

(a) Equipment is open to theft and sabotage during the early phases of deployment. Units must provide guards to secure all unit and personal equipment.

(b) Weapons, ammunition, and sensitive items must be stored in a restricted area and guarded.

(c) Classified documents are open to compromise during deployment. Steps must be set up for maintenance, transportation, turn-in, and destruction of classified material. The most secure facility available must be used.

c. **Rear detachment security.** Security after deployment (maintained by the rear detachment) is as important as security before deployment. No information should be given to the news media from other than official sources. The rear detachment may continue to play a role in the unit cover and deception plan.

d. **Tactical operations center security.** Only personnel listed on TOC access rosters are allowed free entry into the TOC area. Personnel wanting to enter the TOC, who are not listed on the access rosters, are detained until their identity and purpose are proven. Such personnel are escorted at all times while within the TOC area.

e. **Marshalling and out-loading area security.** The ranger unit normally marshalls its personnel and equipment in the unit area, providing its own security force. The USAF security police provide this support at the departure airfield. On Army airfields, Army military police provide security. En route security is provided by the RSE.

f. **Personnel and unit equipment sanitization.**

(1) Before deployment, personnel are relieved of non-mission-essential material having intelligence value. This reduces the intelligence available to hostile forces if they capture a ranger force member.

(2) The ranger force commander sets the degree of sanitization IAW mission requirements.

(3) If strict sanitization is imposed, guidelines are issued as follows:

(a) The only personal items retained are identification card and tags.

(b) Unit insignia and markings are removed from uniforms and equipment.

(c) Personal items, such as wallets and photographs, are secured by the rear detachment commander.

(d) Personnel are briefed on how to compose and retain combat diaries. No personal diaries are allowed.

(4) The commander decides how much US currency is to be taken with the unit. '

g. **Area sanitization.** To avoid the compromise of classified or sensitive mission-related material, searches are conducted before and after leaving the host installation, REMAB, or ISB. These searches are held to locate and remove any material, indicators, litter, or other items that could compromise the ranger unit's mission. Unit commanders ensure that a thorough sanitization of their areas is conducted, to include dismantling rehearsal areas, destroying terrain models, erasing chalkboards, and emptying all classified and nonclassified waste containers. During their searches, commanders must pay close attention to emergency operations centers, tactical operations centers, and marshalling and loading areas. The ranger unit's intelligence officer coordinates support from local counterintelligence elements to search the area again after the ranger unit leaves.

3-5. Operations.

a. **Actions upon receipt of the OPORD.**

(1) Task-organize and equip units for special mission(s) IAW the warning order.

(2) Allocate/reallocate mission-essential equipment.

(3) Distribute augmentee personnel.

(4) Redistribute key personnel,

(5) Prepare a deployment time sequence.

(6) Prepare a movement plan.

(7) Prepare and maintain a current operation estimate.

b. **Advance elements.** A planning element is normally deployed to the REMAB/ISB or JTF headquarters in advance of the main body if the ranger force does not deploy directly to the objective area.

c. **Emergency operations center.** An EOC is setup by the ranger battalion headquarters. The EOC is responsible for relaying all reports to higher headquarters. It remains operational until ordered to cease.

d. **Air movement.** The ranger force always deploys by USAF aircraft. This is normally an administrative air movement to a REMAB or ISB. It may be followed by tactical air deployment from there. For some missions, tactical deployment begins at the unit's home station. Even if the movement is administrative, key personnel are cross-loaded. This is done under the direction of the operations section air officer based on aircraft allocations and load plans.

3-6. Logistics.

The following describes logistics readiness needs and steps for deployment.

a. **Supply.**

(1) **Class I.**

(a) Combat rations are normally palletized when the unit is RRF. These rations are to be used during the tactical operation. If the mission calls for deployment directly into the objective area, the palletized stock of rations is off-loaded in the unit area and issued as needed.

(b) Each ranger deploys with emergency combat rations.

(2) **Classes II and IV.**

(a) **Individual clothing and equipment.**

- Individual A bags are packed IAW a predetermined packing list, tagged, and marked. They are palletized, moved to the departure airfield by the RSE, and loaded IAW the air movement plan.
- Individual B bags are packed, assembled, and forwarded on order by the rear detachment.

- Individual environmental equipment is issued. Equipment shortages are replenished out of stocks maintained by the RSE.

(b) **Air items.**

- While the unit is on RRF, a contingency load of main and reserve parachutes is maintained by the RSE.
- Individual and unit air items are palletized by the RSE and shipped in bulk to the REMAB or ISB.
- For direct-deployment missions, air items and packing material are taken to the unit by the RSE.

(c) **General supplies.** The RSE maintains pallets of common contingency items and special equipment. Pallets are picked up by the out-loading support unit and moved to the unit area for issue or moved with the other pallets to the departure airfield.

(d) **Bulk supplies.** Barrier material is identified, relocated, palletized, and out-loaded IAW the air movement plan.

(3) **Class V.**

(a) The unit basic load of ammunition is secured, palletized, and loaded by the RSE IAW the air movement plan.

(b) For direct deployment missions, selected pallets are taken to the unit area and ammunition is issued. The convoy route is coordinated through the post EOC, and secured by the RSE.

(c) Designated items are rigged by the RSE for airdrop, if enough time exists before movement to the REMAB or ISB begins.

(4) **Class VI.** Personnel deploy with comfort items as needed.

(5) **Class VII.**

(a) All weapons and sensitive items not deployed are assembled and secured by the rear detachment personnel.

(b) Unit vehicles are prepared for airland or airdrop.

(c) Inoperable mission-essential equipment is repaired or replaced.

(d) Special equipment requests are forwarded to the logistics section as they are identified.

(6) **Class VIII.**

(a) Individuals deploy with one set of individual Class VIII items.

(b) Unit medics deploy with modified aidbags.

(c) Controlled medical items (less drugs) are palletized and deployed with section supplies. Controlled drugs are secured and issued to unit aidmen on order.

(d) Medical supply chests are maintained in the unit aid station.

(7) **Class IX.**

(a) Units deploy with initial prescribed load list (PLL) items.

(b) Critical PLL shortages are identified to the RSE for follow-on issue.

b. **Maintenance.**

(1) Upon alert, priority maintenance or replacement of key items of equipment is coordinated by the logistics officer. Only minimum mission-essential equipment is replaced.

(2) Items to be returned from supporting maintenance activities are identified, and pickup is coordinated by the property book officer.

c. **Out-loading coordination and support.**

(1) Upon alert, the RSE provides a liaison officer to the ranger logistics section.

(2) The support platoon leader (SPI,) attends the air movement conference and coordinates the following:

(a) Movement times (for both troops and equipment) to the departure airfield.

(b) Steps for weighing and joint inspection of the pallets and vehicles upon arrival at the departure airfield control group (DACG).

(c) Limitations on the pallet height and weight due to aircraft type, allowable cargo load (ACL), and availability.

(d) Changes in the planned number of pallets due to the mission requirements or airframe availability.

(e) Rigging and loading of the heavy drop and supply bundles.

(3) Empty pallets are placed in position for loading by the RSE in coordination with the SPL.

(a) Companies and staff sections prepare pallets.

(b) For airdrop operations, vehicles are prerigged at the unit location. On order, vehicles are driven to the heavy drop rigging site. The SPL and the officer in charge of the rigger detachment coordinate for a rigging site and the delivery of the required air items to that location. Rigging is done by the RSE, and supervised by a member of the ranger unit. Loading aboard the aircraft is IAW the air movement plan and supervised by RSE and USAF aerial port personnel.

(4) The DACG is provided by the host installation. Upon arrival at the DACG area, unit personnel and equipment are under the control of the DACG. As the deployment begins, a

member of the ranger unit is present at the DACG to ensure an orderly loading of personnel and equipment.

(5) The DACG plans for troop-holding areas in the event of delayed or aborted aircraft.

(6) Unit movement data are maintained by the air officer in the operations section.

3-7. Communications-electronics.

a. **Alert notification communications.** Alert notification of the ranger battalions by the regiment is done by secure communications means. Units in local training areas are informed by radio or through their assigned range field telephone. Units operating out of normal FM radio range during RRF periods set up a high frequency net, or other communications, to their higher headquarters.

b. **Telephone service.** Strict control is maintained over key telephones being used during unit alert. Care is taken to ensure that COMSEC and OPSEC violations do not occur. All telephones are assumed to be monitored; classified, sensitive, or official information is not discussed.

c. **Radio communications.**

(1) **Voice communications.**

(a) A secure radio net is set up early in the deployment. The ranger unit EOC, the installation EOC, and the RSE EOC operate in this net. A liaison officer from the ranger unit is placed at the installation EOC.

(b) The ranger support element liaison officer, attached to the logistics section, operates a secure voice station on a separate logistics net to the RSE EOC and the RSE commander.

(2) **Cryptographic material.** During RRF, COMSEC codes and enabling equipment are kept available to set up the secure voice net.

3-8. Remote marshalling base and intermediate staging base.

a. Ranger units can deploy from their CONUS base directly to the objective area. A more common method would be for the ranger unit to deploy first to either a REMAB or an ISB before insertion into the area of operations. (See Figure 3-1.)

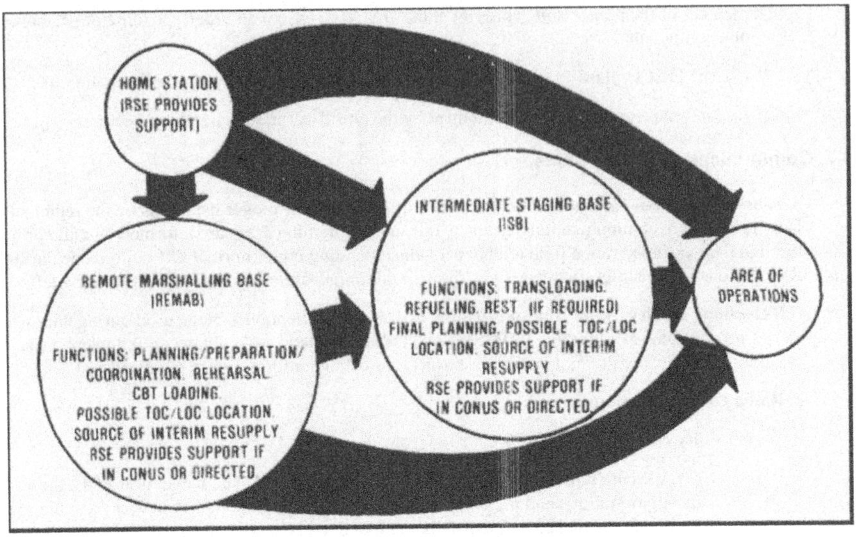

Figure 3-1. Deploying ranger units.

b. The REMAB is a secure base to which the entire ranger unit, including organic and attached support elements, deploys and where mission planning continues. The REMAB is normally within the geographical area encompassed by the command authority of the theater or JTF commander. This ensures that the CSS elements providing support to the ranger unit are operating within their normal area. It prevents or lessens out-of-sector support requirements for CSS elements.

c. The REMAB also provides a secure location for the ranger unit to conduct detailed planning and coordination with the controlling headquarters staff. In the REMAB, the ranger force commander conducts rehearsals, refines and modifies plans, determines PIR, and coordinates with the proper intelligence source to receive that information. It is in the REMAB that specialist augmentees to the ranger force are integrated into the unit, if they have not already joined. Specially trained supporting units, such as aviation and communications elements, also join the ranger force at the REMAB. The REMAB must provide--

(1) Access to the controlling headquarters staff.

(2) Physical security of billeting, planning, maintenance, and communications areas.

(3) Mess, billeting, latrine, and shower facilities for the ranger force and its supporting elements.

(4) Access to a C-141 or C-130 capable airfield, possibly with all-weather operations.

(5) Access to secure communications and processed mission intelligence.

(6) Access to rehearsal areas where sites can be built and live-fire rehearsals can be conducted.

(7) Access to the unit locations of major supporting elements, such as naval landing craft or Army aviation units.

(8) An external security force and an active counterintelligence agency.

(9) Vehicular transport for troop lift, equipment transfer, and administrative use.

(10) Access to maintenance support facilities.

(11) Medical support facilities to augment the ranger force medical personnel.

(12) Covered area to allow for packing of parachutes and rigging of airdrop loads.

d. The ISB is a more austere location. Elements of the ranger force deploy to an ISB to perform final planning, coordination, and task organization. The unit's organization and composition are finalized for movement to the objective area. The ISB is normally not occupied for long periods. Ranger units may deploy to the ISB from the REMAB or CONUS. Deployment to the ISB is more common when terrain or distance precludes insertion to the objective area directly from the REMAB or CONUS. Quick-reaction missions often involve the use of an ISB. Facilities needed at the ISB are limited to the following:

(1) Austere messing arrangements for the combat elements of the ranger force and its attachments.

(2) Austere billets or rest areas.

(3) Water point.

(4) Fuel for aircraft and vehicles.

(5) Areas for test firing of weapons.

(6) Covered and concealed areas for assembly of the ranger force and rigging of parachutes and door bundles.

(7) Austere airfield support facilities, possibly capable of all-weather operations.

(8) Remote location far away from civilians or traffic routes.

(9) Security and counterintelligence elements.

(10) Secure communications.

e. The REMAB and ISB should be located in an area similar in terrain and climate to the objective area. Time spent at the REMAB or ISB lets the ranger force begin acclimatization. Ranger units train yearly in all climates and emphasize physical conditioning. This allows acclimatization quicker than units not as well trained or conditioned. However, ranger units still need an acclimatization period if the terrain and climate of the objective area are different from that of their CONUS base. Not enough acclimatization means that the ranger force could be less efficient, especially when the objective area is hot, dry, or at a great altitude.

CHAPTER 4

Insertion, Extraction, Escape, and Evasion

Insertion is the entrance of a ranger force into a hostile area to execute a particular mission. It is the first critical phase of a ranger operation. These operations often involve deep penetrations of enemy territory by air, water, or land. This chapter emphasizes fundamentals and techniques employed by ranger units when their mission requires insertion and extraction. Extraction is conducted quickly after the mission to avoid casualties.

4-1. Intelligence.

a. Insertion plans are based on timely and accurate intelligence. The headquarters directing the operation must provide current intelligence information to the ranger force. The ranger regiment normally takes action to solicit required information from the controlling headquarters and from other agencies, if security permits. The type of intelligence needed varies depending on the specific mission, but normally includes the following:

(1) Current photographs from aerial or overhead reconnaissance.

(2) Detailed maps.

(3) Data on the population, terrain, and weather from area studies.

(4) Current enemy order of battle, enemy troop locations, and enemy capabilities.

(5) Details on the target area and infiltration routes.

b. To meet its need for current intelligence, the ranger force also relies on aerial reconnaissance and surveillance of the terrain over which it will move. Emphasis is on getting information on the enemy's ability to detect and engage the inserting forces. The location and abilities of air defense radar and weapons systems are critical. Terrain analysis is detailed and focuses on areas suitable for the insertion. Weather information must be detailed and current.

c. In getting information and intelligence data, OPSEC measures are stringent and must be enforced. Coordination with other agencies is kept within security constraints approved by the directing headquarters.

4-2. Deception.

There are always plans to deny the enemy knowledge of the ranger unit's insertion, or to deceive him as to the location or intent of the operation. Feints, false insertions, and other deception operations all add to ranger deception plans. Selection of unexpected means, time, place, and routes of insertion, as well as use of speed and mobility, give less reaction time to the enemy. Planning may include the use of diversionary fires to direct the enemy's attention away from the inserting unit. Deception techniques that may be used include--

a. Multiple airdrops, water landings, or both. This precludes the entire force from becoming engaged at the same time and place, if detected.

b. Dispersion of insertion craft (air or water) both in time and location.

c, Landing a force in an area closer to other potential targets than to the actual targets.

d. Leaks of false information and false messages.

e. False insertions or radio traffic to disguise actual insertions.

f. Diversionary actions such as airstrikes in other than the target areas.

g. Increased reconnaissance flights, either by manned aircraft or remotely piloted vehicles, over false targets.

4-3. Speed and Mobility.

a. Speed of action must be emphasized when the enemy can react quickly to the ranger force presence or when there are limited means of withdrawal. The operation must be conducted swiftly, allowing the ranger force to withdraw before the enemy can react.

b. Major problems facing ranger units are the need for speed and mobility, and the heavy loads carried by soldiers for sustained operations. Ranger units are required to be flexible and to be able to react to any problem within the objective area. However, if equipment is carried to cover every encounter, the load will become too heavy. This will reduce the speed and mobility of the ranger unit. The ranger force commander must always have a balance between the equipment carried and the mobility of his unit. He does this by making three different types of loads: fighting, mission, and existence. He decides what items will make up these loads, where they are carried, and what will be done with them upon contact with the enemy.

(1) The fighting load consists of those items of equipment, weapons, ammunition, food, and water that are common to all soldiers and are necessary for immediate use in combat. These items are normally carried and are retained at all times while in contact with the enemy.

(2) The mission load consists of those items of equipment or ammunition that are required by the mission, but are not normally carried by all rangers in a unit. The load may include spare batteries, demolitions, radios, antitank weapons, or mines. This load is normally spread out among members of the unit and is rotated often. It may be dropped upon enemy contact, but is normally carried to the objective and either used or emplaced there.

(3) The existence load consists of any items designed to provide protection from the elements: sleeping gear, changes of clothes, spare rations, or tentage. This load may or may not be brought into the objective area, depending on the tactical situation. If it is brought, it is normally dropped upon enemy contact or left in an assembly area and retrieved later.

4-4. Stealth.

Methods and steps chosen for insertion must emphasize stealth and try to avoid detection by the enemy. Multiple routes or methods of insertion may be used to preserve the ranger force if some soldiers or units are seen.

4-5. Suppression.

Effort must be made to suppress enemy detection devices, weapons systems, and command and control facilities. This may include electronic countermeasures or direct attack. This helps prevent the enemy from finding and engaging the ranger force en route to the objective area. Deception tactics add to suppression activities. If the ranger force is inserting by infiltration, suppression must not disclose friendly activities.

4-6. Security.

Emphasis on OPSEC prevents compromise of an impending operation. The operation must not be disclosed by overt rehearsals or training activities, the open use or procurement of special items of equipment, or the location of the marshalling area. Several measures can help in maintaining security:

a. Guarded and secure facilities for the headquarters and staff during planning.

b. Issue of operation orders only after units are moved to secure marshalling areas.

c. Alerts, rehearsals, and training to get the local people used to conditions that will exist during actual marshalling.

d. A secure facility set up to isolate units selected for an impending operation.

4-7. Night vision and electronic devices.

a. Rangers use night vision devices to detect and avoid the enemy's forces and his detection devices. These devices also assist in controlling and speeding movement and allow units to traverse almost impassable terrain.

b. The use of electronic homing devices permits precise navigation during reduced visibility. These devices may also be used with visual marking systems.

c. Passive night vision devices are used to assemble rapidly and to reorganize.

d. The use of active night vision devices in assembling, and the degree to which drop zones or assembly areas are "sterilized," depends on the following:

(1) Mission success may depend on not being seen during insertion. When a clandestine insertion is made, the time needed to get to the objective will normally be greater, the use of passive devices will prevail, and the sterilization of the area will be vital. Each ranger must be briefed on sterilization plans and techniques for erasing signs of the insertion.

(2) When detection is likely and the mission depends on getting to the objective before the

enemy can react, speed becomes crucial to the success of the insertion. Insertion may be near or on the objective. Rangers must get in and out quickly. They do minimum sterilization of the area and can use active RSTA devices.

4-8. Rehearsals.

Rehearsals should be close to the actual conditions of insertions and extractions. Security of rehearsals and rehearsal sites is important to the success of the operation. Rehearsals are held on terrain and structures like those in the objective area. If such structures are not available, they are built by the ranger force and the RSE with its supporting engineers.

4-9. Sand tables and terrain models.

Sand tables are useful to acquaint personnel with insertion sites and surrounding terrain. The use of sand tables and terrain models during the issuance of prejump orders and briefings enhances an orderly and speedy tactical assembly.

4-10. Communications.

Reporting during insertion and extraction missions is normally by exception, using an operations schedule.

4-11. Planning insertions.

a. **Centralized planning.** Planning for insertion of ranger units is centralized at the lowest level responsible for the overall operation. This is normally at the joint task force level, with detailed coordination required among all supporting forces. These forces may include other Army elements, USAF, United States Navy (USN), United States Marine Corps (USMC), other US agencies, and allies. Preparation must include detailed contingency planning for use of firepower, shock effect, speed to break contact (if detected), control measures, procedures, and planning. When planning takes a long time, intelligence must be continually updated through all available sources. This includes overhead photography and ground sources to ensure that the initial planning data are still current. Planning stresses the use of deception. Escape and evasion plans, external communication nets, and the location of caches must be known to leaders at all levels.

b. **Reverse planning sequence.** Insertion planning is done in the reverse planning sequence with the ground tactical plan being prepared first. Actions at the objective serve as the basis for the landing and assault plan, the plan for insertion and movement, the loading plan, and the plan for marshalling. All plans are interrelated. Some may be developed at the same time. Consideration of the factors of METT-T guides all planning. Special considerations that apply to these operations are:

(1) **Mission.** The nature of actions at the objective and requirements for special items of equipment and nonorganic personnel impacts on how the unit inserts. When the mission requires rapid deployment, the fastest method of insertion may be needed. In other operations, where success of the mission depends on maintaining secrecy, speed may be less important.

(2) **Enemy situation.** Enemy disposition, strength, and security measures along the route(s) near the objective area affect the method of insertion to be used. Planners must take into consideration whether the enemy:

(a) Uses RSTA detection devices or not.

(b) Has a strong air defense system or possesses air superiority.

(c) Has gaps between his defensive positions.

(3) **Terrain and weather.**

(a) **Topography.** Terrain such as mountains, swamps, or heavy forests favors land infiltration. Deserts, arctic regions, and open prairies favor air or water insertion. Areas not inhabited or with a friendly but scattered population favor land infiltration. A heavily populated area, or one which contains large enemy forces, favors air or water insertion. Hydrography--tides, currents, surf, reefs, and sandbars--must be considered when planning water insertion.

(b) **Weather.** Bad weather with reduced visibility and high winds favors land infiltration, and adversely affects air or water insertion. Limited visibility favors land infiltration; good visibility does not. Adverse weather aerial delivery systems (AWADS) lessens the impact of reduced visibility as a limiting factor on air infiltration. High winds may preclude airborne insertion.

(4) **Troops.** The number of men to be inserted and their level of training may be limiting factors. A need for special skills may call for use of nonranger unit personnel whose physical stamina and ability to perform many infiltration techniques may not equal that of the rangers. Whether supporting forces and special items of equipment are available must also be considered.

(5) **Time.** The main concern is whether the ranger operation is quick-response or deliberate-response. During a quick response mission, planning, preparation, and rehearsal time is short. The timing of the insertion is crucial with respect to weather, the enemy, and if an insertion craft is available.

4-12. Air insertion.

The most rapid form of insertion is by air. Rangers and equipment can be delivered by parachute (static-line or freefall technique), fixed-wing aircraft (airlanded), or helicopter (airlanded, parachute, or rappeling).

a. Suppression of enemy air defense along the insertion corridor may be necessary. The USAF will normally assume the major responsibility for suppressing enemy air defense artillery (ADA) capabilities that may interfere with the inserting forces. This is done by a variety of sophisticated electronic countermeasures applied against enemy equipment and by strikes against known enemy positions. Army fire support elements, special electronic mission aircraft (SEMA), and ground jammers may participate, as well as Navy aircraft, jammers, and gunfire.

(1) A main danger area is the forward edge of the battle area (FEBA) or frontier areas where

the enemy employs most of his sophisticated weapons systems and air defenses. Other dangerous areas are population or troop concentration areas behind enemy lines protected by point air defense systems.

(2) Since the majority of the enemy's detection devices and air defense weapons may be located near the point of entry, fire support, smoke screens (even at night), and suppressive measures may be critical. The use of special items of equipment may be needed to counter the enemy's RSTA effort.

(3) Long-range fires (artillery and naval gunfire) should be planned on known and suspected enemy antiaircraft locations and on prominent terrain features along the route(,s). Once beyond the FEBA, inserting units may be beyond the range of conventional indirect fire. More reliance will then be placed on air assets for fire support and electronic countermeasures.

b. There must be close coordination between the ranger unit and the transporting unit. What the transporting unit does during the insertion has a great impact on the success of the mission. To decrease the chance of detection, maximum use is made of periods of reduced visibility and tactical cover and deception. Drop zones and landing zones are chosen behind tree lines, in small clearings in forests, or on other types of terrain not easily seen.

c. All flight routes over enemy territory should be planned in order to avoid occupied areas and enemy air defenses. Routes should complement the cover and deception plan. They should avoid giving away final ranger objectives.

d. In-flight emergencies, mainly in deep-pedetration flights, should be considered, The ranger force receives a preflight briefing on the route to be flown and is informed of checkpoints along the route. Simple ground assembly plans for contingencies are set up before enplaning. In an emergency, the leader of the unit involved decides whether to continue the mission or abort. This is based on METT-T factors, contingency plans, and distance to the objective as compared to the distance to friendly territory. Contingency provisions for air and water rescue are made.

e. Deciding the DZ location is a joint effort. After receiving input from the air and ground commanders, the joint task force commander makes the final decision. During air assault operations, the ground commander is responsible for the overall process. After reviewing MEYF-T factors, it may be determined that surprise is more important than speed, or vice versa. This may determine the number, size, and locations of DZs and LZs.

f. Planning and preparation continue during the marshalling phase of any operation using air delivery for insertion. A review must be made of items of equipment that are crucial to the ranger force reaching its target area, accomplishing the mission, and returning safely. The need for special supplies and equipment, their transport, the distance to be traversed, the delivery means, and external support are all factors that enhance or reduce the chance for success. Once the means of insertion has been decided, logistics planning continues with emphasis on requesting and coordinating external support. The following must be considered:

(1) Detailed planning for mission logistical needs.

(2) Extra sets of key items of equipment to ensure redundancy and self-sufficiency of

inserting units.

(3) Plans for accompanying, planned, and emergency resupply. The ranger force should prepackage loads to meet specific contingencies. It should consider the use of supplies and equipment already in the operational area.

(4) In quick-response operations, only manportable items of equipment needed to accomplish the mission are carried. Plans and signals must be established for resupply of key items.

(5) Specific plans to evacuate casualties.

(6) All attached personnel and augmentation equipment must be properly prepared for movement.

(7) Provisions must be made for evacuating captured enemy equipment and personnel.

(8) All inserting units should be cross-loaded. Items to be considered in a cross-loading plan are the ground tactical plan, the assembly plan, the configuration of the DZ, the light route and formation of the aircraft, and the number of aircraft available. Cross-loading ensures that key leaders and equipment--for example, door bundles--are evenly distributed throughout the assaulting force. If one or more aircraft abort or are shot down, some key leaders and equipment will still arrive at the DZ. This permits mission accomplishment. Cross-loading aids rapid assembly on the DZ in support of the ground tactical plan. If a ranger is to assemble near the leading edge of a DZ, he is loaded on the aircraft to exit near the front of his stick.

(9) A bump plan is used if an aircraft should abort while still on the ground and there are no spare aircraft. The bump plan ensures that key leaders and equipment are not left behind when the main body takes off. If an aircraft aborts on the ground, key personnel move to another aircraft and replace nonessential personnel. This concept promotes success on the objective.

4-13. Airborne insertion.

a. During the planning phase of an airborne operation, there is no room for error in judgment. It can be planned and conducted well only if there is a constant exchange of information. All personnel involved must be informed of all changes in plans, times, and locations.

b. The reverse planning sequence is the key to the success of an airborne insertion. Most of the planning is based on the ground tactical plan, followed by the air movement plan to support it.

c. The USAF is responsible for flight planning, initial point (IP) selection, and crew procedures throughout the flight. The senior ranger in the aircraft must remain oriented and keep abreast of any last-minute changes. He must coordinate with the aircraft commander.

d. Most airborne insertions are made at either very high altitudes (HALO/HAHO) or very low. For combat operations, drop altitudes may be less than 500 feet above ground level (AGL), depending on the type of parachute used.

e. In airborne insertions, emphasis is on the use of special delivery or navigational techniques since normal insertion is during periods of limited visibility.

(1) **Adverse weather aerial delivery system.**

(a) A self-contained navigational system in some USAF C-130 aircraft is AWADS. The system allows the aircraft crew to fly a specific route to a DZ during darkness or bad weather. The system works with a Doppler radar that determines the ground speed of the aircraft, and an AWADS computer programmed with special flight route information (start point, route, checkpoints, wind speed and direction, and air speed). These two instruments show where the aircraft is in relation to the programmed checkpoints. This eliminates the need for airdrops under visual weather conditions.

(b) Only certain aircraft in formation are equipped with the actual AWADS device. Other aircraft are equipped with station-keeping equipment (SKE). This SKE consists of a computer that tells the aircrew where their aircraft is in relation to other aircraft in the formation.

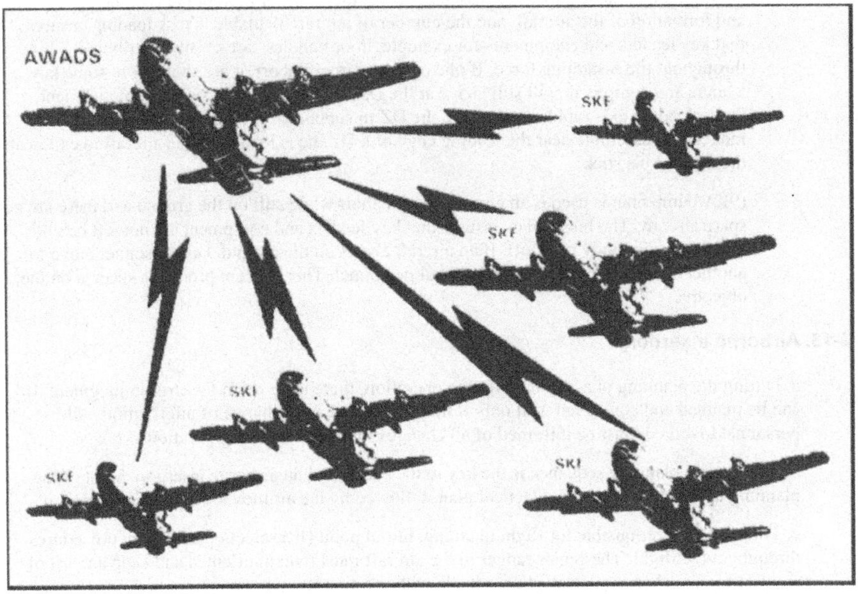

Figure 4-1. AWADS abd SKE aircraft formation.

(c) Airborne insertions using AWADS give the ranger commander the ability to airdrop personnel and equipment during low or zero visibility. This increases the safety of USAF aircraft. It protects them from visually tracked enemy antiaircraft guns and non-all-weather fighters. It also uses adverse weather as a combat force multiplier. Using AWADS, personnel and equipment may be inserted day or night without a pre-positioned USAF combat control team (CCT) or an Army assault team. Time length of the air train is shortened. The air corridor has to be cleared only once. This enhances the element of surprise. The CCTs are brought in with the assault elements. They are able to direct other low altitude parachute extraction systems (LAPES), container delivery systems (CDS), and air-land or heavy drop missions. Use of AWADS enables the ranger commander to conduct rapid, vertical reinforcement of units threatened by an enemy attack during bad weather.

(d) The use of AWADS requires detailed planning, rehearsal, and close coordination among the forces involved. Intense training of both USAF and Army personnel is required.

(e) Personnel must be briefed on terrain through use of photographs, maps, and terrain models. Limited visibility relies mainly on compasses and audible and visual assembly aids.

(f) During reduced visibility, essential heavy-drop serials may precede personnel aircraft. This reduces the time lag between personnel and heavy-drop chalks due to the need to move friendly troops off the DZ. Remaining heavy-drop serials or LAPES can be scheduled following P-hour.

(2) **Reconnaissance teams.**

(a) The operations of the main ranger force may be enhanced by use of reconnaissance teams. These teams are inserted ahead of the main body to provide early warning or to conduct reconnaissance of key objectives. If reconnaissance teams are required at a point along the flight path or near the main DZ, they may be cross-loaded on board any aircraft within the formation. They exit at planned times so as to land as close to their objective as possible. Should such teams be required at a point perpendicular to the axis of the main formation, one of the following is used:

- Another flank aircraft delivers teams to designated points before or with arrival of the main body.
- Selected aircraft deliver the main body, turn, and then deliver the reconnaissance teams.
- The teams are cross-loaded as complete elements on aircraft delivering other units whose flight path crosses the desired locations.

(b) Missions assigned to reconnaissance teams include:

- Providing local security for a USAF CCT, in which case, by joint doctrine, they form the Army assault team (AAT).
- Establishing surveillance or blocking positions on likely avenues of approach into the objective area or DZs.
- Locating and marking targets or assembly areas for the main assault force.

(c) Reconnaissance teams may be inserted by static line or HALO/HAHO operations.

(d) If the reconnaissance element is to locate a specific target or objective, it may be inserted up to 24 hours early. The longer the reconnaissance team remains near the objective area, the greater the danger of it being seen.

(e) In order for the information to be used by the ranger force commander, secure communications and a linkup plan must be prepared and rehearsed.

(3) **Assembly.**

(a) The ranger airborne force can conduct a quick assembly of combat power (weapons, personnel, and command and control elements) on the DZ. Quick assembly and reorganization is crucial because the unit is open to attack. Assembly must be fast and precise. Assembly areas and plans are set forth after considering all the factors of METT-T, mainly the location of the enemy, visibility, terrain, drop information, dispersion pattern, and cross-loading.

(b) The number of assembly areas depends on the location of the DZ, size of the assembly areas, and enemy's detection ability. Unit members move directly to the assembly areas without breaching security. Dissemination of information to all jumpers is the key to a successful assembly plan.

(c) For battalion-size or larger operations, general assembly areas are prescribed in the ranger force OPORD. Company commanders select and designate the exact location of their assembly areas. Although assembly area locations are terrain-dependent, they are generally on the DZ if the drop is at night and off the DZ if the drop is during daylight.

(d) Personnel chosen to carry assembly aids are cross-loaded in the center of their unit sticks. Strong, aggressive leaders should jump first and last in each stick. Upon landing, the leaders roll up their sticks toward the center, locate and move to the proper assembly aid. They may also carry or wear assembly aids. Individual rangers then follow these key personnel. During daylight operations, radiotelephone operators (RATELOS) can display color-coded pennants on their radio antennas to facilitate linkup with leaders. Key personnel may carry and jump with their own radios, and also display color-coded pennants on the radio antennas. Key personnel and RATELOs put their radios into operation before moving to assembly areas. For drops occurring near dawn or dusk, both daylight and night assembly plans are required.

(e) When tactical vehicles or door bundles are dropped, they are normally marked with panels, chemical lights, or strobe lights. The delivery parachutes are often color-coded.

(f) A ranger force should be able to assemble on an unfamiliar DZ, at night, within 45 minutes of the final aircraft pass. Because most ranger operations are time-sensitive, the unit normally moves from the assembly area to the objective when enough of the force has assembled to accomplish the mission. A straggler control point remains at the assembly area under the control of a designated leader. When everyone assembles, or at a specific time, the leader moves to link up with the unit. If available, a radio is left with the straggler control point.

(g) Assembly area security plans are prepared and disseminated. These plans must address hasty defense, sectors of responsibility, observation posts, fire support, and chain of command.

(4) **Contact on the DZ.** During the conduct of parachute insertion, rangers must be ready for enemy engagement at all times. Rangers are open to attack on the DZ before assembling. Just as contingency plans and chance immediate-action drills are developed, briefed, and rehearsed for other chance enemy contacts, rangers must be prepared to execute immediate-action drills for enemy contact on the DZ. The following actions are taken when there is enemy contact on the DZ:

(a) Rangers immediately assemble in fire team-size groups and fight off the DZ to designated assembly areas. Elements attempt to break contact and assemble.

(b) Assembly areas designated by commanders in the mission OPORD serve as primary rally points.

(c) Alternate assembly areas (normally 2 to 5 kilometers from the DZ) are designated in the mission operation order (OPORD). They are used by elements that cannot link up or assemble in the primary assembly area. These elements continue with the mission once assembled.

(d) During limited visibility, personnel should move directly and quickly off the DZ. They must avoid mistaking other friendly elements for enemy, which could cause firefights among friendly forces.

(5) **Reorganization.**

(a) The unit reorganizes according to plans, using designated assembly areas, aids, and identification markings for personnel and equipment. Replacements are named for key leaders or personnel who are lost or injured.

(b) Security teams assemble and move to their positions. Remaining units move quickly to their assembly areas, carrying supplies and equipment.

(c) Subordinate leaders keep the commander informed during assembly and reorganization.

(d) Runners may be used to report assembly status. This maintains radio silence. Designated personnel quickly recover supplies and equipment and move to the assembly areas.

(e) Reorganization is complete when all units have assembled or have been accounted for, and control has been established.

(6) **Pop-up technique.** This involves close terrain flying until the DZ is reached, then the aircraft quickly climbs to jump altitude. Special navigational equipment and aircrew training

are needed but may not always be available.

(7) **HALO/HAHO.** This involves high-altitude, low-opening and high-altitude, high-opening jumps with high performance parachutes designed to maneuver the jumpers to a specific point on the ground. During these operations, midair assembly procedures may be used.

(8) **Arctic.** Airborne insertion in extremely cold conditions is difficult. Times allocated for all parachute and equipment rigging must be increased. Rigging of individual equipment must be IAW FM 57-220. Ahkios can be rigged, packed inside a double A22 container, and dropped in platoon bundles. Because of the increased amount of individual equipment needed, allowable cargo loads for USAF aircraft are reduced.

(9) **Water.** Because of a lack of suitable land DZs, or to avoid detection, rangers may conduct airborne operations onto water DZs close to land. Procedures and techniques are similar to land DZ operations.

(10) **In-flight rigging.** On long-range operations involving extended flight time to the objective area, personnel may rig in flight. In-flight rigging requires detailed planning, coordination, of all jumpers. A rehearsal of these procedures should be conducted before deployment. Aircraft loads must be reduced to carry out in-flight rigging procedures. (See FM 57-220.)

4-14. Air-land insertion (fixed-wing).

a. **Planning factors.** Most of the planning factors that apply to airborne insertion operations also apply to air-land insertion operations. Most USAF aircraft cannot conduct air-land operations in total instrument weather conditions. The AWADS equipment is accurate enough for airdrop navigation, not air-land. Air-land operations are conducted, however, under conditions of extremely reduced visibility. Specially trained USAF crews, using sophisticated navigational techniques, can locate and land on small airfields in bad weather. As in airborne operations, ranger reconnaissance teams, as well as specially trained CCT elements, can jump before the air-land aircraft arrive. They emplace navigational and location markers to assist the pilots. Rangers may rely solely on air-land operations for insertion. This method does not provide for delivery of the force in the shortest time possible. Air-land insertions are normally in concert with an insertion of airborne elements.

b. **Time.** Since turnaround time is crucial, loads in aircraft should be unchained as the aircraft taxies. This also applies to personnel and seat belts. Aircraft should stop as close as possible to the assembly areas where personnel and equipment are off-loaded.

c. **Method.** Air-land insertions normally begin after a ranger Army assault team (AAT) and a USAF CCT have parachuted into the initial objective area and conducted reconnaissance, clearing, marking, and security operations. The ranger ground force commander is the arrival airfield commander. He has control over all arriving, taxiing, and departing aircraft (fixed-wing or rotary) through the CCT, which locates near him. The ranger force command and control element is usually on the first aircraft in and the last aircraft out.

d. **Security.** Air-land operations concentrate large numbers of men and equipment, which presents an open target to the enemy. The ranger ground force commander must ensure rapid off-loading, assembly, dispersion, and movement toward the objective.

e. **Communications.** The ranger ground force commander must have secure radio communications with the ground force, CCT, fire support elements, and higher headquarters. Wire should be laid to connect ground command and control locations.

f. **Operation duration.** Depending on the factors of METT-T, the aircraft may remain on the ground for extraction while the mission is being accomplished or take off at once. They may then orbit, pending recall, or return at a prearranged time.

g. **Extraction.** Extraction of the air-landed ranger force is normally planned and accomplished like an insertion, only in reverse order. A simple, yet effective, way must be found to see that no ranger personnel are left in the objective area.

4-15. Air assault insertion.

Ranger air assault (air-land, rappel, or combination) insertions normally take place during limited visibility. They quickly place the ranger force on, or close to, the objective. The ranger force (depending on the mission) normally plans on fighting when it hits the ground, completing the mission, and then being extracted. Air assault insertions need detailed planning, surprise, flexibility, speed, shock effect, concentration of combat power, and precise timing.

a. **Ground tactical plan.** The reverse planning process is important. The ground tactical plan, developed from the mission assessment, is the first planning area. All other planning begins from this point. The ranger commander and his staff decide stated and implied tasks, and request the proper helicopters for the mission.

b. **Fire support plan.** The fire support available to the ranger commander may be artillery, naval gunfire, mortars, attack helicopters, and USAF or USN tactical aircraft. Because of the timing of ranger air assault operations, preparatory fires are often not used. If they are used, enemy ADA sites normally have priority.

c. **Landing plan.** This plan puts the ranger force into the objective area at the time, place, and proper sequence to support the ground tactical plan. The ranger commander selects primary and alternate LZs and the priority for landing. The exact location of an LZ is determined after a detailed analysis of METT-T. Rehearsals are important. The ranger force must quickly assemble, reorganize, and accomplish the mission. The senior ranger on each aircraft ensures that the personnel on his aircraft offload at the correct LZ.

d. **Air movement plan.** The air movement plan provides for the coordinated movement of the ranger force into the zone of action in support of the 'landing plan. Flight routes, air movement tables, flight formation, altitude, and air speed should be considered. The ranger commander must have secure in-flight communications with his subordinate commanders.

e. **Loading plan.** This plan supports the air movement plan and provides for the pickup zone (PZ) setup and control. It sets priorities for the commander for loading personnel, equipment, supplies, and cross-loading. Air-loading table, aircraft loading formations, bump plan, and rehearsals should be considered.

f. **Staging plan.**

(1) This plan includes PZ selection and operation. The selection of PZs and LZs needs planning and coordination for best use of air assault assets. Site selection in or near the objective must be coordinated in person between the supported ranger commander and the aviation commander. The tactical situation is the key factor in PZ and LZ selection. Other factors are size of landing zones, surface conditions, ground slopes, approach and departure directions, prevailing winds, obstacles, communications, cross-loading, aircraft command and control, PZ and LZ markings, and rehearsals.

(2) The PZs are set up and secured about 15 minutes before touchdown of first aircraft. The fire support plan must support the other plans. All supporting fires must be coordinated with the air mission commander (AMC). Because of the depth of most objective locations, USAF tactical aircraft will be the main fire support method. Orbit locations for fire support aircraft must be planned.

g. **Other planning considerations.** These include escape and evasion, actions at the last LZ, assembly plan, downed aircraft procedures, control measures, forward area rearm and refuel point (FARP), weather delays, and deception plans.

4-16. Amphibious insertion.

Insertion by water may be by SCUBA, surface swimming, small boats, submarines, surface craft, helocasting, or a combination of these methods.

a. **Planning.** Planning for waterborne operations must be thorough. The plan is decided upon between the ranger commander, the transporting unit, and the joint headquarters directing the operation. Plans must be made for all types of enemy action and weather while on board the transporting craft. The transporting unit is given information on a need-to-know basis. Information that could endanger the operation may be withheld until the mission is under way. Initial planning includes:

(1) Time schedule of all events, from the beginning until the end of the joint operation, to be used as a planning guide. Timing for each event must be exact; the success of the operation depends on it. (See Figure 4-2)

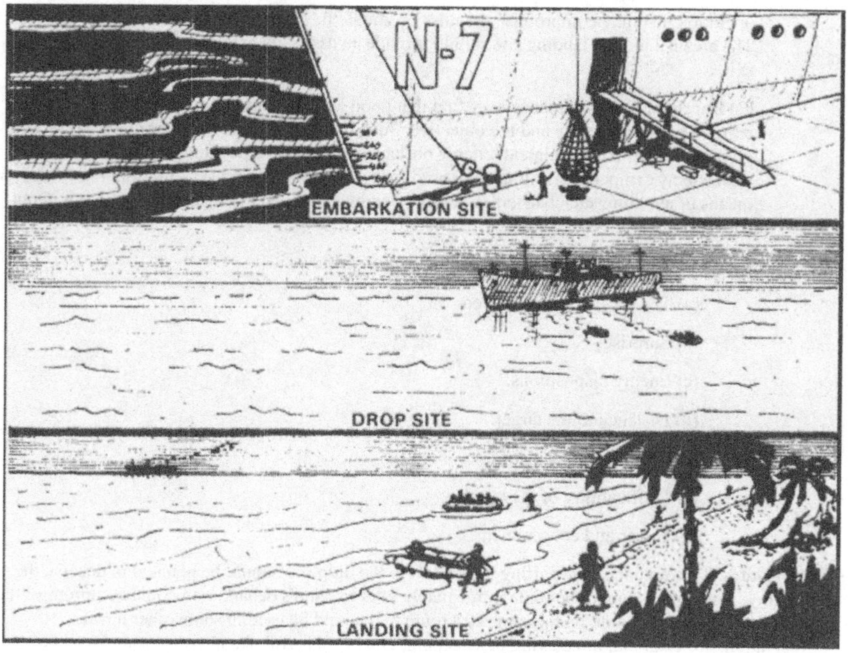

Figure 4-2. Waterborne insertion.

(2) Embarkation site, where troops and equipment are loaded on the transporting craft.

(3) Drop site, where the ranger force will depart the primary craft and load into smaller boats.

(4) Handing site, where the unit will beach its boats, or land directly from amphibious craft.

(5) Loading, when loads are lashed and waterproofed IAW unit standing operating procedures (SOPs). Inspections are important. A bump plan for key personnel and equipment must be developed.

(6) Cross-loading, which will help preserve the ranger force. location of leaders, crew-served weapons, and needed equipment are stressed.

b. **Landing site.**

(1) **Beaches.** Beaches are checked to find one that allows an undetected approach. Landing sites that cannot be approached from different directions are avoided. Stretches of defended or patrolled coastline should be avoided. If sand beaches are used, tracks and other signs of

operation that may compromise the mission should be erased. Rural, isolated areas are best. The area behind the landing site should provide an assembly area and concealed routes for exit.

(2) **Barriers.** Reconnaissance may find that good landing sites are not suitable because of barriers between the site and the objective. Such barriers may be patrolled roads, large areas giving little or no concealment, enemy obstacles or defensive positions, areas populated with enemy sympathizers, or an area used for enemy troop maneuvers and training. Unless a means of avoiding such barriers is devised, landing sites should be limited to those without barriers.

(3) **Other factors.** Beach selection should also consider:

(a) The scheme of maneuver.

(b) Surprise.

(c) Enemy dispositions.

(d) Distance to the target.

(e) Landing and exit sites.

(f) Trafficability of the beach.

(g) Cover and concealment.

c. **Routes.** The route the transporting craft takes to the drop site should be planned to deceive the enemy. A major change in the route immediately after the unit's debarkation could compromise the mission. The route should be the same as a route that would be used in some other naval operation--for example, minelaying or sweeping, or a naval patrol. An alternate route must be planned.

d. **Tactical deception.** There should be other plans to deny the enemy knowledge of the insertion besides the water approach route plan. Plans may include electronic countermeasures (ECM), diversions, feints, ruses, or demonstrations.

e. **Actions at the drop site.** Some operations may permit landing directly from the transporting craft. If not, primary and alternate drop sites must be set. The drop site should be at least 1,500 meters offshore to cover any noise in the loading and launching. If the enemy has a surface radar capability, a drop site should be chosen to avoid detection. The drop site may have to be several miles offshore or may need the use of ECM. The manner in which men, small boats, and equipment are loaded must promote fast action at the drop site.

f. **Navigation.**

(1) Ship-to-shore navigation (to the landing site) may be done by dead reckoning. The course may also be maintained by compass navigation, reference to a shoreline silhouette, radar, or prearranged signals from the shore.

(2) Boat handlers, navigators, and other experts are relied on in various phases of the

transport operation.

(3) Lead reconnaissance teams or scout swimmer teams may be used to plant radar, radio, or infrared buoys, or distinct, shielded lights. Boats may be equipped with radar reflectors or beacons to permit accurate tracking.

g. **Actions on the beach.**

(1) Boats hold offshore while ranger scout swimmers move to a covered and concealed position. After reaching the beach, they make a short listening halt. They check the beach and the area immediately beyond the beach for enemy activity. Swimmers signal the force "all clear," mark the landing site, and provide local security.

(2) When the boats land, chosen personnel move directly into covered and concealed security positions. They prepare to defend the landing site.

(3) If the ranger unit is not carrying equipment, boats may be moved off the beach and buried or camouflaged. If heavy equipment makes it hard to move the boats, chosen personnel start unloading at once. The remainder of the force secures the landing site. Each boatload moves into position under the control of a designated leader. Teams are positioned with the best cover, concealment, and fields of fire available.

(4) The ranger force leader sets up a command post from which he can control the operation. He may establish secure communication with his headquarters.

(5) Boats may be deflated and buried or camouflaged near the landing site or away from it. If the boats are to be hidden near the landing site, a team may be chosen to dig holes or cut brush for camouflage along with the unloading of the boats. A small team may be chosen to sweep the beach to erase tracks and drag marks.

(6) The ranger force should be assembled and prepared to move within minutes after the last boat lands. Every ranger must be briefed on the unit's plan should enemy contact be made upon landing.

h. **Insertion by air from ship.**

(1) Helicopters from a ship can launch the ranger force from over the horizon and move at a low level to the LZ.

(2) Helicopters may be vectored from ships to a landing zone. The aircraft may return to the ship while the ranger force conducts the mission, they may orbit offshore, or they may stay at the LZ. Other aspects of landing and assembling are the same as discussed for helicopter insertions.

i. **Helocasting.** This form of insertion combines helicopters and small boats in the same operation. It is planned and conducted the same as air assault operations, except that the LZ is in the water. While helicopters are moving at low levels and low speeds, rangers drop small boats and jump into the water. They then assemble, climb into the boats, and move to shore. Ranger elements are usually assembled and loaded into the small boats within 20 minutes of the last ranger entering the water. This type of operation is normally conducted during limited visibility.

j. **Fire support.** Fire support may be ship-to-air, ship-to-shore, air-to-ground, or ground-to-ground. Specific missions include:

(1) Suppressing enemy air defense.

(2) Neutralizing enemy field artillery within range of boat or helicopter routes.

(3) Employing on-call fires against shoreline targets.

(4) Destroying enemy aircraft.

k. **Contingency planning.** The following must be planned for:

(1) Actions on enemy contact.

(a) En route.

(b) Helocast site.

(c) Flares.

(d) Aerial attack.

(e) Small arms fire.

(f) Indirect fire.

(2) Bump plan with priority to key leaders, crew-served weapons, and radios.

(3) Downed craft procedures.

(4) Escape and evasion.

(5) High surf.

(6) Adverse weather.

(7) Separation.

l. **Rehearsals.** The ranger force must rehearse all aspects of the amphibious insertion. This includes boat launching, paddling, beaching and assembly, boat commands, and capsize drills.

4-17. Land infiltration.

Land infiltration to the objective may be the best way to complete a mission. This is normal when the enemy has total air control or effective air defenses in the operational area to be penetrated, or when the objective is close to friendly territory. Rangers can infiltrate over any type of terrain, in any climate. However, thick forests, swamps, and broken or steep terrain offer the best chance for success.

a. **Planning.** There are many methods of infiltrating by land, and each has its good and bad points. Plans are designed to get the infiltrating force to the objective area without being seen. Plans should include:

(1) Selecting concealed routes for surprise and protection. Primary and alternate routes are based on detailed map and aerial photo study, ground reconnaissance, and data from agent reports.

(2) Avoiding obstacles, populated areas, silhouetting (when forced to cross obstacles or ridge lines), enemy positions, main avenues of approach, and movement along often used routes and trails.

(3) Selecting the time of infiltration to make use of reduced visibility, such as darkness, rain, snow, and fog, and periods of reduced enemy alertness, such as early morning, holidays, and times following protracted movement or combat. The time of the infiltration is important during critical phases (crossing borders or passing through enemy troop concentrations or populated areas).

(4) Using speed and stealth. All rangers must know routes, linkup points, alternate linkup points, time schedules, danger areas, and actions on enemy contact and at linkup points.

(5) Centralizing planning and decentralizing execution. Centralized coordination must ensure that units are acting IAW cover and deception plans. Control should be given to the leaders of the infiltrating groups. Within their designated zones, these leaders select their own routes and formations. Leaders control the rate of advance. When the enemy has electronic detection devices, radio communications must be kept to a minimum. Key leaders normally go with the group that is using the fastest lane or the lane least likely to be detected.

(6) Seeding infiltration routes to the assembly area with unattended ground sensors. These sensors can help the infiltrating force by reducing the chance of surprise contact during the infiltration, and by helping in controlling the forward progress of the units.

(7) Planning for fire support, which must include coverage of routes, danger areas, assembly areas, objectives, and PZs.

(8) Rehearsing actions at the objective, actions on enemy contact, actions at danger areas, and movement techniques.

b. **Movement.** Land infiltration is conducted using one of three techniques:

(1) **Movement in small groups along one axis.** All members of the force use the best route. Small groups are harder to detect, easier to control, and do not compromise the total force if detected. This technique may take too long and it needs an assembly area or linkup point. If the lead group is detected, other groups may be ambushed.

(2) **Movement in one group.** This technique has no reassembly problem. The same route is used by all, making navigation easier. A large force can fight out of a dangerous situation easier than a small one. It has no coordination problems between other infiltrating units. This form of movement is easier for the enemy to detect. If the force is detected, the mission may be endangered.

(3) **Movement in small groups along several routes at the same time.** This technique avoids putting the total force in danger, is less likely to be seen, forces the enemy to react in

many locations, and makes it harder for the enemy to determine the size of the force or its mission. Some groups travel over poorer routes and have problems of assembly, control, and support; the capture of personnel may endanger the mission. Regardless of which technique is used, ranger units normally move to reduce the chance of enemy contact.

c. **Assembly area.** Following movement into the general area of operations, infiltrating units converge on the assembly area or linkup point. Assembly may begin at once after the passage of lines, or it may take place at any time before reaching the objective. The infiltrating force must remain undetected before assembly. If an infiltrating unit is seen at this time, the enemy may learn the location of the assembly area, and the objective. It may take action to destroy the infiltrating force.

d. **Linkup point(s).**

(1) Selection characteristics include terrain that--

(a) Is free of enemy troops.

(b) Offers cover and concealment from enemy observation and patrols.

(c) Does not restrict movement of friendly units.

(d) Accommodates the unit.

(e) Is easily identified.

(f) Offers escape routes to alternate linkup points.

(g) Offers concealment from air attack.

(2) Action sequence for occupying the linkup point.

(a) Halt short of the point.

(b) Establish local security.

(c) Conduct a reconnaissance of the linkup point.

(d) Occupy the linkup point.

(e) Emplace observation posts and listening posts.

(f) Link up with other groups.

(g) Conduct final preparations.

(h) Continue the mission.

(3) Security may be enhanced with the aid of devices such as night vision goggles and night sights. Unattended ground sensors may be placed on likely avenues of approach into the linkup point. Normally, only passive observation devices are used to assist in avoiding detection.

(4) Plans must set a time schedule for actions at the linkup point. All unit members must move in quickly, consistent with security, so that the unit may proceed to the objective rally point (ORP) and on to the objective.

4-18. Stay-behind technique.

A ranger force can be positioned behind enemy lines by employing the stay-behind technique. In this method, a ranger force allows itself to be bypassed as the enemy advances. This is done to allow the ranger force to perform a specific mission. Stay-behind operations may require the concealment, or cache, of supplies before the enemy bypass. Rangers may need to use captured enemy supplies. The location of the assembly area and the plan for extraction are important parts of stay-behind plans.

4-19. Extraction.

The ranger force is normally extracted after the mission is completed. The enemy may also force the unit to abort its mission and to call for extraction. Contingency plans must be made to extract a force under enemy pressure, as well as one that is not in contact. Primary and alternate extraction sites are planned and coordinated with all ,porting forces. Ranger operations are normally conducted against deep targets. Extraction may begin with land movement but normally it includes the use of boats or aircraft. Plans may include primary I alternate means of extraction.

a. **Terrain.**

(1) The terrain is important in choosing the point of extraction - for example, whether or not the extraction site offers good cover from enemy direct-fire weapons and positions from which the extracting force can suppress the enemy. Tidal conditions and PZ size must be considered.

(2) Unlikely locations such as swamps, jungles, or mountain areas are often used for extraction. Specially trained air and naval crews, using the latest extraction devices, make such terrain useable.

b. **Enemy.**

(1) Extraction is best conducted without enemy pressure. However, such pressure could develop during the operation. Detailed plans are made for emergency or contingency extractions forced by the enemy as well as clandestine extractions not under enemy pressure. If under pressure, units will have to use planned fire support to suppress enemy weapons. At the beginning of the extraction with security posted to provide early warning, the unit may start to thin out the same as with a night withdrawal. Ground security may be relied on until the last unit withdraws under cover of suppressive fires.

(2) If enemy contact is made during an extraction, the commander must decide whether to reinsert units or break contact and move to another extraction point. Fire support and suppression are important and include suppressing enemy air defense by deception, ruses, or decoys.

(3) The ranger unit may have special weapons systems or equipment to aid in the extraction. Shoulder-fired antiaircraft missiles may be used to defend against enemy aircraft threatening

the extraction.

c. **Extraction by air or water.** Extraction by air or water is preferred when the resources are available and their use will not endanger the mission. Other factors that favor their use are when--

(1) Long distances must be covered.

(2) Time of return is essential.

(3) The enemy does not have air and naval superiority.

(4) Heavily populated hostile areas block land movement.

(5) Ranger forces are burdened with prisoners of war,casualties, or critical items of equipment.

d. **Extraction by land exfiltration.** Stealth and evasiveness are key elements of exfiltration. This method of extraction is favored when--

(1) The enemy has air superiority.

(2) The enemy is able to prohibit air or water extraction.

(3) The distance to friendly lines is short.

(4) The terrain provides cover and concealment (for movement on foot) and limits enemy mobile units.

(a) Multiple routes of exfiltration may be used to prevent destruction of the force if soldiers or units are detected. Captured enemy vehicles and equipment may be used to assist in the exfiltration.

(b) The ranger force may exfiltrate in one body or in small groups. Exfiltrating in small groups avoids the time delay of assembly.

(5) The exfiltrating force is lightly equipped and not burdened with captured personnel or material.

(6) The exfiltration route passes through an area occupied by friendly civilians or guerrilla forces who can assist the movement.

(7) Areas along exfiltration routes are not inhabited.

(8) The enemy force is dispersed or is not able to concentrate against the exfiltrating force.

4-20. Escape and evasion.

a. A vital part of all premission planning is the development of en-route and post mission E&E plans. Such plans enhance survival of the ranger force, and the transporting air or boat crews. The ranger force commander is responsible for--

(1) Helping in the development of an E&E plan in coordination with all supporting agencies.

(2) Ensuring that all members of the ranger force and the supporting boat or air crew are briefed on the E&E plan.

b. Each E&E plan is unique. Each situation has unique problems associated with E&E. The E&E plan devised by the ranger force commander must address these problems, while profiting from the individual abilities and training of the rangers and their supporting air and boat crews. The following considerations apply to E&E plans devised during ranger operations:

(1) The plans enhance survival of personnel who can no longer accomplish their assigned mission. The senior ranger in a group must decide if there are any remaining missions that the group can accomplish. If not, then he must try to evade and escape enemy capture if unable to link up and be extracted with the rest of the ranger force.

(2) Because of the depth of penetration behind enemy lines, most successful E&E plans may involve either air or water movement away from enemy-held territory. The air and boat crews supporting ranger operations are highly skilled and dedicated. If there is any chance of a successful rescue, these forces continue to try to retrieve rangers escaping from an enemy-held area.

(3) Rangers possess stamina and endurance. They can move a great distance (especially at night) and over rugged terrain to reach an area where they can rendezvous with search-and-rescue (SAR) aircraft or boats. Rangers are also resourceful and highly trained in communication, stealth, and camouflage. This, coupled with their self-discipline, means that successful E&E can be conducted over a great distance, through heavy concentrations of enemy, and over a long period.

(4) Escape and evasion plans for ranger elements should include avoiding contact with the local civilians. However, the aid of friendly insurgent forces may be enlisted. The ranger E&E plan may also include the use of other special operations forces E&E networks that are in place behind enemy lines. These nets must not be compromised by the volume of evaders.

c. Escape and evasion plans should be based on three phases:

PHASE ONE: That portion of the insertion following entry into the hostile zone where personnel survival and recovery is the only thing considered. Should an aircraft or boat carrying rangers be disabled during phase one, the ranking survivor assumes command. He must consider E&E planning and the experience and skills of other survivors. He should be concerned with the security of the mission, caring for wounded, and moving away from enemy activity while contacting SAR elements for recovery.

PHASE TWO: That portion of the insertion that is near the objective and could permit the ranger element to successfully pursue its mission. Although part of the E&E plan, actions by the ranger force during this phase become a forced land infiltration rather than an E&E. The senior surviving ranger must decide if enough personnel and equipment can be moved to the objective area in time to accomplish the mission. If so, he must proceed, taking along any surviving air or boat crew members. If not, he may proceed directly to the extraction point and wait, or he may contact SAR elements for recovery.

PHASE THREE: That portion of the operation after mission accomplishment. It is a hard time to escape and evade since the enemy is alert and reacting to the ranger forces. The confusion caused by the violence of the attack, coupled with the rangers' ability to move quickly cross-country away from the objective area, offers the best chance of a successful escape and evasion. The surviving ranger elements that cannot be extracted can do one of the following:

(1) Move over land at once to a PZ or beach for extraction by SAR elements.

(2) Move to an area where reentry into friendly lines can be made.

(3) Move to a neutral or nonaligned country where either internment or return to friendly control can be effected.

d. Any plan for E&E should provide for the issue to each ranger of an area map, a covert means of recognition, and a small compass. An emergency ration may also be included and any items needed for the climate in that objective area. The ranger force commander must balance the need for other survival equipment against the added weight and bulk involved.

CHAPTER 5

Strike Operations

The ranger regiment conducts strike operations in support of the Air Land Battle. Ranger strike operations are integrated into an overall plan designed to destroy, delay, and disorganize the enemy. They may also cause him to divert a major part of his combat forces and power to rear area security. Strike operations also help create an environment in which other special operations forces can operate. Strike operations can complement air power in disrupting enemy forces in depth. They allow the theater commander to regain the initiative and lead to decisive action.

5-1. Application.

a. The ranger regiment can conduct strike operations either as deliberate-response or quick-reaction missions. It can be inserted into the area of operations by land, sea, or air.

b. Strike operations can be conducted throughout the depth of the battlefield in support of conventional or unconventional warfare. Multiple strike operations can have a synergistic effect and can create situations that help friendly forces seize and maintain the initiative.

c. Strike operations include raids, personnel and equipment recovery, and interdiction of lines of communication. They are best accomplished by ranger units who use conventional techniques and special skills. Strike operations vary in size. When well augmented and supported, the ranger regiment can conduct strike operations deep in enemy territory using up to three battalions. Strike operations normally are limited in time and may be conducted unilaterally or in conjunction with other special operations forces or allied forces. They are normally ended by the extraction or exfiltration of the ranger force.

5-2. Raid operations.

a. A raid is a strike operation conducted behind enemy lines against strategic objectives, targets of high tactical value, time-sensitive targets, or key personnel and facilities in enemy rear areas. It normally involves a swift penetration of hostile territory to confuse the enemy or destroy his installations. Ranger units can conduct deep penetration raids when given the appropriate assets for insertion.

b. A strategic raid by the ranger regiment would be directed by the NCA and conducted under the OPCOM of a unified or JTF commander. An operational raid would be directed by the unified or JTF commander and conducted by that command directly or under the OPCOM of a corps commander. Typical targets would include:

(1) Command, control, communications, and intelligence centers at front level or combined arms army level.

(2) Nuclear and chemical weapons storage sites and delivery means.

(3) Key logistical centers, warehouses, ammunition complexes, or fuel pumping centers supplying logistical support to a specific front.

(4) Air defense and air traffic control integrating centers and air defense weapons sites.

(5) Key power generating and transmitting facilities, hydroelectric dams, and irrigation systems.

(6) Key ports or rail complexes.

(7) Key installations or facilities, such as airstrips, buildings, bridges, dams, or tunnels.

c. The size of the ranger raid force is determined by a detailed analysis of METT-T. The unit committed to the ground phase of the raid mission is kept as small as possible. It may be as small as a ranger company or it could include all the battalions of the ranger regiment. The force must be tailored to complete the mission quickly, violently, and with few casualties. The size of the force may also be affected by the augmentation of specially trained personnel needed for that mission.

d. Raids are normally conducted in five phases. (See Figure 5-1).

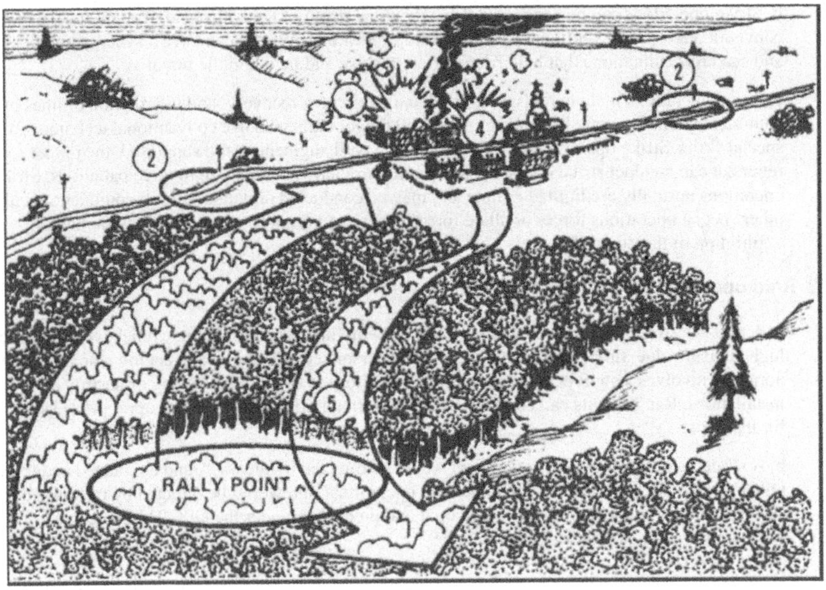

Figure 5-1. Five phases of a raid.

PHASE 1. The ranger force inserts or infiltrates into the objective area.

PHASE 2. The objective area is then sealed off from outside support or reinforcement, to include the enemy air threat.

PHASE 3. Any enemy force at or near the objective is overcome by surprise and violent attack, using all available firepower for shock effect.

PHASE 4. The mission is accomplished quickly before any surviving enemy can recover or be reinforced.

PHASE 5. The ranger force quickly withdraws from the objective area and is extracted.

e. This paragraph discusses the raid after insertion/infiltration. All considerations discussed in Chapter 4 must be applied in order for the unit to get deep behind enemy lines. Upon receipt of a raid mission, the following basics are emphasized:

(1) **Maximum use of intelligence information.** The gathering and dissemination of information must be continuous and provided to the raid force even while en route to the target area. To ensure mission accomplishment, the ranger unit must be kept informed of the latest enemy developments in the objective area to prevent being surprised. Only then can the full combat power of the rangers be concentrated at the decisive time and place.

(2) **Plan development.** The reverse planning sequence and the planning steps set forth in FM 101-5 are followed in the plan development.

(3) **Coordination.** Coordination is through the operational headquarters with friendly units (such as higher and supporting Army or joint headquarters, intelligence agencies) for fire, RSTA, special equipment, and personnel and logistical support.

(4) **Rehearsals.** Rehearsals help validate all aspects of planning for the raid and ensure precision in execution-They allow changes to be made in the plan before it is carried out.

f. A successful raid is ensured by--

(1) Launching the raid at an unexpected time or place by taking advantage of darkness and other periods of limited visibility, and moving over terrain that the enemy may consider impassable.

(2) Avoiding detection through proper movement techniques and skillful camouflage and concealment to include taking advantage of natural cover of the terrain, and using sophisticated equipment to detect and avoid enemy forces.

(3) Timing the operation as close as possible.

(4) Using all available support, both organic and nonorganic, to include use of special weapons such as Air Force "smart bombs" and artillery cannon-launched guided projectiles, with the ranger unit using laser target designators.

(5) Performing quick, violent, precise, and audacious actions that focus full combat power at the decisive time and place.

(6) Disengaging quickly upon mission completion.

(7) Withdrawing swiftly using planned routes and including a deception plan.

g. Four functions are normally performed by ranger tactical formations when conducting a raid. Each subelement is organized and equipped to do a specific part of the overall mission. Depending upon the specific mission, nature of the target, enemy situation, and terrain, these four functions are as follows:

(1) The command group controls movement to and actions at the objective. This unit normally consists of the ground commander, other subordinate leaders in the raid organization, and communications to support these leaders.

(2) The security element, whose organization is determined by the mission of the raid force, size and type of enemy force and its mobility and state of alert, terrain and avenues of approach into the area, and the time needed to seal off the objective area. The security element may perform the following:

(a) Secure the ORP.

(b) Give early warning of enemy approach.

(c) Block avenues of approach into the objective areas.

(d) Prevent enemy escape from the objective.

(e) Provide overmatch for the units at the objective and suppressive fires for their withdrawal.

(f) Provide short-range air defense.

(3) The support element provides the heavy volume of fire needed to neutralize the objective. Because fires from this unit are violent and devastating, they must be closely controlled to ensure the precision needed. On order or as planned, fires are lifted and shifted to cover the maneuver of the assault element by suppressing enemy fire from the objective or aerial fires. The support element may also be given specific locations to cover by fire in support of the security element if an enemy quick-reaction force moves toward the objective area. These may include routes into and out of the objective site, key terrain features, or installations adjacent to the main objective. Once the assault has been completed, or on order from the raid force commander, the support element displaces to the next planned position. Organization of the support element is determined by the following:

(a) Size of the objective, the geography of surrounding area, and the enemy threat (to include air) in the area. This element should be able to neutralize the objective (when supported by air or naval gunfire) and to lift or shift fires either when the assault is launched or when ordered by the raid force commander.

(b) Mission of the assault unit.

(e) Suitable firing positions.

(d) Size and nature of the enemy force in the objective area and those enemy forces capable of reinforcement at the objective.

(e) Fire support from other units (air strikes, naval gunfire, surface-to-surface missiles, and artillery fire).

(4) The assault element seizes and secures the objective and protects demolition teams, search teams, prisoner-of-way teams, or other teams.

(a) The organization of the assault element is always tailored to the mission. Each target area must be examined carefully. To assault, seize, and destroy an installation, position, or equipment, the assault element could be organized into one or more assault teams. The element's mission is to overcome resistance and secure the objective and to destroy the installation or equipment. Other specialized teams may also be needed. For example, sniper teams could be needed to remove key sentries. To capture prisoners, liberate personnel, and seize or destroy equipment, the assault element could be organized into assault teams, prisoner teams, search teams, medical teams, demolition teams, or breach teams.

(b) To destroy a point target or installation in a heavily defended area where the USAF cannot get close en to be effective, the assault element might be organized with one small team equipped with laser target designators. From covered and concealed positions, members of the team could then provide guidance for USAF delivery of laser-guided munitions from a safe distance.

h. The site chosen for the raid LZ or DZ must support the planned actions at the objective. There are two options when choosing sites:

(1) The ranger force can land on or near the objective and seize it before the enemy can react. (See Figure 5-2.) This avoids forced marches over land carrying heavy combat loads. If there is no suitable landing area near the objective, or the enemy has a strong reaction force nearby, this option is not favored.

Figure 5-2. Airborne landing on the objective.

(2) The ranger force can land unseen far from the objective. It then assembles, reorganizes, and moves into an ORP near the objective. (See Figure 5-3.) The objective is seized after security and support elements are in place. This option makes coordinated action easier by setting up control of small units before engaging the enemy.

Figure 5-3. Airborne landing distant from the objective.

5-3. Interdiction operations.

a. AirLand Battle doctrine set forth in FM 100-5 is a shift in emphasis toward keeping the initiative in battle. The deep battle seeks out enemy elements separated from the main battle area and attacks them before they can be deployed. Interdiction of selected enemy forces by attacking them or blocking chokepoints hinders, delays, or interrupts the enemy's use of his lines of communication; it denies him the use of certain areas; it destroys industrial facilities, military installations, and equipment; and it stops or slows his advance along a particular route of approach. The ranger regiment's ability to conduct interdiction is vital to execution of the AirLand Battle doctrine.

b. Interdiction by ranger forces includes blocking or channeling approaches, holding key terrain for a limited time. and destroying facilities.

c. Ranger forces conduct interdiction missions that supplement those conducted by long-range Army and USAF weapon systems. Large-size ranger forces are normally used only when other systems fail or are not available, and the target is of such priority that it must be attacked. Small ranger laser target designator teams may work in concert with USAF or USN strike aircraft to direct terminally guided weapons systems. Rangers should be directed against crucial interdiction targets that disrupt, confuse, or delay the enemy and allow friendly forces to gain the initiative at the forward line of own troops (FLOT).

d. Typical interdiction operations are--

(1) Destroying rail lines, switches, rail yards, and rolling stock.

(2) Destroying bridges, overpasses, tunnels, and truck parks.

(3) Closing mountain passes or routes in restricted terrain.

(4) Cutting oil or natural gas pipelines and destroying pumping and distribution systems.

(5) Destroying dams, locks, spillways, and flood control gates.

(6) Destroying rail transload points, switches, and yard engines.

(7) Destroying shipyards, dry docks, wharfs, lighters, cranes, and stevedore facilities.

(8) Destroying electrical generation facilities, power lines, transformers, and grid monitoring centers.

(9) Destroying water pumping stations and filtration systems.

(10) Destroying radio and TV stations, phone lines, microwave terminals, satellite receiving stations, and telephone exchanges.

(11) Ambushing enemy forces and personnel moving along lines of communication.

e. Considerations when using rangers for interdiction are:

(1) Insertion and extraction.

(2) Support resources available.

(3) Ranger force vulnerability.

(4) Target identification.

(5) Threat environment.

(6) Other or supporting means of interdiction.

(7) Command and control.

(8) Time and magnitude of the effect on the enemy forces at the forward edge of the battle area.

(9) Collateral effects of target destruction.

f. Interdiction targets are not identified and attacked at random. They are part of an overall deep attack plan to destroy or cripple an entire system. Interdiction is based on the assigned deep campaign plan that directs which targets are to be attacked, by whom, the results desired, and the priorities of attack for specific systems. Based on his plan, the ranger force commander selects those ranger elements best suited to conduct the attack. Target selection, which requires detailed intelligence and planning, is based on:

(1) **Target criticality.** A target is critical when its destruction or damage would have a major influence on the enemy's ability to conduct or support operations. Each target is considered in relation to other elements of the particular target system designated for interdiction. The criticality of a target changes with the situation. For example, when the enemy has few locomotives, a railroad bridge is less critical.

(2) **Target accessibility.** The ranger force must be able to insert into the target area. Terrain, security controls, location, and means of insertion are considered when studying a target for accessibility. systems. Based on his plan, the ranger force commander selects those ranger elements best suited to

(3) **Target restorability.** The enemy's ability to restore a damaged facility to normal operating capacity affects target selection. The enemy's repair and maintenance ability and his reserve stocks must be considered.

(4) **Target vulnerability.** A target is vulnerable when it is open to attack by means available to ranger forces. Vulnerability is affected by the nature of the target such as type, size, disposition, and security. It is also affected by the means available to attack it such as explosives, incendiaries, and special devices.

g. Key to the interdiction process is the ability to locate and evaluate target systems for attack and the timing of the attack. A target system is a series of related functions that serve a common purpose.

(1) A target is one element, installation, or activity identified for attack such as a locomotive, train, bridge, or prison.

(2) A target complex is several targets in the same general area, such as a railway marshalling yard, an airfield, or port and dock facilities.

(3) A target system consists of an industrial system and its sources of raw materials; the rail, highway, waterway, or airway systems over which these materials are transported; the source of power and method of transmission; the factory complex; and the means by which the finished product is transported to the user. The targeting process is usually accomplished in the special operations force cell at EAC and the battle coordination element (BCE) at corps level.

h. Interdiction techniques include ambush, emplacement of conventional demolitions, target designation for laser-guided munitions, and emplacement of beacons or sensors.

NOTE: The enemy surface-to-air threat normally precludes the use of an airborne platform for direct observation and designation. This may be accomplished by the on-ground ranger force.

i. Through interdiction, ranger units are an ideal force to contribute to the Army's AirLand Battle operational concept. This includes taking the initiative, moving quickly to strike deep in the rear of the enemy, and defeating the enemy force beyond the FEBA. Planners and commanders must ensure that the ranger units are used on missions that are synchronized to contribute to the overall deep battle campaign. Strikes against critical targets must combine with efforts of other assets to take the initiative from the enemy, limiting his flexibility, initiative, and momentum.

5-4. Personnel and equipment recovery operations.

Recovery operations are specialized raids to liberate imprisoned or detained personnel or to return certain items of equipment to friendly control. These operations include recovering and extracting downed or hijacked aircrews and political or military leaders. Rangers use surprise and combat power to overwhelm resistance before harm can come to the detainees or prisoners.

a. **Planning.** The planning of personnel and equipment recovery operations is the same as for the raid. The main difference is that the ranger force commander must plan for the extraction of the recovered personnel and the loading and extraction of sensitive equipment. Personnel and equipment recovery raids are often executed as quick-response missions. They normally require the ranger force commander to plan and execute the recovery quickly.

b. **Augmentations.** This type of strike operation often requires that the ranger force be augmented with nonranger personnel with special skills or abilities. Examples are medical personnel, technical experts, mechanics or crew members trained in repair and retrieval operations, linguists or translators, and other special operations force elements.

c. **Organization.** The organization of the recovery force is the same as for a raid. Sometimes special teams are needed to perform certain missions involved in the recovery. This mission is conducted aggressively, usually ending in close combat during darkness or other limited visibility conditions. Reconnaissance teams may be inserted ahead of the ranger force to reconnoiter the objective and to locate the detained personnel.

(1) All planning and execution takes place as described for the raid. The emphasis must be on detailed, timely intelligence.

(2) Medical teams must be available to care for the detainees.

(3) The ranger personnel must be adept at searching and clearing of buildings.

(4) The ranger force must--

(a) Locate the specific building or area where the friendly detainees are being held.

(b) Gain entry into buildings, use stun grenades, and avoid obstacles and booby traps.

(c) Engage all enemy personnel using selective-kill techniques.

(d) Locate, identify, and secure detained personnel.

(e) Move personnel to a secured extraction site.

(f) Leave no detainees or rangers behind in the objective area.

d. **Execution.** The key to success is surprise and speed. The enemy must be assaulted when least expected. The ranger force must ensure that friendly detainees are not harmed during the assault - either by rangers or enemy personnel.

e. **Extraction.** The ranger force commander should plan for the extraction point to be as close as possible to the point of recovery. This prevents the detainees from having to move cross-country or for the ranger force to transport bulky equipment long distances. For small groups of detainees or small items of equipment, the extraction point can be further away. Recovery operations may use any method or combination of methods of insertion and extraction (see Chapter 4). Close planning and coordination is normally required with US Army, USAF, or USN aviation for both delivery and evacuation of the target area.

CHAPTER 6

Special Light Infantry Operations

Ranger units are normally employed on strike operations. However, broad US strategic needs demand a wide range of other military abilities. The ranger regiment may conduct special light infantry operations when larger conventional forces could not be used. The regiment conducts these operations using light infantry doctrine, methods, and techniques.

These operations include, but are not limited to, movement on foot, military operations on urbanized terrain (MOUT), passage of lines, ambush, and reconnaissance. Ranger units do not normally conduct planned operations against enemy armored forces. Battlefield surprises or METT-T considerations may cause ranger units to engage armored forces for short periods. The ranger regiment can conduct a conventional light infantry defense for short periods. A defense is not normally conducted adjacent to other friendly forces. It is normally a perimeter defense from which the unit then breaks out from encirclement and links up with friendly forces.

6-1. Movement.

a. Most missions involve movement on foot in tactical operations. Ranger units infiltrating behind enemy lines seek to avoid chance enemy contact. They choose the movement that allows them to retain security and control. To avoid loss of surprise and initiative, casualties, and mission failure, ranger units normally--

(1) Avoid chance enemy contact, if possible.

(2) Move on covered and concealed routes.

(3) Avoid likely ambush sites and other danger areas.

(4) Practice camouflage, noise, and light discipline.

(5) Maintain all-round security.

(6) Make unavoidable enemy contact with the smallest element possible.

(7) Retain the initiative to attack at the time and place of the ranger unit's choice.

b. In planning the movement of his unit, the ranger commander must consider all the elements of METT-T and determine how they will affect the movement. He must determine the route, formations, security measures, methods of command and control, location of key leaders and weapons, and action to be taken on chance enemy contact.

(1) **Routes.** The ranger force commander conducts a careful route analysis, using the factors of observation, concealment, obstacles, key terrain, and avenues of approach (OCOKA). He chooses a route that maintains the security of the force while ensuring surprise. He tries to choose a route that will avoid enemy contact and speed movement. The commander makes a detailed route reconnaissance using maps, aerial photographs, and reports from persons familiar with the area. If possible, he reconnoiters the route himself or uses a reconnaissance team. He may have that team act as guides or mark the route using covert marking devices.

(2) **Formations.** Ranger units use the movement techniques found in current light infantry doctrinal publications. The traveling, traveling overmatch, or bounding overmatch technique is chosen based on the likelihood of enemy contact. The ranger force normally tries to avoid detection and contact with the enemy while moving. The commander chooses a movement technique that ensures that if unexpected contact were made, it would be with the smallest element possible and not the entire formation. The distance between moving ranger elements depends on the terrain, visibility, and enemy situation. March interval is normally great enough to allow each succeeding element either to deploy or change direction if enemy contact is made. Distances are close enough that each element could quickly assist the element to its front.

(3) **Security.** Security is to the front, rear, and flanks when enemy contact is likely. A front security element is placed well forward with adequate communications with the main body. The security element is used to detect the enemy and warn the main body. The main body then either changes direction or hides until the enemy is past. If the enemy does detect the main body, the security element may be used to attack the enemy from the flanks or rear. Flank security elements are placed next to the route of march. They move forward either by alternate or successive bounds. An alternative is to have the security element moving adjacent to the column along routes paralleling the direction of march. Rear security is handled the same as front security. Plans are made for the rear guard to support the main body during chance enemy contact. This is done either by maneuver or by furnishing supporting fire.

(4) **Command, control, and communications.** All available communication is used, consistent with OPSEC and movement security, to assist in maintaining control during movement. March objectives, checkpoints, and phase lines may be used to aid in control. The number of reports is reduced, normally only exception reports are needed. The ranger force commander should be well forward in the formation but may move throughout as the situation demands. Communications with security elements are mandatory. Operations security often precludes the use of radios. Therefore, connecting files, runners, and visual signals are normally used. Detailed planning, briefing, rehearsals, and control are valuable if there is enemy contact. Alternate plans are made to cover all possible situations.

(5) **Weapons.** Rangers are placed within the formation to allow a mix of their individual weapons to cover the front, flanks, and rear of the formation. Rifles, grenade launchers, and light automatic weapons cover all sectors. Crew-served weapons such as machine guns, antitank weapons, and mortars should be given designated locations within the formation. This allows the element leader to employ them quickly. They should be placed near the lead element, but not so close as to be engaged along with it. Crew-served weapons are used to

provide overmatch, especially when employing the bounding overmatch technique of movement.

(6) **Rehearsals.** The ranger force, expecting movement as part of an operation, conducts the following combat drill rehearsals:

(a) Conduct movement.

(b) React to enemy contact.

(c) Maneuver under fire (fire and movement).

(d) Cross danger areas.

(e) Break contact.

(f) Conduct a hasty ambush.

(g) Make a temporary halt.

(h) Establish a perimeter.

(i) Establish a rally point.

(j) Take actions upon consolidation.

(k) Use arm-and-hand signals.

(l) React to flares and indirect fire.

c. After a ranger force is inside enemy territory, it must be alert to avoid detection before reaching the objective.

(1) Should the ranger force security element become aware of enemy presence, it alerts the main force. The main force tries to move away without alerting the enemy and continues the mission.

(2) If the enemy does make contact, the security element then engages the enemy, trying to deceive him as to the size of the ranger force and its direction of movement. After breaking contact, the security force divides into small groups and moves to a rally point to rejoin the main force.

(3) The ranger force does not fight en route to the objective unless there is no alternative, then it quickly breaks contact and moves away.

(4) Ranger units may need to set up a temporary defensive position for resupply, evacuation of wounded, planning, or extraction. One technique that can be used is the temporary perimeter defense.

(5) If the moving ranger force comes under indirect fire, it moves quickly out of the impact area and continues the mission.

(6) The most effective means of combating an ambush is through the immediate return of a

high volume of fire by all weapons. This requires alertness, discipline, dynamic leadership, and rehearsed plans.

(a) Regardless of the method of movement, all weapons should be positioned for immediate use. A well-executed ambush is hard to counteract. However, the effectiveness of an ambush can be reduced by the immediate return of fire. Enemy weapons may be silenced, thus creating weak points in the ambush position. Smoke and high-explosive (HE) grenades are effective in executing counterambush actions. They create confusion, provide a screen for movement, and disrupt the ambush plan.

(b) When all elements of a ranger force are not trapped within the ambush, the elements that are free to maneuver should initiate an immediate flank or rear assault against the ambush force. A flank assault permits better coordination with personnel in the ambush. Supporting fires are provided by those weapons that cannot be hand carried and fired in the assault.

(c) Specific actions against a near and far ambush are as follows:

- **Near ambush.** Elements caught in the kill zone return fire at once and move out of the kill zone. Elements not in the kill zone help by maneuvering and conducting a flanking attack. The unit assaults the ambush position.
- **Far ambush.** Those ranger elements not in the kill zone set up an overmatch position and lay down a base of fire for the elements in contact. The elements in contact move out of the kill zone and prepare to assault the enemy, using individual movement techniques and fire and movement.

(7) The reorganization after enemy contact involves the use of rally points, plans for local security, reorganization of the force, evacuation of dead and wounded, and movement based on the unit mission.

6-2. Ambush operations.

a. **Definition.** An ambush is a surprise attack, from a concealed position. It is used on a moving or temporarily halted enemy to destroy or capture him and his equipment. An ambush is a useful tactic because--

(1) Small, well-trained, disciplined forces, with limited weapons and equipment, can destroy much larger enemy forces.

(2) It reduces the enemy's overall combat effectiveness by destruction and harassment of his forces.

(3) Enemy morale and effectiveness suffer heavily at little cost to the force executing the ambush.

b. **Execution.** A successful ambush must be executed with precision, violence, speed, and audacity of execution. For success, ambush operations must emphasize the following:

(1) **Surprise.** Surprise, more than any other single aspect, enhances the value of an ambush. Surprise increases the potential for inflicting damage on the enemy with less danger to the

ranger force.

(2) **Coordinated firepower and shock effect.**

(a) Coordinated firepower is used for maximum shock effect. Massive volumes of accurate fire, explosives, and mines, coupled with an aggressive attack, breaks the enemy's spirit to fight back. Surprise increases shock effect and the chances for success. Shock effect can cover unexpected defects in an ambush--for example, ambushing a much larger force than expected.

(b) All weapons must be sited with interlocking fires in the kill zone and along likely avenues of entrance or exit. Mortars should be used if the terrain permits. Tripods and traversing and elevating mechanisms are normally used with machine guns to lock in fires. All riflemen use firing stakes to mark left and right limits, and elevation stakes. There is a tendency to shoot high in an ambush--especially at night. The M203 grenade launchers are sited to cover the dead space and routes of escape. When the rules of engagement permit, plastic bags containing CS crystals may be taped to the front side of Claymore mines.

(3) **Control.** Control is essential.

(a) Leaders must have contact with all members of their unit to alert them to the oncoming enemy. Leaders should not move around the ambush site during this crucial period. A method used to alert members can be to tie strings or vines to rangers' legs or arms. By a series of light tugs, all members of the ambush can be alerted to enemy presence.

(b) The leader must initiate the ambush with a casualty-producing device. A bank of Claymore mines on a double-ring main is an excellent device to spring an ambush. Other good techniques are to use a 90-mm recoilless rifle firing antipersonnel (APERS) rounds or a machine gun. All must be under the direct control of the ranger leader. Whistles or pyrotechnics must not be used. They will allow the enemy time to react.

- As soon as the enemy is hit, he reacts. The ambush force has only a few seconds to destroy the enemy in the kill zone. He will quickly recover from the initial shock and leave the kill zone--either with a direct counterattack or withdrawal, Subsequent fires and other banks of Claymore mines must be planned.
- The leader always executes the ambush, except when a member of the ambush knows he has been discovered. He then has the authority to execute--with killing fire, not by yelling.
- The cease-fire must be controlled by the ranger leader. A whistle or other device may be used to get attention, and then cease-fire is signaled.

(4) **Security.** The flanks and rear of an ambush site are open to counterattack. Flank and rear security may be enhanced by -

(a) Echeloning in depth.

(b) Designating sectors of observation.

(c) Positioning of RSTA devices.

(d) Enforcing noise and light discipline.

(e) Having a good withdrawal plan.

(f) Securing routes of withdrawal.

(g) Executing with speed and violence.

(h) Positioning security forces to sea] off the ambush area.

(i) Having good camouflage.

(5) **Simplicity.** A simple, direct plan improves the chance of success. The ambush plan must be clear yet concise to offer the greatest likelihood of success. For example:

(a) Mission statements for security, support, and assault elements must be clear, concise, and direct.

(b) Tasks to be performed by the ambush elements should be easy to understand.

(c) Contingency plans should be simple.

(d) Routes into positions and withdrawal routes should not cross. They should be the shortest, most secure routes.

(6) **Training and self-discipline.** Precise and violent execution is how ranger units win against enemy forces. All advantages must be exploited. Discipline must be strict. There must be no sleeping, talking, eating, or smoking in the ambush site. If an ambush is to be set up for long periods, then the elements of the ambush must be pulled back at set times to the ORP for rest. Extended ambushes of 24, 36, or 48 hours require six- or eight-hour shifts. It may take a company to man an extended platoon ambush position. Tired troops cannot man an ambush well. Other vigorous operations cannot be performed all day and have an alert ambush all night.

(7) **Maximum use of specialized equipment.** This includes the most sophisticated weapons, communications equipment, and RSTA devices. The effectiveness of any mission is enhanced by the use of this equipment.

(8) **METT-T.** Every ambush mission must be evaluated by the ranger commander IAW METT-T to design an ambush that works.

(a) **Mission.** Clearly state the mission--for example, stop enemy resupply, ambush enemy armor columns, or capture personnel or equipment.

(b) **Enemy.** State the type of enemy units--for example, guerrillas, main force, logistics carrying parties, or armor units. What is the size of the force, its habits when attacked, and type of support? What is the enemy night vision ability and expected time of movement?

(c) **Terrain.** Is it wooded, mountain, jungle, or urban? Some good ambushes can be set in a city or suburb. Are the routes available to the enemy suitable to his activity?

(d) **Troops.** Squads and platoons are the basic ranger ambush forces. An ambushing force should be able to engage a force two to three times its size--but evaluate enemy and terrain, A platoon ambushing an enemy tank company in daylight in open terrain is risky. But engaging that same column at night or in a forest may give the ranger platoon an advantage.

(e) **Time.** How long will it take to emplace the ambush? A complex, lethal ambush with plenty of "dirty tricks" takes hours to set up; a hasty ambush on a trail takes 15 to 30 minutes. How long is the ambush to be in place? What hours of the day, or at nighttime only? How soon must it be set up? Do not set up an ambush in darkness--even if it is to be executed in darkness. Try to allow every ranger to '6see" and "lay" in his weapon in daylight. Time is important.

c. **Organization of the ambush force.** A ranger unit conducting an ambush must be task-organized to perform the following functions: assault, security, and support. The ambush forces should be task-organized according to the TOE--by platoons, squads, and fire teams. The TOE should not be changed to create smaller elements for an ambush. The TOE formations may be reinforced with machine gun, recoilless rifle teams, or a 60-mm mortar squad.

(1) **Assault.** The elements assigned the assault mission either move directly into their positions or through a release point. The mission may include any combination of the following:

(a) Conduct the main assault.

(b) Halt an enemy's motorized column or any moving target,

(e) Kill or capture personnel.

(d) Recover supplies and equipment.

(e) Destroy vehicles and supplies.

Search teams are not always used. The ranger leader must decide how and when to use search teams. When rangers leave the security of their well-chosen, concealed ambush position, they are subject to the fires of the enemy who may also be hidden and ready. If it is at night, do not throw tripflares or shoot illumination to light the search area as this will illuminate and expose rangers to the enemy. Always assume there is hidden enemy--the ambush will not kill them all. Night observation devices (NODs) or a red filtered flashlight taped to M16s should be used to quickly search in ranger buddy teams. If the return fire from the enemy is great or if the ambush missed the main body, then the leader may choose to break contact and leave without searching the kill zone.

(2) **Security.** The elements assigned the mission of security may move to their positions

directly or by way of a release point. Their missions may include any or all of the following:

(a) Secure flanks, rear, or ORP.

(b) Provide early warning.

(c) Seal off the kill zone to prevent the enemy from escaping or reinforcing.

(d) Assist in executing the ambush.

(e) Cover withdrawal of main ambush force.

(3) **Support.** The units assigned a support mission provide fires that may include employment of--

- Heavy automatic weapon fires.
- Antitank fires.
- Mortar fires.
- Mines.
- Flame munitions.

d. **Ambush site.**

(1) **Choose an ambush site.** When choosing an ambush site, all sources of information must be used to enhance surprise, exploit the enemy's weak points, and take advantage of the terrain. An analysis of the terrain is also based on the other factors of METT-T to include maps, aerial photos, and, if possible, personal reconnaissance. Emphasis is on--

(a) Natural cover and concealment for the ambush force.

(b) Routes of entry and withdrawal (at least two) that should be direct and easy to reach.

(c) Good observation and fields of fire.

(d) Harmless-looking terrain.

(e) Few enemy escape routes.

(f) Limited enemy reinforcement ability.

(g) Nearby assembly or rendezvous area.

(h) Terrain that will canalize enemy into kill zones, and natural obstacles to keep him there.

NOTE: Try to select a site covered by friendly supporting indirect fires.

(2) **Take advantage of the terrain.** Emphasis must be on exploiting all natural cover and concealment afforded by the terrain. Site the ambush and individual positions based on the terrain rather than trying to adapt the terrain to a fixed geometric design.

(3) **Restrict enemy movement.** Restricting enemy movement by natural or man-made

obstacles should also be planned.

e. **Types of ambushes.**

(1) Ambushes have two categories for ranger purposes--area ambush and point ambush.

(a) Area ambushes may be set up by ranger platoons, companies, or battalions. They are used to interdict enemy movement in a given area or inflict casualties on his forces. Area ambushes consist of a series of point ambushes. The size and location of the ambushes are dictated by the METT-T analysis.

(b) Point ambushes are set at the most ideal location to inflict damage on the enemy. Such ambushes must be able to accept the enemy force from more than one direction. The ranger force levels may enable it to execute an ambush in two or three main directions. The other directions must be covered by security that gives early warning of enemy attack.

(2) The basic ambush is linear (Figure 6-1). An ambush is set on an expected avenue of approach. This ambush can accept contact from three basic directions--left, right, and front. The rear is secured by a security team.

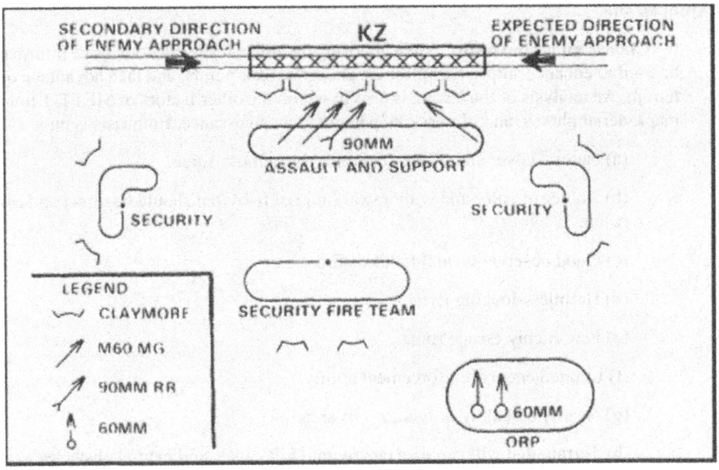

Figure 6-1. The linear ambush.

(3) A variation of the linear ambush is the L-shaped ambush (Figure 6-2). The L-shaped ambush is formed with the base (bottom) of the L perpendicular to the expected enemy direction of advance. This is a good ambush for a road, jungle trail, or an area where the enemy is canalized and his approach route is known. The L-shaped ambush can handle an enemy coming from the expected direction (toward the base of the L) from the front (stem

of the L). It is less effective against an enemy formation that comes from the opposite direction. The ambush must be executed when the enemy main body has cleared the base of the L. The flank security must protect the rear of the base of the L. In the L-shaped ambush, weapons must be sited to avoid direct or ricochet fire into friendly forces. The security elements must be considered as fire fans are being planned.

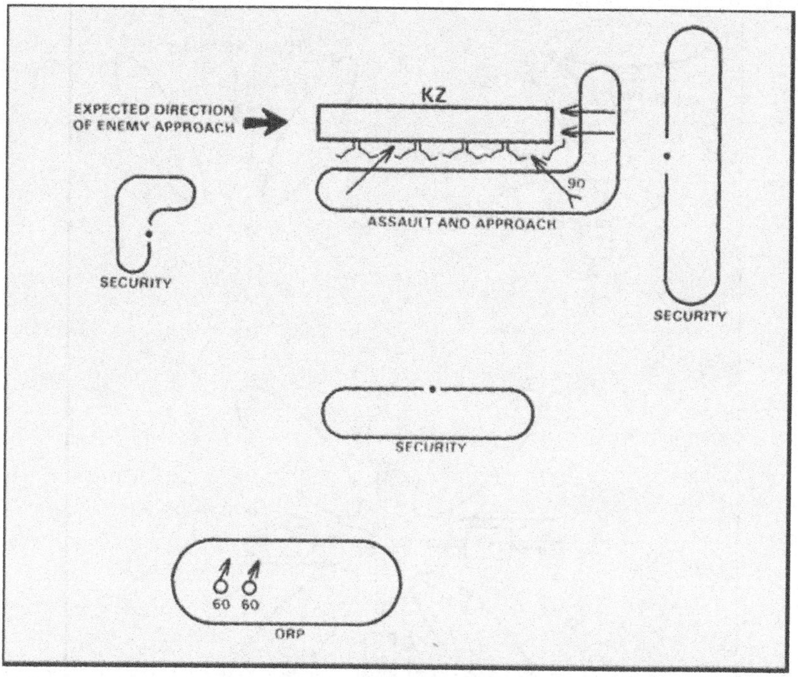

Figure 6-2. The L-shaped ambush.

(4) Other ambushes (the Z, the X, the V, and so on) are all variations of the linear and L formations (Figure 6-3). The more advanced ambush formations are difficult. Units must be expert in the basics before they are ready to employ such techniques.

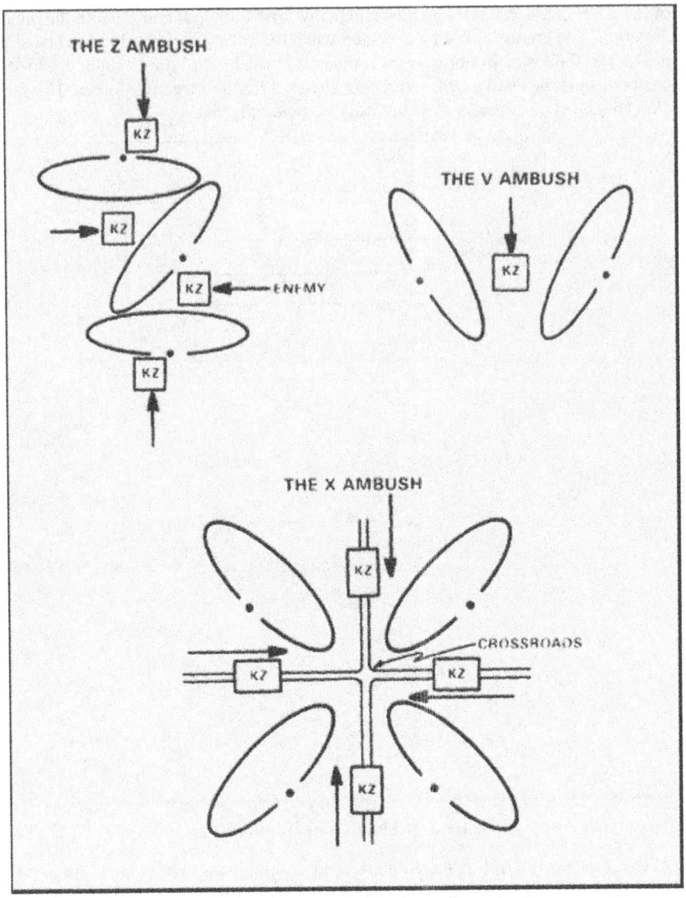

Figure 6-3. Other ambush variations.

(5) Another type of point ambush is the mechanical ambush (Figure 6-4). It consists of Claymore mines set in series with a double-ring main. It is command detonated or detonated by a triggering device activated by the enemy. Mechanical ambushes are normally manned. Rangers prepare to engage the enemy with direct fire if the mechanical ambush does not detonate or if it is wholly or partially ineffective. Mechanical ambushes are an effective way to interdict a large area using a small ranger force. If the mechanical ambush is effective and

rangers do not reveal their presence, the enemy is confused. This has a devastating effect on his morale and effectiveness.

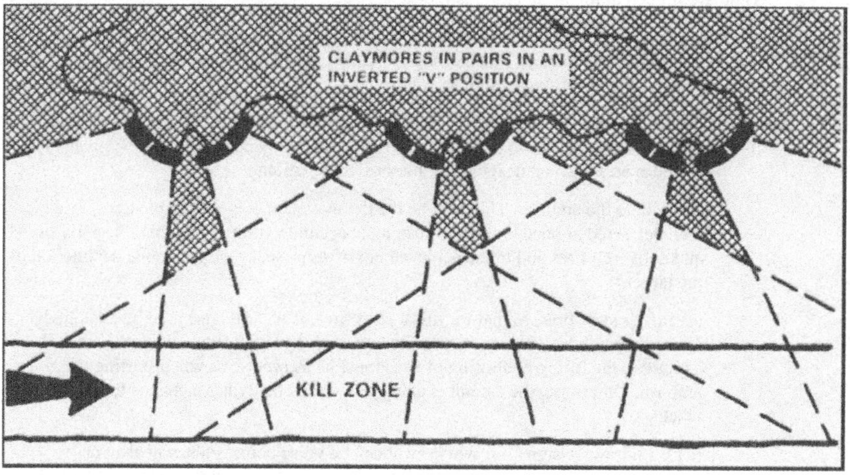

Figure 6-4. The mechanical ambush.

f. **Execution of the ambush.**

(1) **Occupation.** Stealth and security are important factors.

(a) Position security teams and early warning detection devices first.

(b) Use the rest route to main ambush position consistent with security.

(c) Quickly occupy the ambush position and set up communications and signaling devices.

(d) Position key weapons (automatic and antiarmor).

(e) Rig Claymore mines, tripflares, and booby traps.

(f) Ensure that all weapons are correctly positioned. Assign sectors of fire to provide mutual support and cover dead space.

(2) **Camouflage.** Each man must be hidden from the enemy, During preparation for the mission, each man camouflages himself and his equipment, and secures his equipment to prevent noise. At the ambush site, positions are prepared with minimum change in the natural appearance of the site. All debris resulting from preparation of positions is concealed to prevent evidence of occupation.

(3) **Movement.** Movement is kept to a minimum. The number of men moving at a time is closely controlled. Every man is as quiet as possible, especially at night. Light discipline is rigidly enforced at night. Smoking is forbidden.

(4) **Signals.** Audible and visual signals, such as whistles or pyrotechnics, must be changed often to avoid setting patterns and alerting the enemy. Three or four simple signals are needed to execute the ambush. Signals are used to--

(a) Provide early warning of an enemy approach. A signal by the security force to alert the patrol leader to the correct direction of enemy approach may be given. This includes arm-and-band signals, radio, or field telephone.

(b) Initiate the ambush. This may be the detonation of mines or explosives, Fire is then delivered at once in the heaviest, most accurate volume possible. Properly timed and delivered fires add to the achievement of surprise, as well as to the destruction of the target.

(c) Lift or shift fires, Signal for lift or shift fires, if the kill zone is to be assaulted. Voice commands, whistles, or pyrotechnics may be used. When the kill zone is assaulted, the lifting or shifting of fires must be as precise as when starting the ambush. Otherwise, the assault is delayed and the enemy has a chance to recover and react.

(d) Withdraw. A signal for withdrawal can be voice commands, whistles, or pyrotechnics.

(5) **Objective rally point.** The ORP is located far enough from the ambush site so that it will not be overrun if the enemy manages to attack the ambush unit(s). Existence loads are normally left in the ORP. Routes of withdrawal to the ORP are scouted (when possible by each man), Withdrawal routes should provide cover and concealment for the unit and hinder enemy pursuit. Withdrawal routes are a main consideration in the selection of the ambush site. They may be the key to survival after execution of the ambush. On signal, the ambush force quickly (but quietly) withdraws to the ORP. If pursued, the withdrawal may be by bounds, with mines or hasty ambushes used to delay pursuing forces.

(6) **Ambush variety.** If one method of ambush is used predominantly, the enemy will develop an effective defense against it and will be affected less by the shock of the ambush since he knows what to expect. More than one method of ambush should be used. No single method will fit all combinations of terrain, equipment, weather, and enemy capabilities. Variety should also be used when signaling - for example, use multiple signals, both audible and visual. Use weapons fire, mines, and RSTA when possible and vary signals to avoid compromise.

(7) **Swift action.** Speed in the execution of the ambush and the withdrawal should prevent enemy reaction forces from engaging the ambush force. Speed is often a shield against casualties and failure. When there is contact with reaction forces, speed may enhance quick disengagement. Ranger units must not only perform better than the enemy but also move faster than the enemy can react.

g. **Successful ambush.** Emphasize the following:

(1) Intelligence to ensure the enemy is ambushed at a time and place when he least expects or is least prepared to fight. This is preferred during reduced visibility to achieve surprise.

(2) Detailed planning and thorough training and rehearsing of all elements in all phases of the ambush. This ensures maximum shock effect through swift, precise execution.

(3) All available RSTA devices to permit daytime effectiveness at night when moving, shooting, or detecting enemy movement. To avoid detection, active RSTA devices should not be used until after the ambush has been triggered.

(4) All available firepower with emphasis on antiarmor, area and automatic weapons, and precision-guided munitions.

(5) Speed to achieve surprise and enhance security of the force.

(6) Cover, concealment, and overall protection afforded by the terrain when moving or when occupying ambush positions.

6-3. Antiarmor operations.

a. Antiarmor operations are not primary missions for ranger forces. They lack antiarmor firepower and have limited antiarmor ability. Ranger units can, however, perform limited antiarmor missions on close terrain by interdiction or ambush. The antiarmor weapons they do have are man-portable, but the ammunition is too heavy to carry many rounds cross-country. A ranger battalion has ten Dragon missile launchers and nine 90-mm recoilless rifles. It is also equipped with light antitank weapons (LAWs) and laser target designators (LTDs) that can guide Copperhead rounds.

b. Every ranger must be skilled with the limited antiarmor systems available. Proper selection, fortification, and camouflage of antiarmor ambush sites, route selections, withdrawal procedures, and initiative are important. Armor kill zones (AKZs) are normally set up after infiltrating the ranger force behind enemy lines under limited visibility.

c. A ranger antiarmor force is organized into four elements: command and control, support, security, and armor-killer teams (Dragons, LAWs, 90-mm recoilless rifles). These elements function the same as a raid force. The exception is that the assault force is the armor-killer team with the mission of destroying enemy armor, and it will probably not enter the AKZ. After the ambush occurs, the ranger force withdraws to its ORP. Indirect fires (if available) are then brought to bear on the AKZ.

d. Ranger units may conduct antiarmor ambushes and interdiction operation by use of laser-guided Copperhead rounds or air-delivered precision-guided munitions. Air-delivered family of scatterable mines (FASCAM) can be used to provide the ranger force security against attack by armored forces.

6-4. Military operations on urbanized terrain.

a. The ranger regiment normally tries to avoid the manpower-intensive, high-casualty battles associated with MOUT. However, due to the nature of strike operations and the types of targets selected, the regiment may conduct MOUT for limited periods. As in all ranger operations, the ranger force commander tries to surprise the enemy, attack him from unexpected directions, and destroy him with a violent, precise attack. The maze of cellars, rooms, sewer systems, rooftops, and alleys common to urban areas presents many opportunities for the ranger force.

b. Offensive operations by a ranger force exploit the centralized planning and decentralized execution aspects of ranger unit training. Individual action and initiative are always channeled to accomplish the commander's intent. The ranger unit's offensive tactics are based on quick action and boldness. Attacks by the ranger force are short and sharp. Execution is marked by swiftness, daring, timing, and surprise. The enemy's sleeping habits, eating habits, and relief times are watched. launched almost always at night or under smoke screens, attacks are often made without mortars or artillery, relying on surprise for success. Snipers and sharpshooters are used to shift the balance of battle.

c. Deliberate defensive operations are not good ranger unit missions. However, if a ranger unit must defend for a short period, the close terrain of urban areas allows it to function well against a stronger force. The buildings in a city break up advancing enemy formations and force them to move along streets. The ranger force executes an active defense anchored on strongpoints, such as a building or a group of buildings. It launches surprise counterattacks against the enemy's flank and rear. The aim of ranger defense is to give the enemy no respite and to make him feel threatened by a sudden, deadly attack. Ranger units should be withdrawn from defensive positions in urbanized areas as soon as possible. Prolonged MOUT and the associated casualties quickly degrade a ranger unit's combat effectiveness.

6-5. Perimeter defense.

The Perimeter defense may be conducted to control key terrain, gain time or economize forces.

a. **Planning.**

(1) **Warning order.** The order must be issued soon after mission receipt. It should contain the mission, time schedule (including rehearsal and inspection times, mission time, and place for OPORD), uniform and equipment, and special instructions.

(2) **Operation order.** The leader normally gives the OPORD on the ground to be defended. Locations and sectors of fire for squads and crew-served weapons should be designated. If time permits, leaders walk the terrain with subordinates as the OPORD is issued. The OPORD covers security, communications, signals, target reference points (TRPs), armor kill zones, and priority of engagement rules for antiarmor weapons. The OPORD also states the priority of work.

(a) Set up security and communications.

(b) Position crew-served weapons and personnel.

(c) Clear fields of fire.

(d) Prepare fighting positions.

b. **Organizing.**

(1) Squads are positioned with overlapping and interlocking sectors of fire and observation.

(2) The M60 machine guns are positioned either with an assigned final protective line (FPL) of fire or with a primary direction of fire (PDF). The PI)Fs are employed on covered avenues of approach such as gullies. The M60 FPLs should interlock across the unit front.

(3) The 90-mm recoilless rifles and Dragons are positioned with frontal cover to engage attacking armored vehicles from the flank.

(4) Riflemen are positioned to support and protect crew-served weapons. Riflemen are assigned FPLs that interlock at a point just beyond hand-grenade range.

(5) Dead space is covered with Claymore mines, antipersonnel mines, and M203 fires.

(6) Antitank mines, positioned to provide close-in protection from armored vehicles, are covered by LAW fires.

(7) Leaders plan indirect fire for long-range suppression of armored vehicles and short-range protection from dismounted assault. Mortar or artillery final protective fires (FPF) are placed on the most dangerous dismounted avenue of approach.

(8) Range cards are prepared for all crew-served weapons. Squad and platoon leaders prepare sector sketches and fire plans.

(9) The ranger force camouflages everything from ground and aerial observation.

(10) Leaders designate alternate and supplementary positions.

c. **Executing.** Leaders control the distribution and concentration of fires in their sectors. They constantly consider the overlap of fires with units to their left and right.

(1) Squads and crew-served weapons engage targets in sectors assigned by platoon leaders.

(2) Each weapon is used in its best role:

(a) Rifle, against enemy infantry within 300 meters.

(b) Machine gun, against enemy infantry within 1,000 meters.

(c) Grenade launcher, against infantry within 350 meters and armored vehicles within 200 meters.

(d) Light antitank weapon, against armored vehicles within 200 meters.

(e) Recoilless rifle, against armored vehicles within 400 meters.

(f) Dragon, against armored vehicles within 1,000 meters.

(3) Most dangerous targets are engaged first.

(4) Squads shift fires as needed or as directed by platoon leaders.

(5) The platoon masses fire on selected targets while maintaining coverage of the entire platoon sector.

(6) Platoon leaders control fires.

(7) After repelling an enemy assault, the platoon consolidates and reorganizes the defense.

6-6. Breakout from encirclement.

A breakout from encirclement is conducted when units, operating behind enemy lines, find themselves cut off from friendly forces and surrounded by superior forces. Given ranger unit missions, the chance of operating as an isolated force behind enemy lines is great. The breakout is characterized by determining enemy weak points, deception, massing of combat power, and a direct attack for a violent and timely breakout.

a. **Characteristics and fundamentals.** The breakout from encirclement involves the following characteristics and fundamentals:

(1) Set up a hasty defense. Adjust the perimeter as needed and make contact between subunits. Assign each a specific sector.

(2) Plan fire support. Plan FPFs on likely dismounted avenues of approach. Mass antiarmor weapons on the most dangerous mounted avenue of approach.

(3) Conduct reconnaissance. Identify enemy weak points. Locate enemy armored forces and the routes they must use. Locate enemy automatic weapons.

b. **Planning.**

(1) Organize all support personnel into maneuver elements--for example, company headquarters, mortars, and antitank. Organize the ranger force into a rupture force, reserve, support force, and a rear guard.

(2) Issue a fragmentary order. It should contain current local enemy situation and cover all pertinent areas of the five-paragraph field order.

(3) Select rally points.

c. **Execution.**

(1) Reduce the defense perimeter. This is done slowly until the unit is massed for the breakout attempt.

(2) Create a diversion. This is done either by fire from the rear guard force or by a limited attack in a direction other than that of the main breakout attack.

(3) Break through the defense with the rupture force. Hold the shoulders of the breakthrough.

(4) Move the reserve through to exploit the breakthrough.

(5) Move the support troops through with casualties. Have them collect casualties en route.

(6) Have the rear guard follow the support troops. It may have to fight a delaying action and use fire and movement to break contact with the enemy.

(7) Consolidate and reorganize following the breakout.

(8) Have the rupture force break contact with the enemy once the rear guard joins.

6-7. Linkup operations.

During operations behind enemy lines, ranger units may be required to link up with other friendly ground forces. Linkup may occur in airborne, air assault, amphibious, or shore-to-shore operations. It may occur during the breakout of an encircled force or during the convergence of separate forces. It may involve conventional or unconventional US forces, allied forces, or friendly guerrillas. Rangers may be required to link up with other rangers, partisans, or friendly ground forces that are conducting offensive operations.

a. **Planning.** Ranger commanders plan detailed linkup operations to avoid friendly forces firing on each other. Coordination and planning of communications, recognition signals, fires, command relationships and responsibilities, and control measures are essential. Coordination should be done before the start of the operation.

(1) Site. The linkup site should be easy to recognize, have cover and concealment, be located away from prominent terrain the enemy might use, be defensible, and provide multiple access and escape routes. An alternate site should be planned in case of compromise.

(2) **Communications and recognition signals.** Frequencies, call signs, codes, visual signals, and alternates for each of these should be planned before departure of friendly lines. Radios may be used to report the location of each unit to the site, and occupation and securing of the site. A system of mutual recognition must be devised to keep friendly patrols from firing on each other. This may include visual signals such as arm bands, panels, colored lights, or RSTA devices.

(3) **Fires.** The headquarters directing the linkup sets up fire control measures and priority of fires. The most commonly used control measure is the restrictive fire line (RFL), beyond which one force may not fire without coordination with another force. The RFL may be adjusted as two units draw closer together. Such an adjustment needs close coordination and should be planned before the operation. Successive phase lines between the two units can serve as on-order RFLS, provided that neither unit comes too close to the RFL before it is shifted.

(4) **Command relationships and responsibilities.** These are set up by the headquarters directing the linkup. Ideally, liaison personnel are exchanged before the operation. The key to success is liaison and guides.

b. **Execution.** Each member of the ranger force must understand the plan. Rehearsals are necessary in order to avoid confusion during the linkup. One uninformed ranger can start firing and cause enough confusion that friendly forces could be identified as enemy. Actions must be quick since units are open to enemy attack while they try to link up.

(1) **Moving units.** A linkup between moving units is hard. As the units move closer to each other, the chance of them engaging one another increases. Therefore, the linkup units must adjust their movements to each other and continuously coordinate on a designated secure radio net. If possible, one or both units should come to a short halt before linkup.

(2) **Stationary units.** Moving units must know the positions of the stationary units. They must keep stationary units advised of locations. Stationary units guide the moving units to the linkup point by radio. The stationary units must be ready to accept and position the moving units smoothly and quickly.

6-8. Passage of lines.

A passage of lines is an operation in which one unit moves through another unit that is stationary and disposed in a tactical formation on a FEBA. Ranger forces may conduct a passage of lines to get behind the enemy. Movement in forward unit areas must be controlled, coordinated, and kept to a minimum. This avoids conflict with friendly troops or the activation of RSTA devices. Ranger units treat the positions of forward units as danger areas. They must be assumed to be under enemy surveillance in all weather or visibility. The ranger force is open to attack during the passage. Personnel and units may be more concentrated than is desired. Fires of the stationary unit may be masked for a limited time. The passing ranger unit may not be in a good position to react to enemy contact. Detailed reconnaissance and coordination are crucial to ensure that the passage is conducted quickly and smoothly.

a. **Principles.**

(1) **Coordination.** The ranger force commander, or his representative, must coordinate departure and reentry of friendly front lines (FFL). This is done either directly with the units through which the rangers pass or through designated staff agencies. Items to be coordinated include fire support, passage lanes, control measures, guides, communications, intelligence, CSS, and actions on enemy contact.

(2) **Reconnaissance.** The ranger force commander should make a ground reconnaissance of the area through which the unit will pass. He should pick the initial rally point (IRP) at this time. The reconnaissance should include observing the area just forward of the friendly unit for routes, danger areas, and obstacles. During reconnaissance or coordination, rangers should wear the same uniform as the troops in the forward unit. This will avoid attracting the enemy's attention.

(3) **Security.** Maintain security to avoid contact with the enemy while departing or reentering. It is important that the ranger force make no enemy contact at this time. Control is hard if a fire fight starts just forward of an FFL.

b. **Departure of friendly front lines.**

(1) Designate an IRP.

(2) Maintain security. Use an appropriate movement formation when departing the FFL.

(3) Do not move within the friendly unit's area without a guide.

(4) Make final coordination with the friendly unit commander.

(5) Count the ranger force members as they depart.

(6) Make a security and listening halt.

c. **Departure of friendly units.**

(1) The ranger force arrives at the forward unit and is met by a guide. The guide leads the rangers to an assembly area chosen by the ranger commander during initial coordination with the forward unit commander.

(2) No one should move, either singly or as a unit, anywhere in the FFL without a guide.

(3) The ranger commander makes final coordination with the forward unit commander. He determines if any changes have taken place since initial coordination and learns of any recent enemy activity.

(4) Before leaving to coordinate, the ranger commander gives instructions (called a contingency plan) for what should be done while he is gone. These instructions state where he is going, who is going with him, how long he will be gone, what to do if he does not return, and actions to be taken if there is enemy contact.

(5) On the commander's return from final coordination, he may issue a fragmentary order (FRAGO) to cover any changes.

(6) Three common threats and techniques for countering them during departure from the FFL are:

(a) **Ambush and chance contact.**

- The ranger force commander must avoid enemy contact while departing.
- The ranger force goes to the friendly side of the departure point where a point team is dispatched to see that the area forward of the barriers is clear.
- The point element should check the area out to the first covered position or an area large enough to allow the unit room to maneuver if it is hit while departing.
- The distance varies with the size of the ranger force and the terrain.
- The point team notifies the ranger commander when the area is clear. The rest of the unit is brought forward.

(b) **Indirect fire.** If enemy indirect fire is falling, the ranger commander should not halt his unit after final coordination at the forward command post. Instead, he should move it quickly through the FFL. The point element should be far enough ahead of the main body to provide security from ambush or chance contact.

(c) **RSTA.** If the enemy has radar, sensors, or night vision devices, they may be offset by the following countermeasures:

- Use a well-hidden departure point, such as a reverse slope or a dense woods.

- Infiltrate the unit through the departure area and have the rangers rendezvous in a defilade.
- Pass through during rain, fog, snow, or darkness to help conceal the passage.
- Employ electronic countermeasures.

(7) Have a security and listening halt after the ranger force has moved out of sight and sound of the FFL. This is a short halt to see if enemy is in the area, and to accustom the unit to the sights and sounds of the battlefield. Halt in a position that covers the unit from chance friendly small-arms fire. If the security and listening halt must be in an exposed position, have each man lie flat on the ground.

(8) During movement, have frequent halts to see that the unit is not being followed and that no enemy is in the area.

d. **Reentry of friendly front lines.**

(1) Set up and occupy a reentry rally point (RRP).

(2) Maintain security at the RRP and at the reentry point.

(3) Use as few personnel as possible to locate the reentry point while the bulk of the unit remains in the RRP.

(4) Meet a guide at the reentry point. Normally, a forward-of FFL password is used since it may be overheard by enemy.

(5) Count the ranger force members to ensure no infiltrators follow the unit, especially during reduced visibility.

(6) Give the friendly unit commander a spot report, providing combat information of immediate tactical value.

e. **Reentry of friendly units.**

(1) The unit is moved into a rally point near the reentry point. This rally point should be on a prominent terrain feature where the ranger commander can pinpoint his location, with respect to the reentry point.

(2) The forward unit is alerted by radio that the ranger unit is ready to reenter. A code word is used for security and brevity. The code word must be acknowledged by the forward unit before the ranger force reconnoiters the reentry point. This shows that a guide has been sent to the reentry point and is waiting for the unit.

(3) If the ranger commander is certain of the reentry point location, he moves the entire unit directly to the reentry point. He has the point element coordinate the reentry.

(4) An effective method of locating the reentry point is through the use of a thermal imagery device. It detects the body beat of the reentry guide, even though he may be well camouflaged or his location masked by smoke. If ground surveillance radar is available, it can be used to vector the ranger unit to the reentry point.

(5) The reconnaissance party should never move parallel to friendly barriers or move around the wire. If the reentry point cannot be found by the initial reconnaissance or if the reconnaissance party finds only the barrier wire, the unit should notify higher headquarters and move to another rally point to wait until daylight. The unit should not stay in a rally point from which a radio transmission has been made.

(6) When the reentry point is found, the ranger commander can bring the unit forward. If the reentry point is easy to find, he can call by radio for the unit to come forward.

(7) The guide leads the unit through the barriers to the security position coordinated with the forward unit commander. The unit halts in the security position, remaining in movement formation. The ranger commander then gives a spot report to the forward unit commander. The ranger commander tells him anything of intelligence value or of immediate tactical use to the forward unit.

6-9. Reconnaissance platoon operations.

a. The reconnaissance platoon organic to the ranger regiment consists of a platoon leader, a senior reconnaissance sergeant, and three five-man reconnaissance teams that include a communications specialist in each team (see Appendix A). This platoon is responsible for short-duration reconnaissance of objectives for the ranger force commander. This reconnaissance is routinely in support of future strike or special light infantry operations. The teams can also be used by the ranger force commander to conduct deep target acquisition and designation, or to confirm or disprove information reported by other sources. Rarely will teams be used to attack enemy targets.

b. The reconnaissance teams of the ranger regiment do not replace (or perform the same reconnaissance tasks associated with) long-range reconnaissance patrol (LRRP) units or long-range surveillance units (LRSUs) that may be organic to the corps or division. These units are organized, trained, and equipped for long-term passive surveillance and reconnaissance in enemy territory. The ranger reconnaissance teams are trained and organized to support the immediate intelligence-gathering needs of the ranger force commander. They are not to act as a human intelligence (HUMINT) collection asset for other operational headquarters.

c. The reconnaissance platoon headquarters trains, equips, and controls the employment of the three reconnaissance teams. It operates the communications stations needed to receive reports from deployed teams. In the absence of the S2, it debriefs team members upon extraction or exfiltration. It can provide one or more teams to support each ranger battalion on independent operations. The reconnaissance teams can--

(1) Infiltrate the objective area by parachute (HALO, HAHO, or static lines), helicopter, fixed-wing aircraft, SCUBA, small boat, foot, or other means.

(2) Remain undetected in the vicinity of the objective area up to five days.

(3) Perform reconnaissance operations employing a full range of night observation devices, infrared detection devices, unattended sensors, and photographic equipment.

(4) Perform demolition target analysis.

(5) Operate small watercraft and inflatable boats.

(6) Emplace unattended ground sensors, omni-directional navigational beacons, hand-emplaced expendable jammers, and electronic target designation devices.

(7) Collect combat information to satisfy priority information requirements and mission-essential elements of information. Teams report that information by use of long-range, secure, burst-transmission communications equipment.

(8) Perform DZ selection, marking and reception duties.

(9) Report objective area weather conditions.

(10) Perform highly selective, limited attacks or ambushes when so tasked.

(11) Link up with the main body of the ranger force in the objective area, or escape and evade the enemy in order to return to friendly lines.

(12) Act as part of the AAT during airborne operations.

d. There are many limits of the reconnaissance teams.

(1) Mobility is normally restricted to foot movement in the area of operations.

(2) Use of radio and other active electronic or optical devices makes the teams open to enemy detection.

(3) All supplies and equipment needed by the team must be carried on the initial insertion, since resupply increases the probability of detection by the enemy.

(4) Conduct of insertion and extraction operations deep in enemy-held territory is difficult due to the need for secrecy, timeliness, security, and accuracy of location.

(5) Medical treatment of team casualties is limited to individual first aid. Casualty evacuation is difficult unless linkup with the main ranger force has been achieved.

e. The training of reconnaissance team members parallels that given to all members of ranger units with emphasis on infiltration, communications, and reconnaissance skills. Extensive training is also given in survival, evasion, resistance, and escape (SERE) and advanced first aid. Training emphasis is placed on actions at the reconnaissance site. The reconnaissance teams must conduct their mission and remain unseen. Every team member is trained to consider the need for information balanced by the likelihood of detection. Detection by the enemy compromises future operations.

6-10. Reconnaissance planning.

a. Detailed planning ensures that the reconnaissance mission is successful. It is also vital to the survival of the reconnaissance team. This planning is conducted by the ranger force planning group assisted by elements of the JTF staff. The supporting air or naval commander, navigator, and electronic warfare officer may aid the ranger force staff in the planning.

b. Upon receipt of the mission, the team selected begins an intense preparatory phase at the REMAB or the ISB. There the team receives its initial mission briefing. The items covered allow the team to complete its detailed planning. The reconnaissance platoon leader reviews the team plan. He ensures compatibility and integration into the ranger force commander's plan for employment of the main force. Staff specialists at ranger regiment and battalion--such as those concerned with intelligence, weather, or communications--provide help.

c. Planning for the use of reconnaissance teams is conducted within the framework of the ranger planning sequence for the entire mission. There are five planning considerations for the employment of reconnaissance teams:

(1) Operations security must be considered in all staff efforts to provide maximum protection for the reconnaissance teams. This includes intelligence, CE, logistics, administration, and maintenance. Operations security must be integrated throughout every ranger force operation. It consists of four main categories of security measures: signal security, physical security, information security, and deception. All are related and must be considered at the same time for each operation.

(2) Some missions may require rapid deployment to the operational area, thereby dictating the quickest method of insertion. However, others may depend on maintaining total secrecy, regardless of time.

(3) The transportation means selected for the delivery and recovery of the teams depends on the mission. This selection is also based on the abilities, limits, and availability of mission support platforms.

(4) The distance to and from the objective area must be considered, as well as the distance from the insertion point to the actual target. Reconnaissance teams can move great distances over rugged terrain. However, this movement does require time and limits the team's ability to carry bulky or sensitive equipment.

(5) Consideration must also be given to intelligence information available concerning the objective area which includes, but is not limited to--

(a) **Enemy situation.** The enemy threat--his abilities, disposition, security measures, detection, and air defense systems - affects the selection of the means for delivery or recovery.

(b) **Terrain.** Land formations must be considered in selecting the method of infiltration and exfiltration. Terrain affects the selection of altitudes, approach and exit routes, landing areas, drop zones, and beach landing sites. Air infiltration routes that provide terrain masking are desirable in static line parachute operations.

(c) **Weather.** Seasonal weather conditions affect team delivery and recovery. Factors to be considered are temperature, precipitation, visibility, clouds, and wind. If para-SCUBA techniques are used, high surface winds and their effect upon surf conditions or period of reduced visibility may preclude the use of parachutes, inflatable boats, or swimming as entry or recovery techniques. These same conditions favor land infiltration and exfiltration, The AWADS reduces the impact of visibility as a limiting factor for air infiltrations.

(d) **Astronomical conditions.** Periods of sunrise and sunset, moon phase, moonrise and moonset, and twilight must be considered.

d. A reconnaissance team mission must be specific and support the mission. Information concerning the target may not be gathered because the risk of detection by the enemy is so great. Detection may preclude the conduct of the planned operation. If more than one mission is assigned, priorities are set, All team missions must be coordinated to avoid other friendly forces operating near the target. Ranger forces conduct operations at such depth behind enemy lines that no other US or allied forces are normally present. There is a chance that a friendly indigenous guerrilla force may be operating near the objective, but ranger reconnaissance teams do not normally make contact with them.

e. The reconnaissance platoon leader joins with the intelligence and operations sections of the ranger force headquarters in the initial planning for reconnaissance team operations. The method of operation while moving, communication procedures, reporting, surveillance procedures, and actions on enemy contact should be included in the platoon's SOPS.

f. The ranger force intelligence officer normally prepares the reconnaissance plan IAW the guidance provided by the controlling headquarters. He is helped by the operations officer and the reconnaissance platoon leader. Selected team members and a member of the unit providing the insertion and extraction means are briefed on the mission early in the planning phase. They should participate in the detailed planning to follow. The essential details of a reconnaissance plan normally include--

(1) Areas to be observed and recommended areas or positions from which this can be done. These positions are determined in advance of the insertions of teams. They are based on a study of terrain, road and rail nets, enemy order of battle, delivery means, operations plans of the controlling headquarters, and the desires of the ranger force commander.

(2) Load plans and procedures. This includes both insertion and extraction. Alternate sites are chosen for possible use. Primary and alternate DZs are selected, if applicable.

(3) The flight plan. This includes insertion and extraction flight routes and alternate routes.

(4) The movement plan to and from a patrol base. This includes the times, routes,. and destinations when individuals or small groups are required to operate away from the patrol base.

(5) The fire support plan. This includes the location and times of planned air strikes in support of the main body, and all planned fires by the main assault force. The reconnaissance teams should know the frequencies and call signs for tactical air support.

(6) The timing and ground tactical plan for execution of the main force operation. This includes disposition of the main force, special mission teams from the ranger force, or any other special operations force elements.

(7) The communications plan. This includes frequencies, reporting schedule, emergency reporting procedures, and alternate communications plans. It also includes the ground-to-air

communications plan and the friendly recognition signal to be used during linkup with the main assault force or other friendly forces.

(8) Plans for use of technical specialists or special equipment.

(9) The plan for the treatment and evacuation of sick or wounded team members from the objective area.

(10) The E&E plan. This includes routes out of the objective area, passwords, and alternate extraction zones or methods far from the objective area.

(11) Mission-specific elements of information desired by the ranger force commander, and any special or on-order missions the team must conduct.

g. The team leader uses specific steps in planning, preparing, and executing missions. These procedures are comprehensive, yet adaptable to any situation. Three of the most important steps are:

(1) **Issuance of warning order.** This action gives the team maximum time to prepare for the mission. It should be issued as soon as the team leader has been notified of a mission and updated after he has made his tentative plan.

(2) **Issuance of OPORD.** The team leader conveys the information and instructions needed to accomplish the mission. He requires a briefback from each team member.

(3) **Rehearsals.**

h. The specific steps used when planning missions are as follows:

Step 1. Study the mission.

Step 2. Plan use of time.

Step 3. Issue initial warning order.

Step 4. Study terrain and situation.

Step 5. Make tentative plan.

Step 6. Organize.

Step 7. Update warning order.

Step 8. Coordinate (continuous throughout).

Step 9. Consult aerial photographs and map updates.

Step 10. Complete detailed plan. (To be combined with paragraphs 3, 4, and 5 of the operation order.)

Step 11. Issue operation order.

Step 12. Supervise (throughout), inspect, rehearse.

Step 13. Execute mission.

i. A workable E&E plan must be developed. Such a plan enhances the survival of the reconnaissance team members. It ensures that they are psychologically prepared to conduct the mission, by knowing there is a system and plan for returning them to friendly lines. The E&E plan must include a section describing the actions of the reconnaissance team members should their insertion craft be disabled en route to the objective area. It should also include what to do if the team is unable to link up with the main ranger force, or if the main ranger force has not been inserted into the objective area.

j. Rehearsals and briefbacks are best for finding flaws in procedures or planning. Thorough coordination of all procedures is essential. The exact type of aircraft or other means that will be used for insertion should be used for training. Rehearsals should be conducted under terrain, astronomical, hydrographical, and meteorological conditions close to those to be met on the operation. The team leader conducts a briefback with the platoon leader. This lets the platoon leader decide if the team is ready for the mission. It also allows him to make changes.

6-11. Execution of insertion.

a. Reconnaissance teams normally conduct blind insertions. Blind insertions are amphibious, heliborne, or parachute drops onto unmarked DZs without help from a DZ-marking party or a CCT. Blind insertions often depend on favorable light and weather conditions in the objective area. Favorable conditions for the reconnaissance team often means low-lying cloud cover, fog, and darkness. The teams are able to operate during reduced visibility by the extensive use of night observation devices. The team's extensive training and high level of land navigation skills allow it to quickly traverse rugged terrain and not be detected.

b. Team insertion into the objective area often needs a cover and deception plan. Team members can conduct HALO and HAHO operations. Teams can be inserted into areas not normally acceptable as a DZ for conventionally trained parachutists. By the use of HAHO techniques, reconnaissance teams can assemble in the air and glide to a landing at a location far from the drop point.

c. If reconnaissance teams need special or bulky equipment that cannot be carried on the individual team member, it can be airdropped, hidden near the DZ, and retrieved later. Such container drops are often used during low-level static line insertions. The high-altitude airdrop resupply system (HAARS) can be used to deliver such items at the same time as reconnaissance teams are inserted by HALO.

CHAPTER 7

Combat Support

Combat support is used to enhance the combat power of the maneuver companies. Combat support elements may be organic or nonorganic. They can conduct a variety of missions in providing indirect and direct fire support, security, intelligence, and maneuver and fire control.

7-1. Intelligence support.

a. The success or failure of a ranger mission often depends on the accuracy, detail, and timeliness of intelligence information. Both deliberate and quick-response missions need specific information that can be obtained only by a multidiscipline collection and analysis agency. An active interface with all levels of the intelligence system helps get the needed information quickly.

b. The theater or JTF commander is responsible for providing intelligence support. The 1st SOCOM also provides information and analyzed intelligence to the ranger force commander. The main source of processed intelligence is the Echelons Above Corps Intelligence Center (EACIC), similar to the division level military intelligence (MI) all-source production center (ASPC). The EACIC provides intelligence, security, and electronic warfare (EW) interfaces (see Figure 7-1). It is through the EACIC that ranger commanders can request support from national, joint, combined, or CONUS intelligence analysis systems.

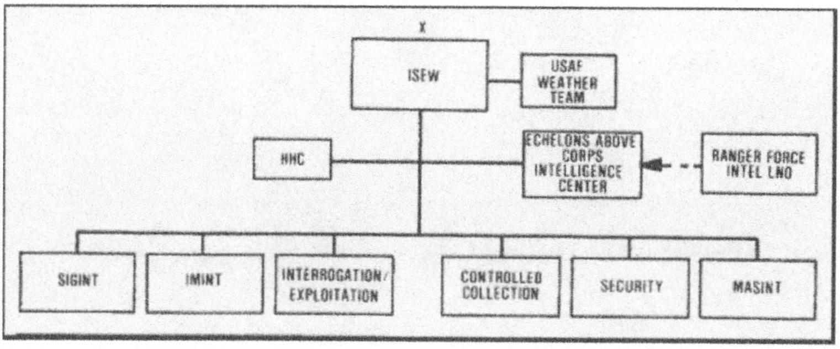

Figure 7-1. Intelligence at echelons above corps.

c. Intelligence requirements for ranger operations are unique, highly sensitive, and compartmented. The intelligence officer of the 1st SOCOM and the appropriate 1st SOCOM regional planning cell coordinate with theater intelligence agencies to provide needed information.

d. The ranger regimental staff and the ranger battalion staff have personnel in their intelligence sections to correlate and analyze information from many sources.

e. The reconnaissance platoon, organic to the ranger regiment, maintains surveillance of the objective area before the insertion of the main ranger force. It reports any last-minute developments that would affect the ground tactical plan. The platoon can also be tasked to determine any mission-specific items of essential information that the ranger force commander needs to complete his plan. This platoon is a highly specialized unit that is reserved for use by the ranger force commander. It is not a HUMINT collection agency for use by the theater or JTF commander. Reconnaissance teams normally link up with the ranger force in the objective area and are extracted with the main body. If this linkup cannot be made, they will exfiltrate enemy territory. They may use escape and evasion nets set up by other special operations forces or Department of Defense (DOD).

f. The ranger regiment normally places a liaison team at the headquarters of the command having OPCOM of a ranger force. This team provides command and control communications and liaison. It puts an intelligence liaison officer (LNO) at the EACIC or the appropriate ASPC (see Figure 7-2). This provides coordination to ensure that the intelligence needs of the ranger force are relayed to the intelligence processing center, and that the analysis is on the specific needs of the ranger force commander. This LNO coordinates with another located at the corps or EAC targeting center. Close coordination between the regiment, the intelligence analysis center, and the targeting center is essential to effective employment.

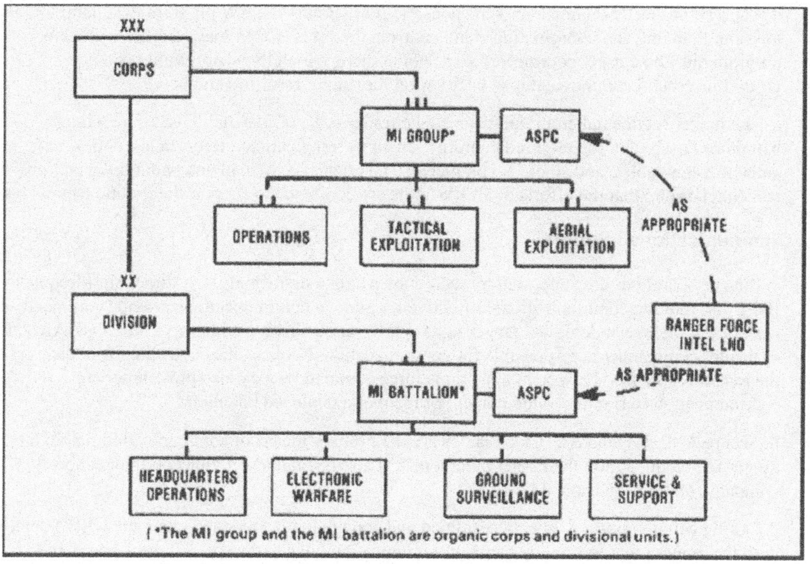

Figure 7-2. Intelligence source at corps and division.

7-2. Electronic warfare support.

a. Depending on the nature of the target and enemy abilities, a ranger mission may need support from EW units. This is mainly true during insertion. Electronic warfare operations disrupt or destroy the enemy's command and control of his forces and weapons systems, and retain friendly use of the electromagnetic spectrum. Electronic warfare is also used to support deception operations. They mislead the enemy by manipulation, distortion, or falsification of indicators, and persuade him to react in a manner prejudicial to his own interests.

b. Active jamming and chaff dispersal can prevent enemy early-warning radar from detecting the airborne force or determining its route. False transmissions can aid deception plans or feints used to help the ranger force's entry behind enemy lines. Selective jamming and imitative deception can disrupt the enemy command and control nets. This prevents the enemy from reacting to the presence of the ranger force in time to prevent mission accomplishment, or slows and interrupts his deployment of reaction forces.

c. Ranger elements, or other military forces, may conduct closely coordinated direct attacks against enemy EW sites during a ranger force's insertion or extraction. Not only does this reduce the enemy's ability to direct EW against the ranger force command, control, and communication system, but it misleads and confuses the enemy as to the true ranger objective.

d. The JTF or theater commander is responsible for planning the EW program to support a ranger mission. Planning and coordinating elements from the 1st SOCOM may also help. The EW program must be closely coordinated so as not to disrupt friendly air-to-ground or ground-to-ground communications, or to reveal the ranger force intentions.

e. The ranger regimental communications-electronics (CE) officer and those of the ranger battalions ensure that the required coordination concerning jammed frequencies, codes, authentication tables, and so on, is completed. The ranger regimental and battalion operations personnel ensure that the efforts of all EW units are combined to support the ground tactical plan.

7-3. Communications support.

a. Ranger operations are supported by secure long-range, lightweight, real-time, high-frequency (HF), and satellite communications. Effective long-range communications provide command and control links between deployed ranger units and the controlling headquarters. The JTF, ARFOR, or theater commander is responsible for communications between the controlling headquarters and the ranger regiment. The special operations force commander may also provide secure communications terminals to the ranger regiment or a deployed battalion.

b. Secure AM, FM, and SATCOM radios are the primary means of communication within the ranger regiment. Within the ranger battalion, AM and FM radios provide communications to company, platoon, and squad level.

c. During certain missions, specially trained and equipped quickreaction elements (QRE) deploy with the ranger force to provide secure communication links to the special operations force commander. These JCSE elements operate on either SATCOM or tactical satellite terminal (TACSAT) channels or through an airborne communications relay platform (see Figure 7-3). Depending on the mission, an airborne battlefield command and control center (ABCCC) aircraft or a joint airborne communications center/command post (JACC/CP) may be used. Their communications systems can operate at all levels of the national chain of command to permit a quick response to the tasking authority.

Figure 7-3. Airborne communications relay platform.

d. When only one ranger battalion is employed and the regimental headquarters is not the controlling headquarters, the ranger regiment provides a liaison cell to the controlling higher headquarters. This liaison cell includes a communications element from the regimental communications platoon. It can provide secure SATCOM, teletype, and facsimile transmission support to the ranger force in the objective area.

e. If two or more ranger battalions are employed, the ranger regimental headquarters deploys and acts as the command and control headquarters. The regimental communications platoon would then provide another communications link to the special operations force commander.

f. The ranger regimental CE officer ensures that the necessary communications links are set up and coordinated. The many communication means and channels available provide for effective control of a deployed ranger force. However, communication means must be closely coordinated at all levels of command to control the complex operations of a ranger mission. Planning and coordination with supporting aviation, transport, fire support, medical, and logistical elements before an operation are vital to efficient communications. The regimental CE officer must consider the communication systems linking the ranger force and other services. Air-to-ground and ship-to-shore communications are vital and must be set up early in ranger operations.

7-4. Fire support.

a. **Planning.**

(1) The ranger regiment has limited organic fire support assets. The six 60-mm mortars in each ranger battalion (two in each rifle company) provide the only organic indirect fire

available to the battalion commander. Each rifle company also has three 90-mm recoilless rifles and three Dragon missile systems that provide a small measure of organic direct fire support.

(2) The missions assigned to the ranger regiment and its subelements normally require fire support from organizations outside of the regiment. To plan, coordinate, and control these fires, the ranger regiment is organized with personnel dedicated to these tasks.

(3) The ranger regimental headquarters has a fire support element consisting of one fire support officer (FSO), two noncommissioned officers, and two fire support specialists. They prepare the fire support annex to the regiment's operation order. They ensure that all available fire support is planned to best support the regimental commander's ground tactical plan. They act as the primary coordinator for all external fire support, such as artillery, armed helicopter, close air support, naval gunfire, and short-range air defense. They also coordinate with each ranger battalion FSO to ensure unity of effort. The fire support element coordinates the efforts of air force, naval, and marine liaison officers when they are attached.

(4) The regimental commander may assign part of the fire support element to a committed ranger battalion as an attachment to help in the integration and planning of fires. The ranger battalions request through the fire support element that more or different fire support means be integrated into the ground tactical plan.

(5) The ranger battalion headquarters has a fire support team headquarters consisting of the battalion's FSO, two noncommissioned officers, and two fire support specialists. The FSO is the prime fire support coordinator for the ranger battalion commander. He maintains coordination with the regimental FSO and any attached fire support coordination elements, such as the USAF tactical air control party or naval shore fire control party. The FSO performs the following:

(a) Advises the ranger battalion commander on all fire support matters.

(b) Recommends allocation of fire support.

(c) Prepares fire support plans.

(d) Assigns target numbers.

(e) Processes target lists from the ranger company fire support teams to eliminate duplication.

(f) Monitors and functions as net control on the ranger battalion fire support coordination (FSCOORD) net.

(g) Reports changes in the status of fire support units to the ranger battalion commander, staff, and FISTS.

(6) When the ranger battalion is operating a single TOC, the FSO and his element are normally located there. When the ranger battalion is operating two TOCS, one TOC is normally manned by the FSO, the fire support sergeant, and a fire support specialist. The

senior fire support sergeant and the other fire support specialist would then monitor the FSCOORD net at the other TOC. The FSO ensures that any attached fire support elements also divide when needed to operate in a two-TOC configuration.

(7) The ranger battalion fire support team headquarters has three fire support teams assigned to it. These are normally allocated one to each of the ranger rifle companies. The company FSO and his team--

(a) Locate targets and request and adjust surface-to-surface fire support (mortar, field artillery, and naval gunfire).

(b) Plan fires to support the company ground tactical plan and prepare target lists.

(c) Coordinate fire support requests through the FSO for surface-to-surface and air-to-surface fires.

(d) Report battlefield information.

(e) Control air-to-surface fires of helicopters and prepare to assume control of fires from fixed-wing aircraft.

(f) Coordinate airspace use within the ranger rifle company operational area.

(g) Advise the ranger company commander on all fire support capabilities, limitations, and methods of employment.

(h) Inform all fire support units of target priority changes.

(i) Designate targets for laser-guided munitions.

(j) Assume operational control of, and administrative responsibility for, all augmentation fire support personnel (TACP, shore fire control party, and attack helicopter liaison teams).

(8) The company FSO is normally with the ranger rifle company commander. The FOs and their RATELOs are normally with the ranger rifle platoon leaders.

b. **Coordination.**

(1) Coordination of fires is a continuous process. It involves implementing the fire support plans of the ranger force commander and reacting to changing tactical situations. The success of a mission often hinges on how well fires are coordinated during the operation.

(2) The fire support planners within the many echelons of the ranger force become fire support coordinators during the actual execution of the operation. They use standard fire control coordination measures.

c. **Organic fire support.**

(1) **Indirect fires.**

(a) The 60-mortars constitute the only organic indirect fire support available within the ranger regiment. The mortar squads in each ranger rifle company normally operate as a section under the control of the mortar section sergeant. However, they can operate alone for short periods.

(b) The 60-mm mortar is used to destroy enemy equipment and light vehicles and to attack enemy personnel. Its main advantage is the ease with which it can be brought into operation against the enemy. This ease of employment allows the ranger force commander to concentrate combat power at the critical time and place. This helps to surprise and destroy the enemy.

(c) Mortar fires are employed by the ranger rifle company commander and coordinated by the company FSO. They are normally requested and adjusted by the FIST observers. The mortars are normally emplaced in a central position, and their fires are controlled by the fire direction center (FDC) using the indirect-fire method. The mortars may also use the direct-alignment or the direct-lay method.

(2) **Direct fires.**

(a) Each ranger battalion is authorized 90-mm recoilless rifles and Dragon missile launchers. These weapons constitute the heaviest organic direct-fire ability in the regiment. The antitank section in each ranger rifle company may operate as a complete section, but normally individual weapons are assigned throughout the company sector wherever their fires can best support the mission. These weapons can destroy most enemy armored vehicles and can penetrate the flank and rear armor of all enemy main battle tanks. The 90-mm recoilless rifle has a multiple projectile round that is effective against exposed enemy personnel. The HE round is highly effective against buildings and bunkers.

(b) The main disadvantage of direct-fire antiarmor weapons is their weight. These weapons, and their ammunition, are man-portable but place a strain on the unit when moving across rugged terrain.

(c) The ranger battalion is trained in the use of light antiarmor weapons, such as the US M72A2 LAW and the Warsaw Pact rocket-propelled grenade (RPG). The following provide protection for the ranger force against attack by enemy armored forces:

- The employment of light antiarmor weapons at all organizational levels.
- The placement of the 90-mm recoilless rifles or guided-missile systems on the most likely avenues of enemy-armor approach.
- An aggressive, offensively oriented employment.
- The ability to move into and across terrain impassable to armored vehicles.

d. **Field artillery fire support.**

(1) Many ranger unit operations are out of the range of supporting field artillery fires. Whenever such fires are available, they are planned for and integrated into the ranger force ground tactical plan.

(2) Coordination and exchange of call signs, frequencies, and target lists occur before insertion of the ranger force. Unless required by the tactical situation, ranger units do not normally receive augmentation forward observers from supporting field artillery units. The degree of mental and physical training necessary to execute certain ranger techniques requires more preparation than field artillery units are able to provide. The FIST from the ranger battalion requests and adjusts field artillery fires for the ranger force.

(3) Field artillery fires can be used to support the ranger force even if the objective area is out of range. Field artillery cannon fire and multiple rocket launch system (MRLS) fire can be used to suppress enemy air defenses. This helps the ranger force as it crosses the FEBA during airborne or air assault operations.

(4) Field artillery fires can be used to support the exflltration of ranger elements as they approach friendly lines. These fires can contribute to the deception plan and add combat power to feints used to support ranger operations.

(5) When in range of the objective area, field artillery units can be used to emplace field artillery delivered FASCAM to enhance the security of the ranger force.

(6) Copperhead rounds fired by 155-mm field artillery units can be terminally guided by the ranger FIST forward observers (FOs). They can attack hardened point targets or enemy armored vehicles by using a man-portable laser target designator (see Figure 7-4).

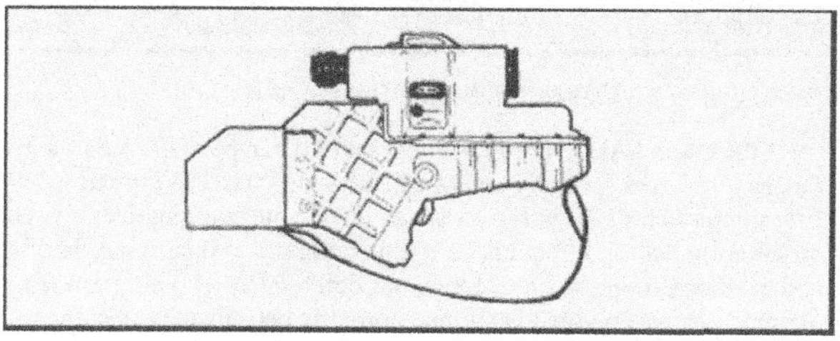

Figure 7-4. AN/PAQ-1 laser target designator.

e. **Aerial fires.**

(1) Aerial fire support is usually the prime means supporting the ranger force due to the distance behind enemy lines at which most ranger operations take place. Aerial fire support can be provided by either fixed-wing aircraft or helicopters.

(a) Fixed-wing aerial fire support may come from USAF, USN, or USMC units. The type of unit providing support, the aircraft, and the mix of ordnance carried all affect the fire support planning and coordination process. Some aircraft have a night and all-weather strike ability enabling them to support the ranger force during any level of visibility. Operations during poor weather that limits visibility to less than 3 nautical miles are still somewhat restricted. The ranger fire support coordinators must ensure that the correct aircraft are requested and employed effectively on the enemy. The

TACP directs and adjusts aerial fires in the objective area.

(b) The ranger force can use ground laser target designators to pinpoint targets for air strikes, as well as electronic navigation aids to permit nonvisual air strikes (beacon bombing). The ranger rifle company FIST or the TACP can control a laser-designated standoff air strike (see Figure 7-5).

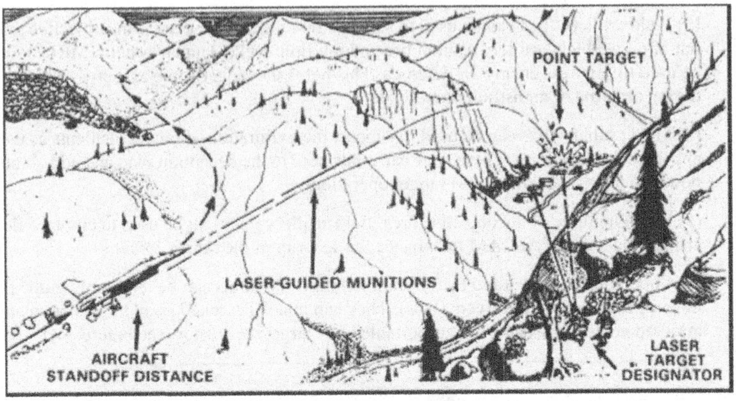

Figure 7-5. Standoff air strike.

(c) If the enemy ADA ability is not great or it can be degraded to a low level, the ranger force uses specially equipped and armed AC-130 aircraft (see Figure 7-6) for fire support. These aircraft provide an invaluable combination of firepower, night observation and illumination, communications, and long loiter time. A well-planned and executed suppression of enemy air defense (SEAD) program, coupled with ECM directed against enemy ADA units, normally permits the use of AC-130 aircraft.

Figure 7-6. AC-130 aircraft.

(2) The attack helicopter, armed with a mix of antitank guided missiles (ATGM), 2.75-inch rockets, 20-mm cannon, and 40mm grenade launchers, is accurate and responsive aerial fire support. The ranger force commander plans to use all sources of aerial fire to help accomplish his mission. However, the nature of ranger operations may preclude the extensive use of armed helicopters due to their limitations. Some attack helicopters are limited in range and lack an all-weather ability. They are restricted during operations at night by a lack of sophisticated night vision devices. They may lack the speed to go with the ranger force on some heliborne insertions. Attack helicopters may be used to escort and assist the ranger force as it crosses the FEBA. They may also be used to conduct feints and demonstrations to cover the insertion of the ranger force. As the MV-22 Osprey and the AH-64 Apache enter service, more use will be made of these aircraft to support ranger operations beyond the FERA, (See Figures 7-7 and 7-8).

Figure 7-7. MV-22 Osprey.

Figure 7-8. AH-64 Apache attack helicopter armed with Hellfire missiles.

(a) If attack helicopters are used to support a ranger operation, planned indirect fires are normally delivered along entry and exit corridors. Attack helicopters approach and depart the objective area using nap-of-the-earth flight profiles.

(b) Fires from armed helicopters are normally requested and controlled by the company FSO or one of his FOs, operating on a special ground-to-air net. The laser target designator may be used to precisely identify targets for the AH-64 Apache. Friendly unit locations may be marked by smoke, panels, lights, mirrors, or infrared sources.

f. **Naval gunfire.**

(1) During amphibious assaults or operations near a coastline, the ranger force may receive indirect fire support from naval gunfire. Normally, a ranger battalion would be supported by either a destroyer or a cruiser in a direct support role. However, the type and importance of the mission, the type of targets, the ships available, the hydrographic conditions, and the enemy capability determine how many and which type of ships are provided to support the ranger force.

(2) Naval gunfire is characterized by large volumes of highly destructive, flat-trajectory fire. Planned strikes in support of the ranger force may also include surface-to-surface missile fire. Some naval guns can fire a laser-guided projectile much like the Copperhead. This projectile, called a Deadeye round, is terminally guided the same as the Copperhead. Deadeyes can be guided either by an attached spotting party from the ANGLICO or by ranger LTD teams.

(3) When ranger units are being supported by naval gunfire, a shore fire control party is attached to the ranger force, This party normally consists of a liaison team and several spotter teams (forward observers). The liaison team is integrated into the operations of the fire support element (FSE) at the ranger battalion TOC. The spotter teams are attached to the ranger rifle companies. The shore fire control party LNO is the ship's representative to the ranger force commander, through the FSO. The spotter teams request and adjust fires from surface vessels. They can also request and control air strikes by carrier-based aircraft.

(4) The LNO and the spotter teams operate in the ground spot net, communicating with the ship by high frequency (HF) radio to request and adjust naval gunfire. The spotter teams communicate with the LNO using very high frequency (VHF) radios. The LNO also can communicate with aircraft using ultrahigh frequency (UHF) radios.

(5) Coordination and control measures that apply to naval gunfire are the same as for field artillery, with the addition of two terms, which are:

(a) Fire support area (FSA)--a sea area within which a ship may position or cruise while firing in support. It is labeled as FSA and numbered by a Roman numeral - for example, FSA VII.

(b) Fire support station (FSS)--a specified position at sea from which a ship must fire. This is very restrictive positioning guidance. It is labeled as FSS with a Roman numeral--for example, FSS VII.

g. **Air defense artillery.**

(1) Army air defense artillery units are rarely used to directly support the ranger force since

ranger operations normally take place deep behind enemy lines. However, it is possible that during the insertion or extraction phase of an operation, ADA units may support the ranger force as it crosses the FEBA.

(2) The ranger regiment's fire coordination element and the FSO in the ranger battalion are mainly concerned with adequate coordination of ADA fires. This prevents friendly units from being engaged. It allows the integration of ADA fires into the scheme of maneuver. The fire support planners ensure that enough coordination has been made so that ADA units are aware of the ranger force and know enough about its flight paths, routes, and altitudes to prevent engagement by friendly units. When conducting this coordination, care must be taken to pass only needed information to the ADA unit. To aid coordination, an ADA officer may be attached to the ranger force.

(3) The plan for a heliborne or airborne insertion and extraction of the ranger force must include the correct identification-friend-or-foe (IFF) codes, the flight profiles, the times of FEBA crossings, and the estimated number of aircraft making the crossing. There must also be a plan for crossing by straggler aircraft and battle-damaged aircraft that cannot reply to IFF interrogation. If possible, the routes selected should avoid concentrations of friendly ADA units.

(4) During an aerial crossing of the FEBA by a ranger force, the regimental headquarters may assign a liaison officer to the headquarters controlling the ADA fires in the crossing area. The LNO is then responsible for close and continual coordination. This ensures that the crossing is made without friendly ADA units engaging the ranger force while it is exiting or entering friendly lines.

(5) During ranger operations, the primary source of air defense fires is shoulder-fired antiaircraft missiles carried by specially trained individual rangers. These weapons provide the ranger force commander a limited defense against attack by enemy aircraft. The decision to take these weapons into the objective area depends on the factors of METT-T. Ranger operations should be completed before the enemy can react with ground forces. However, enemy aircraft may be able to react in time to threaten the ranger force. These lightweight missiles offer a highly effective, short-range counter to enemy air attack.

(6) The ranger force commander must define the weapons control status and rules of engagement for all short-range air defense (SHORAD) teams under his control. Priorities of target engagement must also be determined to prevent multiple engagements of a single enemy aircraft. The ranger battalion FSO and the attached ADA officer are responsible for the coordination of ADA fires within the objective area. Standard ADA weapons-control status and rules of engagement may be modified by the ranger force command.

7-5. Aviation support.

a. **Planning.**

(1) The ranger regiment may need extensive aviation support on some missions. The ranger battalion is organized and trained to move quickly to the objective area by air and to be resupplied by airdrop for short periods.

(2) Often the decision to execute a mission depends on the amount and type of available insertion and extraction aviation support. Normally, the theater commander or JTF commander allocates the type and amount of aviation assets to support a ranger operation. The theater or JTF commander's staff is responsible for the planning and coordination concerning aviation support. The ranger force commander must play an active part to ensure mission success.

(3) Due to the extensive training requirements and specialized operational techniques needed to support special operations, special air operations units in the Army and the USAF have been equipped and trained for the mission. These forces train with the ranger regiment and jointly develop tactics, techniques, and procedures of employment. These special units operate diverse aircraft that have unique capabilities. These help the ranger force commander to effectively employ his units. Command of these special units is maintained at a very high level in the chain of command. They are assigned to support a specific operation. Then they return to their parent unit's control.

b. **Coordination.** In addition to the staff of the theater commander or JTF commander, the coordination of aviation support occurs at all levels in the ranger force. These operations often are conducted in phases. A separate air mission commander is then designated. The AMC is responsible for the completion of each phase of the aviation mission, as well as the close and continuous coordination with the ranger force, commander. The ranger regimental air operations officer and the movements control officer are responsible for the planning, integrating, and controlling of aviation support.

c. **US Air Force.**

(1) The USAF provides special aviation support to the ranger regiment and other elements of the special operations force. The USAF has elements that are equipped with MC-130 Combat Talon aircraft (see Figure 7-9) and AC-130 Spectre gunships, as well as HH-53 Pave Low, UH-60, and UH-1 helicopters. These aircraft are specially equipped to conduct long-range infiltration, exfiltration, attack, and search-and rescue operations. The USAF also provides C-141B and C-130 aircraft for strategic and tactical airlift of the ranger force. There are specially trained USAF crews that can operate using AWADS and SKE. This gives them an excellent means of inserting the ranger force during reduced visibility.

Figure 7-9. MC-130 Combat Talon aircraft.

(2) The AFFOR commander provides the airlift control element (ALCE) support at the departure airfield. The ranger air operations officer ensures that coordination of aircraft load limits, air movement times, specific load plans, and manifests is conducted in an effective and timely manner. The air operations officer and the movements control officer ensure that a departure airfield control officer (DACO) from the deploying unit is designated to provide face-to-face coordination at the departure airfield.

(3) The USAF normally provides an airborne battle command and control craft. This is a specially equipped C-130-type aircraft that provides communications and data system down-links to ground stations.

(4) The terminal guidance needed for assault landings and air traffic control in the objective area is provided by a specially trained and equipped CCT. This element has habitual training and operations relationships with the ranger units. The USAF is also responsible for the specially trained and equipped weather teams that can be sent into the objective area with the CCT.

(5) The USAF Tactical Air Command (TAC) includes fighter, interceptor, ground attack, electronic countermeasure, early warning, reconnaissance, and airborne command and control squadrons, any of which may be used to support the ranger force. The Tactical Air Command provides the results of aerial reconnaissance to the AFFOR, who then processes it for use by the theater or JTF commander's staff. This reconnaissance may be by low-level photo, imaging infrared, side-looking airborne radar (SLAR), low-light-level television, or overhead photography.

(6) The Military Airlift Command (MAC) equips and trains special long-range SAR aircraft and crews. These aircraft, mainly helicopters, are equipped with sophisticated navigation, observation, and communications equipment, as well as extended-range fuel tanks and in-flight refueling capabilities (see Figure 7-10). The ranger regiment uses these aircraft for certain missions.

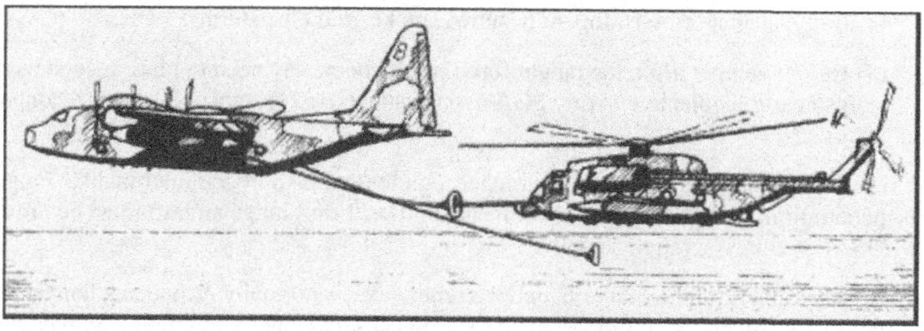

Figure 7-10. In-flight refueling.

d. **US Navy and US Marine Corps.** Both the USN and the USMC have units equipped and trained to support ground forces. The type of aviation support normally provided to the ranger regiment is close air support against targets in the objective area, or as suppression missions against enemy air defense installations. Long-range helicopters from the USMC may be used to support an amphibious operation or to insert special teams during ranger operations. The USN also has a search-and-rescue ability that may be used to support the ranger regiment.

e. **Special operations aviation units.**

(1) The US Army has specially equipped and trained SOA units tasked to support the ranger regiment, as well as other elements of the special operations force. These SOA units have a habitual training association with the ranger regiment and use dedicated aviation assets to -

(a) Insert, extract, and resupply the special operations force.

(b) Conduct armed escort, reconnaissance, surveillance, and electronic warfare in support of the special operations force missions.

(c) Provide airborne command, control, and communications for special operations force elements.

(d) Provide general support aviation during peacetime and contingency operations.

(2) Penetration of enemy airspace is normally done during limited visibility at very low altitudes. Penetrations are planned to maximize the use of existing terrain and exploit blind spots in enemy air defenses. On-board navigational aids are crucial to mission success. There normally is not any terminal guidance to the selected LZ or DZ. Deception operations of a tactical or operational nature help SOA units penetrate to the objective and reenter friendly lines. Advanced helicopters and short-takeoff-and-landing aircraft provide the help needed to do these missions.

(3) To survive enemy air defense systems and prevent mission compromise, SOA units avoid detection. Detected aircraft may reveal the location of supported special operations force ground forces or may be destroyed by enemy ADA fires. The SOA normally avoids detection by operating at night without illumination. This condition requires flight using sophisticated night vision devices. The SOA also operates in other conditions of reduced

visibility, such as clouds, fog, rain, snow, smoke, and dust storms.

(4) Before an operation, the ranger force commander may need to place reconnaissance teams near the objective. Army SOA assets can be used to transport ranger reconnaissance teams.

(5) Elements of SOA provide the ranger regiment with short and intermediate range penetration, resupply, and extraction capabilities. Long-range aircraft must be provided from USAF special operations assets.

(6) The ranger regimental or battalion commander is normally provided a liaison aircraft while the ranger force is in the REMAB. This ensures that he can travel quickly to the dispersed headquarters of the agencies providing support for the mission. This aircraft and crew are provided by the theater or JTF commander from EAC general support aviation assets. It may be a helicopter or fixed-wing aircraft, depending on the location of the REMAB to a suitable landing field.

(7) Army helicopters and fixed-wing aircraft may be used to provide SEMA aerial reconnaissance; radar, infrared and photographic imagery; and electronic intelligence support to the ranger force commander. This support is normally provided by the aerial exploitation battalion of the theater Army military intelligence group.

7-6. Engineer support.

a. The ranger regimental and battalion operations officers plan and request engineer support for ranger operations.

b. The most common engineer-type tasks needed during a ranger operation are the destruction of enemy facilities, equipment, and material by use of conventional explosives; and the interdiction of major lines of communication. To reduce the need for external support, the ranger regiment trains selected individuals in advanced demolition techniques. All members of the ranger regiment are well trained in the basic employment of conventional demolitions. Therefore, there is no need to attach combat engineers to conduct normal demolition missions. If specialized skill or knowledge is needed for a specific mission, qualified personnel can be attached.

c. During the occupation of the REMAB and the rehearsal for an operation, the ranger regiment needs construction engineer support to build simulated target areas. The simulated target areas should be camouflaged to prevent the enemy from recognizing the target. The construction engineers augment the ranger unit's self-help construction efforts. Limited-facilities construction and improvement may also be needed at the REMAB to provide security and isolation. The theater commander or JTF commander provides the construction engineer support.

d. Certain missions, such as those of interdiction, may need engineer special weapons teams. The teams are provided by the theater commander or JTF commander, or they may come from other special operations force assets. They are attached to the ranger force for a specific mission and released at the completion of the extraction and debriefing phase.

e. Some operations may need the ranger force to use captured enemy heavy construction equipment. Selected members of the ranger regiment are trained to operate such equipment.

However, construction engineer personnel may be attached to the ranger force for this mission.

CHAPTER 8

Combat Service Support

Combat service support consists of the logistical and administrative effort to maintain the battalion's ability to fight. The commander forecasts his needs for CSS. He employs CSS units to ensure accomplishment of the mission.

8-1. Ranger support element.

a. The ranger support element is a non-TOE organization formed from elements of CSS units collocated with CONUS-based ranger units augmented by selected personnel from the host installation. The mission of the RSE is to provide all the CSS needed to outload and deploy a ranger unit from CONUS. The personnel designated as part of the RSE have the same alert notification, reaction, and security requirements as members of the ranger unit they support. The RSE is normally built around a command and control group provided by a major CSS unit on the ranger unit's host installation. The remaining CSS elements and teams are then designated. An RSE consists of--

(1) An emergency operations center liaison group that provides timely coordination between the ranger unit and the functional elements of the RSE.

(2) A command and control group that coordinates and directs the actions of the functional elements.

(3) A supply section responsible for the timely issue of all Class II, III, IV, and VII items. This section maintains a stockage of contingency items marked for use by the resident ranger unit. It also issues items out of common stocks.

(4) A transportation section moves equipment and personnel from the unit area to the departure airfield and picks up equipment from local storage sites. It may also operate material-handling equipment and trailers to support loading of aircraft.

(5) An ammunition section responsible for the requisition, storage, issue, loading, and security of the ranger unit's basic load(s) of conventional munitions.

(6) A mess section provides all Class I support to the ranger unit. This includes operating a dining facility, providing food to remote sites, and drawing and issuing operational rations.

(7) A load team loads and unloads all equipment and supplies for the ranger unit. This team prepares pallets for loading, documents hazardous or dangerous cargo, and loads equipment pallets on USAF aircraft. It can also provide limited security for loads.

(8) A communications and electronics section provides secure communications among the RSE subelements, ranger unit, EOC, and DACG.

(9) An engineer platoon that is responsible for the construction of targets and rehearsal areas. This element of the RSE may or may not be activated, depending on the mission given the ranger unit. If the ranger unit deploys from CONUS to the objective area, this platoon has a key role to play in the premission preparation.

(10) A maintenance team whose primary mission is to ensure that the vehicles used by the RSE remain operational. It also provides predeployment maintenance and Class IX support for ranger unit vehicles and equipment.

(11) A medical team that provides unit-level medical care while the organic ranger medical section completes its outloading preparations. This team also provides any Class VIII supplies that the ranger unit may need.

(12) A security team responsible for security at the EOC, ranger unit location, and departure airfield. This team enforces stringent physical security precautions and maintains a constant counterintelligence (CI) surveillance of the outloading procedure. The security team provides armed guards for the protection of ammunition convoys, equipment movement, and vehicles.

(13) A rigger detachment that maintains, packs, rotates, and issues the personnel and equipment parachutes required by the ranger unit. It maintains enough stock of personnel parachutes to conduct a battalion-level parachute assault. The riggers lend technical skill to the rigging and aerial delivery of bulky, unusual, or delicate equipment.

b. The primary mission of the RSE is to provide the support necessary to quickly load and deploy the ranger units. The theater commander provides CSS to the ranger force once it has arrived at the OCONUS REMAB or ISB. The transition of support from the RSE to the CSS assets of the theater commander may be delayed for certain subelements in order to ensure a smooth deployment and enhance mission accomplishment. These subelements are released when the in-theater CSS assets can provide the support. Retained subelements of the RSE may be released from support requirements all at once, released in increments, directed to continue to provide support until mission completion, or augmented by in-theater CSS assets. The maintenance, medical, security, and rigger elements are those most likely to continue providing support OCONUS. This is likely if deployment is to an austere ISB pending immediate employment.

c. Once a ranger unit has been deployed OCONUS and has closed to a REMAB or ISB, the theater commander's staff coordinates CSS. If 1st SOCOM elements have deployed to the theater or operational area, the special operations support battalion (SOSB) of the 1st SOCOM coordinates CSS. Ranger logisticians at regiment and battalion level are mainly planners with limited logistics operations ability. Ranger logistical and personnel officers coordinate with CSS elements who provide such services.

d. The Theater Army Materiel Management Center (TAMMC) is the point of contact for information concerning in-theater logistical assets. In a developed theater of operations, a series of theater Army support groups will have been set up. One of these groups is assigned the mission of supporting the ranger force. The 1st SOCOM may also deploy certain assets available to provide CSS to the ranger force. The many CSS sources, the rapidly changing nature of SOF operations, and the unique CSS requirements of ranger units combine to make providing logistical and personnel support a challenge. The ranger unit's logistics officer must be included in the early phases of planning and deployment.

8-2. Logistics and personnel.

a. The echelons-above-corps CSS structure provides support to the ranger regiment as directed by the theater Army (TA) commander. In a low-intensity contingency involving a corps or less, there may be no echelons-above-corps logistics and personnel support elements. The corps support command (COSCOM) involved may be able to meet all needs from its own sources. However, in a combined operation in a high-intensity environment, the theater commander assembles a large CSS organization to provide support to the TA. This CSS base normally provides for all the ranger unit's requirements. During contingency operations in an area where there is no TA, the JTF commander provides the CSS elements. Until the buildup of these elements is complete, the ranger regiment must be prepared to operate from austere facilities with only the supplies it brings from CONUS.

b. The ranger regiment is normally assigned to the theater command for CSS. The TA commander, as directed by the theater commander, provides CSS to the ranger regiment and sets up support priorities. The EAC CSS organizations are under the command of the TA commander and consist of subordinate area and functional commands. The TA plans, coordinates, and executes its CSS functions through area-oriented support commands and specialized subordinate organizations. During the early stages of conflict, the TA may be organized as shown in Figure 8-1.

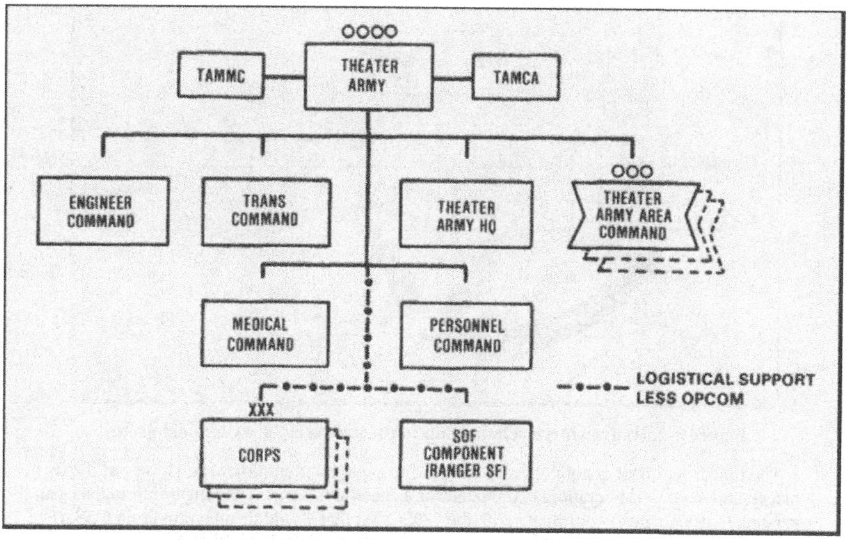

Figure 8-1. Ranger force relationship to theater Army.

c. The Theater Army Area Command (TAACOM) is the key CSS operator in the communications zone. It functions as a major subordinate command under the theater Army. The primary mission of the TAACOM is to provide direct support CSS to units such as the ranger regiment located in or passing through its assigned area. While located in an OCONUS REMAB, the ranger regiment receives most of its logistical support from the TAACOM.

d. The TAACOM area support group (ASG), which is subordinate to the TAACOM, is structured with operating units to provide direct support CSS to units located in or passing through its assigned geographical area of responsibility (see Figure 8-2). Area support groups are usually located along the lines of communication. The ASG is the prime source of CSS for the ranger regiment. It contains the units that actually provide services requested by the ranger unit logistics officer.

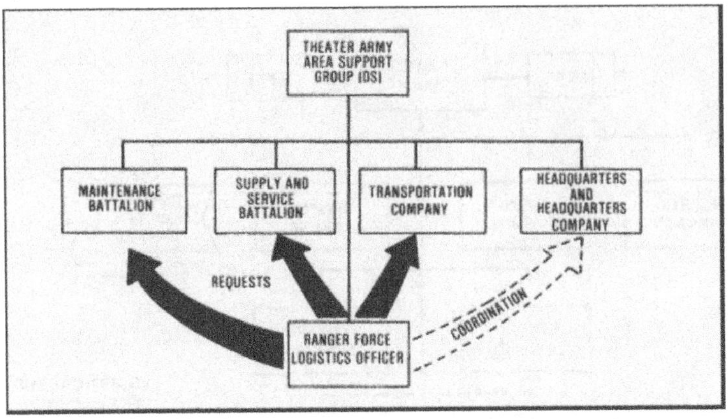

Figure 8-2. Ranger force relationship to theater Army area support group.

e. The ranger regiment is not normally assigned to a corps for employment. However, it may be placed under a corps' operational command for a specific mission. The ranger unit logistics and personnel officers must coordinate with the COSCOM that would then be providing CSS. (See Figure 8-3.) Corps support groups are major subordinate elements of the COSCOM. They provide command and control of COSCOM companies and battalions, which provide direct support (DS) and general support (GS) supply and field services to nondivisional units, such as the ranger battalion. (See Figure 8-4.)

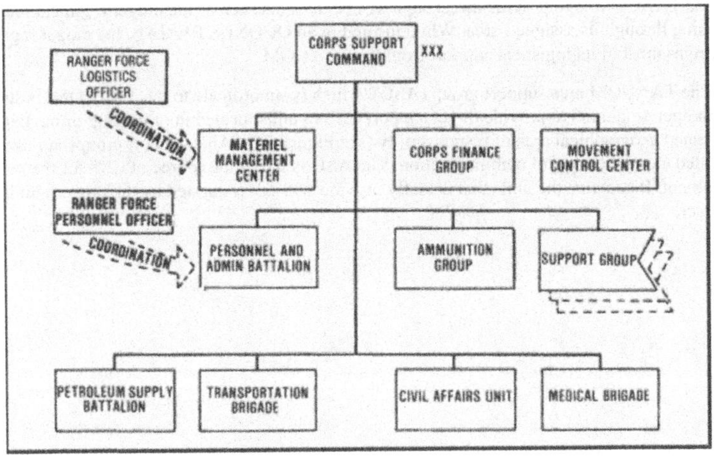

Figure 8-3. Ranger force relationship to corps support command.

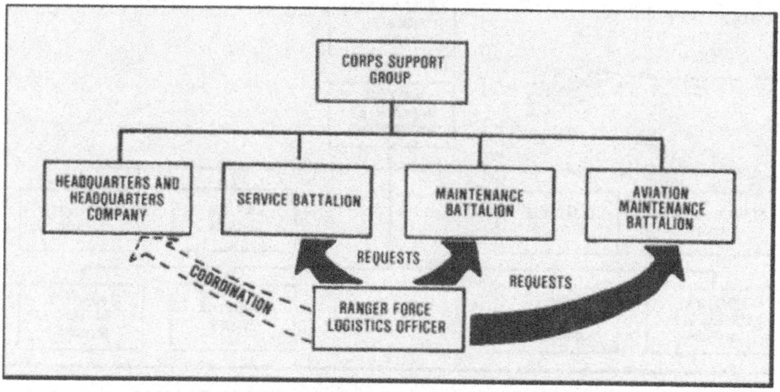

Figure 8-4. Ranger force relationship to corps support group.

f. Normally, the ranger regiment receives its logistical support from the corps support group responsible for the geographical area in which the REMAB or ISB is located. The corps support group provides all supplies except Classes V and VIII. The medical brigade provides Class VIII supply and medical maintenance support. The ammunition group provides Class V support. Financial and personnel services are provided by the corps finance group and the personnel and administration battalion, respectively.

g. Transportation needs are met by the transportation brigade. The ranger unit places ground transportation requests with either the Movement Control Center (MCC) or a designated movement control team. All airlift supply requirements are passed IAW FM 100-27. Standard and nonstandard supplies are normally requested from the supporting direct support unit (DSU). If authorized by the corps commander, the ranger regiment can place requirements for nonstandard items on the appropriate division within the corps MMC.

h. The COSCOM MMC (see Figure 8-5) performs supply and maintenance management for all classes of supply. The ranger unit logistics officer may coordinate with the MMC division chiefs or operations section.

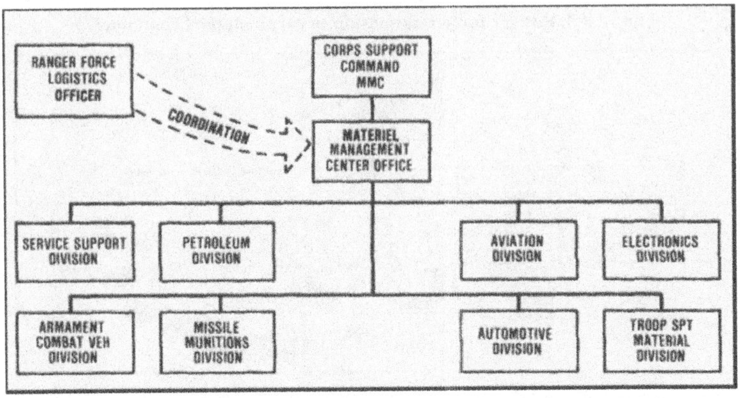

Figure 8-5. Ranger force relationship to COSCOM MMC.

i. Critical logistics operations, such as those supporting ranger and other SOF elements, need intensive management. They may be assigned to special staff sections that act as special operations force logistics managers. If designated, a logistics manager for special operations force elements would be a major point of contact for logistical support for the ranger regiment.

j. The ranger regiment may also receive logistical support from the SOSB. Normally, the SOSB provides administrative and logistical support to the Headquarters, Army Special Operations Forces (ARSOF). It complements the support provided by the Army component in the theater of operations by coordinating, augmenting, or assisting CSS forces. The SOSB provides unique quick-reaction support to the ARSOF or deploying SOF elements.

k. The ranger regiment does not make inordinate demands on the logistical system. However, a ranger unit may need certain nonstandard or low-density equipment. The concurrent need for nonstandard Class IX items requires management by logisticians to ensure support.

(1) Ranger battalion mess sections are responsible for Class I operations at the REMAB. The ranger regimental headquarters, which has no mess section, either collocates with one of the battalions or is supported by a mess facility operated by a nearby unit. It may share this facility on an area basis. The ranger unit personnel officer coordinates with the supporting DSU, which initiates the issue of Class I supplies.

(2) Class II, IV, and IX support to the ranger force is normally not of a great volume. It may consist of some nonstandard items. Depending on the need, the item, and the security, the SOSB may get some nonstandard items directly.

(3) Class III support needed by the ranger regiment is small and limited to standard items. The normal source of supply for Class III is the supply and service company providing area support.

(4) Class V support needed by the ranger regiment is small. It may include requests for nonstandard items of ammunition and explosives. The ranger logistics officer passes requests for standard items to the ordnance company (conventional ammunition) of the ammunition battalion assigned to the TAACOM ammunition group. Requests for nonstocked items are passed through the TAACOM Materiel Management Center or 1st SOCOM channels. The ranger unit must contact the ammunition company early to ensure a continuous supply of ammunition.

(5) Class VII supplies needed for the ranger regiment are often nonstandard or in excess of TOE. (See Appendix C for an equipment density listing.) All nonstandard or excess requests must be coordinated through the TAMMC or 1st SOCOM channels.

(6) Class VIII supplies for the ranger regiment are normally standard items supplied by the area support Class VIII point. The ranger regiment personnel officer sets up the Class VIII supply points of contact through the ranger battalion's medical officer.

(7) Aerial delivery equipment is provided by the airdrop equipment repair and supply company of the theater Army support group (TASG) (GS). There is normally one such company for each TAACOM. This unit supplies and repairs aerial delivery equipment. It does not pack and rig supplies for aerial delivery. The ranger regiment's requirements for aerial delivery supplies normally consist of standard items. Certain low-density items, such as HALO or HAHO parachute equipment, may also be provided through 1st SOCOM channels.

(8) The supporting motor transportation group of the transportation command is responsible for unit movement of the ranger regiment. The ranger regiment does not have any organic transportation assets. It depends on outside support whenever motor movement is needed.

8-3. Resupply.

a. Resupply operations for ranger units are normally covered during the planning and coordination phase of the mission. Ranger units try to complete their mission quickly and then be extracted or exfiltrate from the objective area. Ranger units carry into the objective area the equipment and munitions they need. They do not normally remain long enough to need extensive resupply. Certain operations may require the use of bulky or heavy items of equipment that cannot be carried by personnel. If the ranger force is airlanded on or near the objective, one or more of the aircraft may contain palletized supplies. These can be quickly off-loaded for future use.

b. If the ranger force is inserted into the objective area by parachute assault, aircraft can also drop resupply loads either just before or just after the personnel drop. These loads might include extra water, ammunition, demolitions, or barrier material. During peacetime, preloaded airdrop pallets are stored at CONUS depots for such contingency use. These mixed loads can be quickly moved to a departure airfield and inserted along with, or following, the ranger force. By dropping a resupply of water, food, equipment, and munitions with the ranger force insertion, the heavy-drop aircraft do not have to conduct a separate penetration of the enemy air defense zone. The ECM and SEAD programs used in the insertion of the personnel aircraft mask the penetration of the heavy-drop aircraft.

c. The aircraft dropping ranger personnel can also carry door bundles to deliver bulky or heavy items to the objective area. These door bundles, normally made up of A-7 and A-21 containers, may be color-coded to identify different loads or units. The most common door bundle loads are such items as 90-mm recoilless rifles and ammunition, 60-mm mortar ammunition, Dragon missiles, SHORAD missiles, or more demolition material. These bundles are pushed from the personnel aircraft just before the lead jumper's exit. They may be marked with lights or infrared sources for locating them at night.

d. If the tactical situation requires that a ranger force receive a resupply during the course of an operation, the most common method would be by airdrop. Airdrop could be needed if the operation takes longer than planned, initial supplies are lost, or changes occur in the operational plan. The ranger unit logistics officer and operations officers normally plan for resupply needs and schedule follow up loads of Classes I, II, IV, V, and VIII. There are two different airdrop requests channels for the ranger force to use:

(1) Nonemergency airdrop requests are sent through logistics channels. The ranger LOC sends requests for the airdrop of supplies and equipment to its supporting supply unit, and the supporting unit forwards the requests to the COSCOM or TAACOM MMC. The MMC coordinates with the movement control center (MCC). Jointly they direct the actions of the supply and transportation units. The MCC sends the request to the joint transportation board (JTB). The board validates it and assigns it a priority. The JTB then sends the request to the airlift control center for action. If the items requested are not on hand in the airdrop company, the MMC directs the proper supply activity to prepare the items for shipment to the airdrop unit. The MMC coordinates with the MCC, and the MCC ships the items. The airdrop unit prepares the supplies for the mode of delivery to be used. It also makes sure the supplies are delivered to the air terminal at the right time. (See Figure 8-6.)

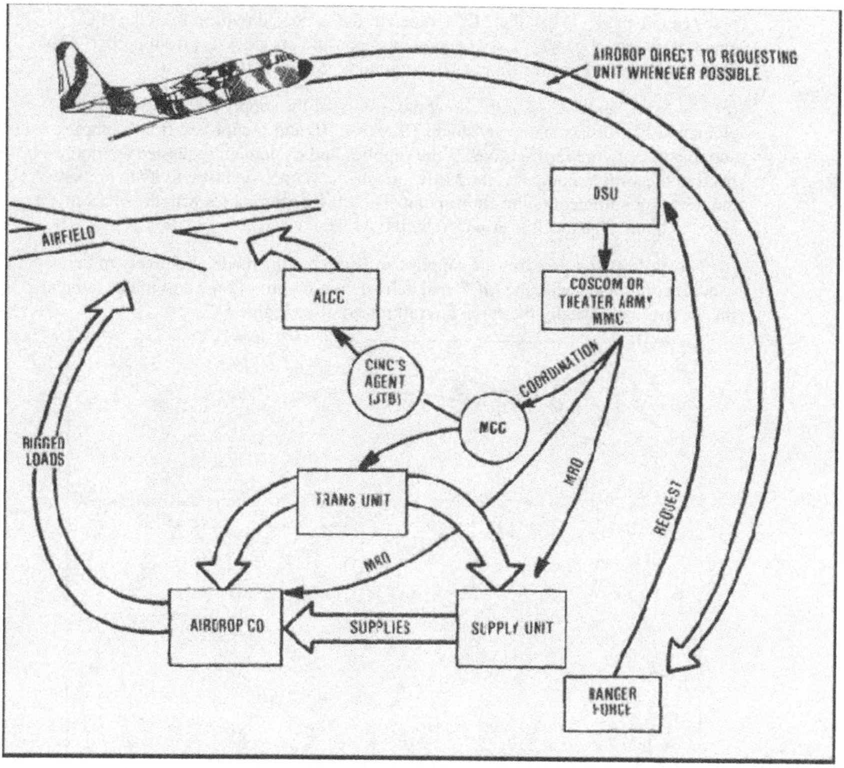

Figure 8-6. Ranger requests for nonemergency airdrop.

(2) Emergency requests for airdrop are passed through command channels in the quickest, most secure way.

(a) The request is usually sent to the command level that has authority to approve it, usually the tactical operations center at corps or above. The Air Force airlift officer at the direct air support center (DASC) coordinates with the Assistant Chief of Staff, Operations (G3) at the tactical operations center to determine whether airlift resources can be used for the mission.

(b) The TOC sends a request for an immediate airlift to the joint force commander's (JFC) designated agent for validation. At the same time, the airlift liaison officer advises the airlift control center (ALCC) that a request is coming through channels so plans can be made. When the ALCC receives the validated request from the JFC agent, it diverts or cancels less essential missions and provides airlift support. The TOC sends the request for supplies or equipment to the MMC.

(c) The MMC and MCC jointly direct the actions of the supply and transportation elements. The airdrop company stocks Classes I, III, and V supplies. It holds these supplies for emergency missions. If the supplies and equipment requested are not on hand in the airdrop company, the MMC directs the proper supply activity to prepare the items for shipment to the airdrop unit. The MMC coordinates with the MCC to ship the items. The airlift is arranged by the ALCC.

(d) The airdrop unit prepares the supplies according to the mode of delivery to be used. It coordinates with the MCC and delivers the supplies to the designated point at the air terminal (usually the cargo aircraft ramp). (See Figure 8-7.)

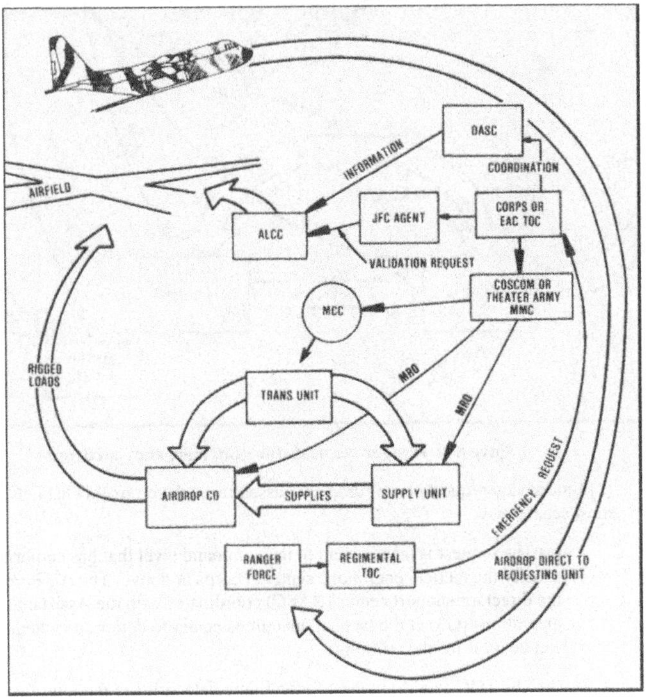

Figure 8-7. Request channels for emergency airdrop.

e. Aircraft delivering airdrop resupply deep behind enemy lines need to take precautions to avoid enemy detection and antiaircraft fire. The USAF is responsible for most of these precautions, such as ECM, suppression of enemy air defense (SEAD), and WILD WEASAL support. These do not concern the ranger commander. The choice of the type of parachute delivery system to use, however, may affect the ground commander's plan. The ranger force commander must specify any tactical restrictions that would require a specific type of delivery system. The ranger LOC and regimental logistics officer must advise the ranger force commander on the type of parachute system to be used. Conflicts must be quickly resolved so as not to interfere with the tactical operations.

(1) The safest way for the airdrop aircraft to penetrate enemy air defense and remain undetected is often by use of very low-level flight. There are four low-level parachute delivery systems that can be used without the aircraft being required to climb to a higher drop altitude and risk detection by enemy ADA.

(a) The low-altitude parachute extraction system (see Figure 8-8) needs a length of level ground with enough clearance for the delivery aircraft to make a very low pass over the extraction zone. The load is extracted by parachute from the aircraft and slides to a stop on a rugged, skid-type pallet. Heavy loads can be delivered using this system.

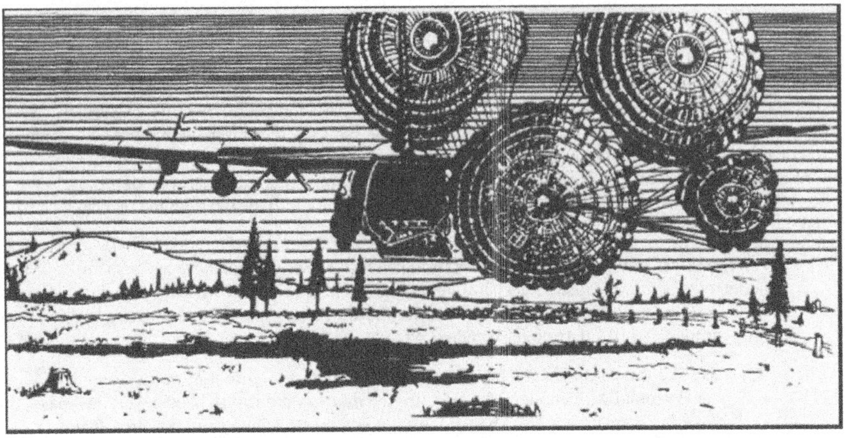

Figure 8-8. Low-altitude parachute extraction system.

(a) The containerized delivery system (CDS) (see Figure 8-9) consists of multiple, individually rigged A-22 containers, each with its own parachute, weighing up to 2,000 pounds each. The system provides single-pass delivery of up to 16 containers by C-130 aircraft and up to 28 containers by C-141 aircraft. The CDS loads may be delivered into drop zones using multiple points of impact to allow for tactical separation. The ranger force commander can direct the use of multiple points of impact, noting the advantages to be gained against the requirement for multiple passes.

Figure 8-9. Containerized delivery system.

(c) The high-speed, low-level, airdrop system (HSLLADS) (see Figure 8-10) is a single A-21 container specially rigged to withstand the parachute opening shock when airdropped from C-130 aircraft at high speed. This system can be used to deliver up to 600 pounds for each container, with a maximum of four containers for each pass, at speeds up to 250 knots.

Figure 8-10. High-speed, low-level, airdrop system.

(d) The CTU-2/A is a high-speed, aerial-delivery container (see Figure 8-11) that can be used to deliver supplies from high-performance aircraft flying at a minimum altitude of 300 feet above ground level and a maximum airspeed of 425 knots. The CTU-2/A is carried on the bomb racks of fighter or bomber-type aircraft. Upon release, a pilot parachute deploys the main chute and the container descends slowly.

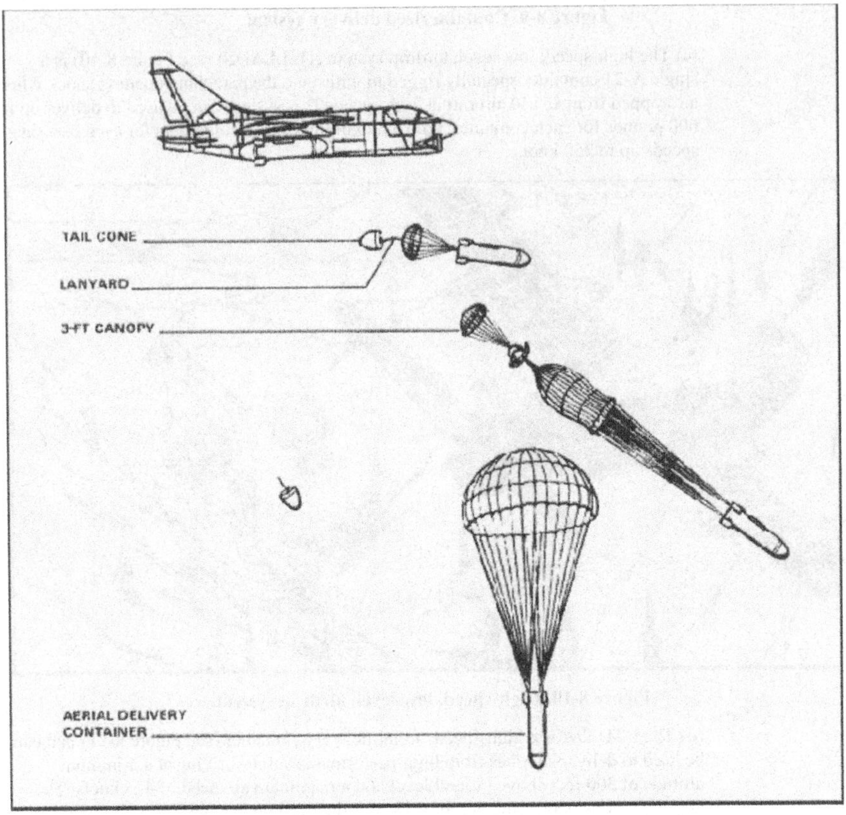

Figure 8-11. CTU-2/A high-speed, aerial-delivery container.

The CTU-2/A can be used to deliver up to 500 pounds of supplies such as weapons, water, food, or munitions. The container can be destroyed by burning. The main advantage of this system is that it can be delivered by high-performance aircraft deep behind enemy lines and in a dense air defense environment. The accuracy of this system is equal to that of a conventional bomb strike.

(2) Sometimes it may be better to drop resupply loads to the ranger force from a high altitude. The HAARS (see Figure 8-12) permits containerized unit loads weighing from 200 to 2,000 pounds to be delivered from aircraft at speeds up to 150 knots from up to 25,000 feet AGL. The HAARS consists of a cargo parachute, an airdrop container, an altitude

sensor, and a pilot chute. The pilot chute gives the descending bundle a speed slightly greater than an accompanying HALO parachutist. The system provides for the steady free-fall descent of loads from altitudes between 2,000 and 25,000 feet to an altitude at which a barometric sensor actuates deployment of the main parachute. This allows safe and accurate delivery of loads onto unprepared drop zones. The HAARS can deliver a payload to within 260 meters from a target impact point from a 10,000-foot altitude, with a proportional degree of accuracy from 25,000 feet. Ranger units can use the system for resupply of battalions and smaller units with rations, ammunitions, and medical supplies, breaking down the containerized material into man-packed loads.

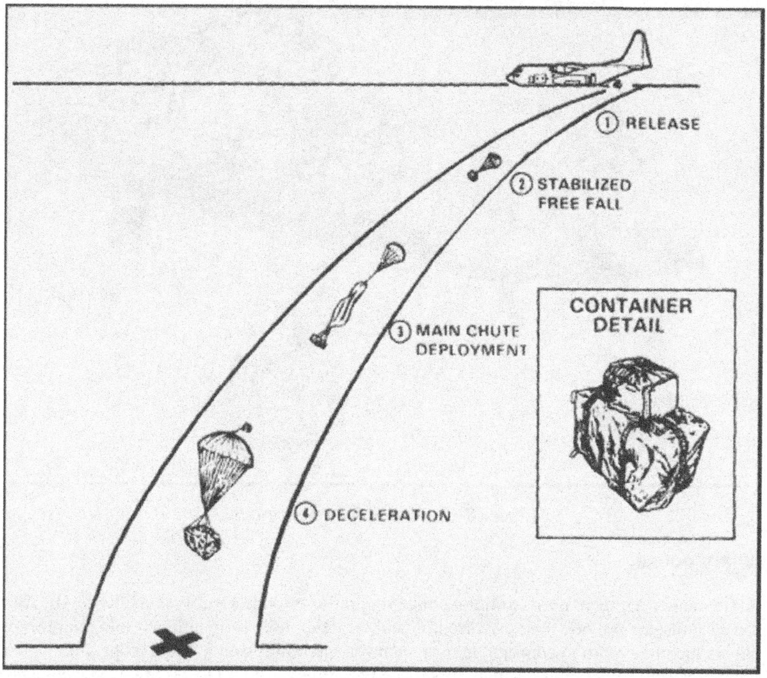

Figure 8-12. High-altitude, airdrop resupply system.

(3) The ranger force may need aircraft loads delivered during periods of poor visibility. The adverse weather aerial delivery system allows aircraft to drop resupply bundles accurately in adverse weather or darkness. These AWADS operations can be conducted safely and effectively with a minimum 300-foot ceiling AGL and a minimum visibility of less than half a mile. The minimum-drop altitude is 500 feet above the highest obstruction. These minimum ceiling or visibility restrictions do not apply to the Combat Talon aircraft.

f. Resupply to the ranger force can be provided by Army helicopters (see Figure 8-13). Normally this is done by SOA units. The danger to helicopters during long-range, low-level penetration of enemy rear areas requires that resupply be conducted when there is limited visibility.

Figure 8-13. Resupply bt helicopter.

8-4. Military police.

a. The ranger regiment needs military police support to provide security at the REMAB. Although ranger units can provide their own security and local defense, using military police for this purpose allows the ranger unit to concentrate its manpower and training on mission preparation. A military police security company is normally designated to provide support to 1st SOCOM elements located in the communications zone. Military police provide security patrols of the REMAB, convoy security, and vehicular search. They provide entry and exit control of the sensitive compartmented information facility (SCIF).

b. A preemptive strike by enemy special operations forces is a threat to the ranger regiment while located in the REMAB; either by conventional attack or by terrorist-type bombing. The preparation and location of the headquarters, the type of buildings, the location of landing zones, the traffic flow, and the size of the local population all influence the size of the military police unit. A military police platoon, augmented with bomb and guard dog teams, should be enough security at the REMAB. These security forces would not normally deploy from the REMAB to an ISB with the ranger unit.

8-5. Counterintelligence.

Counterintelligence support to the ranger regiment is normally provided by a CI team from the military intelligence group at TA level. The CI team provides a liaison and informational flow from the CI agencies operating within the theater. It also acts as advisor to the ranger unit commander on OPSEC while performing CI surveillance operations in the vicinity of the REMAB. The CI support may also be provided through 1st SOCOM channels. It is through 1st SOCOM that liaison with national intelligence and counterintelligence agencies is achieved.

8-6. Personnel service support.

The personnel service support needs of a ranger unit are much like that of any other unit of the same size. To receive the support his unit needs, the ranger personnel officer must meet with the proper agencies within the theater of operations.

a. **Strength accounting.** Within the ranger regiment, strength accounting is conducted through normal personnel channels. Regimental strength figures are passed quickly to the TAACOM or COSCOM. These figures are crucial to the theater commander for planning future operations. When matched with casualty reports, they provide the information needed for replacements.

b. **Replacement operations.** Ranger units need exceptionally well-trained and physically conditioned personnel as replacements. Ranger operations require that replacements train with the unit before they can become an effective member of the team. There are five sources of replacements for ranger units:

(1) **In-theater volunteers.** Use of in-theater volunteers, preferably with experience in ranger units, is the quickest way to replace personnel. These volunteers have a knowledge of the basic skills required of rangers and have some operational experience. The ranger regimental personnel officer coordinates with the theater personnel officer to screen and select those volunteers with the required MOSs and grades. Volunteers from other special operations force components, or from other theaters of war, are identified by DA and screened for assignment to the ranger regiment. Even though these volunteers have individual training and experience, it takes several weeks to become fully integrated into the ranger unit's operations and to achieve the needed skill level.

(2) **Reorganization of units.** This method of reconstitution is used when two or more ranger units have high personnel losses. The remaining members of these units are reorganized into a single unit. This method provides the ranger regimental commander with enough ready-trained, physically fit personnel to form a full-strength unit. Extensive unit training must be conducted before such a reorganization regains the unity and esprit of its component parts. The high standards found among the subelements of the ranger regiment makes this a good method of reconstituting a ranger unit. The units that have lost all their personnel may be reconstituted later, using some other method, or left unreconstituted,

depending on the situation.

(3) **Hospital returnees.** These personnel, former members of the ranger regiment who were injured, evacuated, treated, and have now recovered, are valuable assets. They provide the unit with a major reconstitution ability. These personnel should be returned quickly to their original battalion. This enhances unity and increases morale and esprit. Returnees may not be physically able to return to full duty right away. While recuperating, they may be assigned to limited duty as instructors. The ranger unit uses the returnee's combat experience and individual skills to screen and train newly assigned volunteers. Lessons learned in combat can help prepare the unit for future operations.

(4) **Replacement volunteers from CONUS.** These personnel have training in the basic skills and may have completed the ranger course, but they do not have any ranger unit experience. They need several weeks of unit training to become prepared for combat operations.

(5) **Ranger indoctrination program.** During wartime, this program provides screening and training of newly assigned personnel. Volunteers who have neither served in a ranger unit before nor have graduated from the ranger course require extensive individual training before they are ready to begin unit training with a ranger battalion. This method of reconstitution of ranger units requires the longest training period. During this reconstitution phase, a ranger battalion needs extensive use of ranges, maneuver areas, and support by aviation and other special operations-type units. The ranger regimental commander may choose to move the program from CONUS to a location near the regiment's REMAB.

c. **Casualty reporting.** Casualty reports are handled through regimental personnel channels and must be timely and accurate. The reports are used to check strength accounting data, to assist graves registration personnel in the recovery and identification of remains, and to notify next of kin.

d. **Health service.** The health service support needed by the ranger units deployed OCONUS is usually small. Area medical support provided by host country or theater medical command assets usually meets their needs. Evacuation of wounded members of the ranger force begins upon extraction from the objective area. Personnel wounded in action are evacuated through USAF or Army channels to the nearest medical facility that can provide treatment. The ranger force medical officer and personnel officer coordinate with the area medical support facility providing preventive and routine medical care to the unit while at the REMAB. Vaccine needed for innoculation of the ranger force members and any special Class VIII supplies needed for the mission are drawn through normal medical channels.

e. **Financial service.** Financial support varies with the situation. In mid- and high-intensity conflicts, regular paydays are suspended in the theater of operations. As a nondivisional unit, the ranger regiment coordinates with the theater finance center or corps finance group for financial support. In the REMAB, ranger units are provided financial support from the closest finance support unit. The ranger unit personnel officer appoints Class A agents to help with the disbursement of funds. The ranger force may be given a specific amount of local currency or military scrip depending on the length of time in the REMAB and the operational situation. The ranger unit personnel officer also coordinates with the theater finance center to get any third-country currency that may be included as part of an escape-and-evasion kit provided to ranger elements.

f. **Postal service.** Postal service to the ranger unit is through normal channels. Mail for members of the deployed force is forwarded by the rear detachment commander to an established theater Army general support postal activity. This activity forwards the mail to the postal unit designated to support the REMAB site. The ranger unit commander appoints subordinate unit officers as censors. The exact location of the REMAB, and all aspects of the mission and its location, remains classified until released by an official Army spokesman. The ranger unit commander must ensure that the forwarding of mail, or the writing home of personal letters, does not endanger the ranger force mission. All outgoing mail is censored during wartime. For security reasons, outgoing mail may be held for a short time before it is forwarded. Mail from the ranger unit is never sent by host country or other nation's postal services.

g. **Chaplain activities.** Chaplain activities are provided by the chaplain assigned to the ranger battalion. He provides unit coverage through services, rites, and sacraments regardless of his or the unit member's religious affiliation.

h. **Legal service.** Ranger units do not need extensive legal service support. Legal service support is obtained through coordination with the theater Army staff judge advocate.

APPENDIX A

Organization and Equipment Introduction of the Ranger Regiment

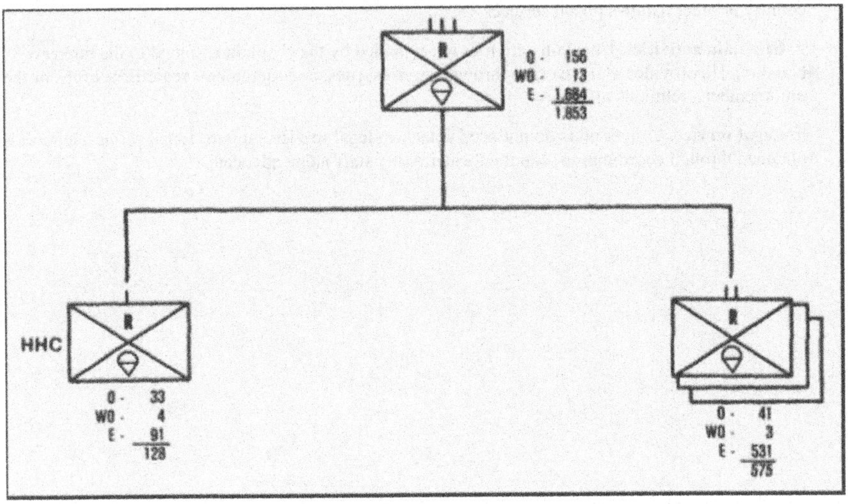

Figure A-1. Ranger regiment.

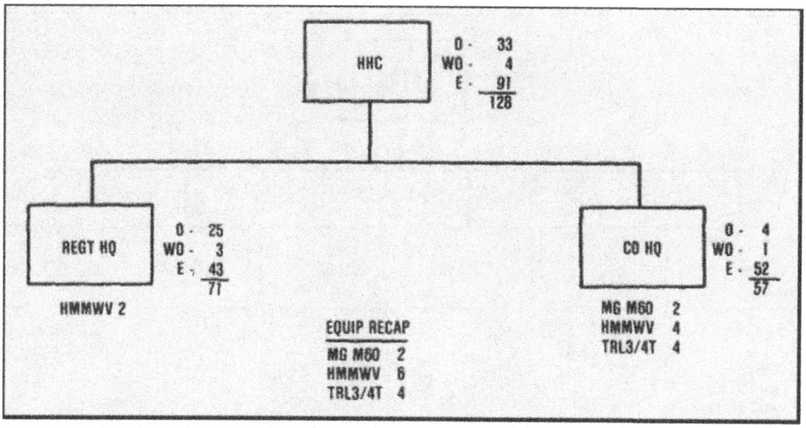

Figure A-2. Ranger regimental HHC.

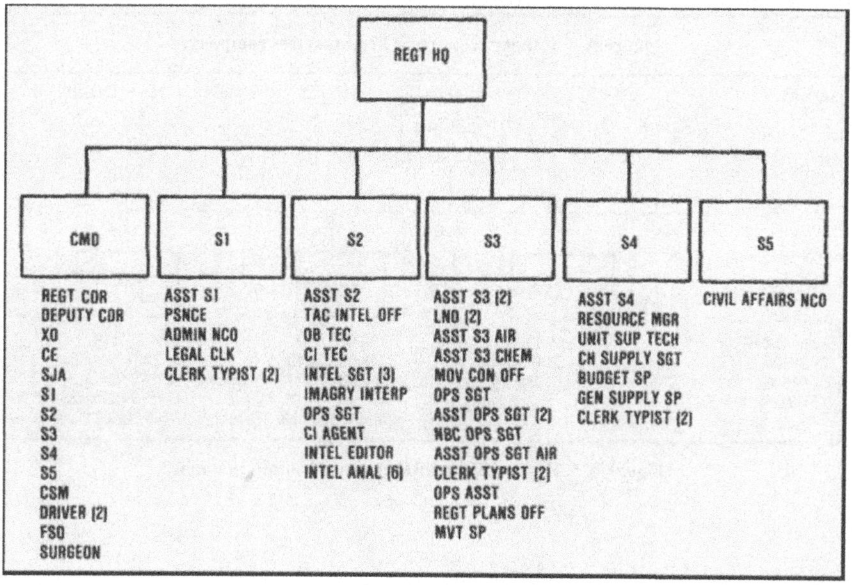

Figure A-3. Ranger regimental staff.

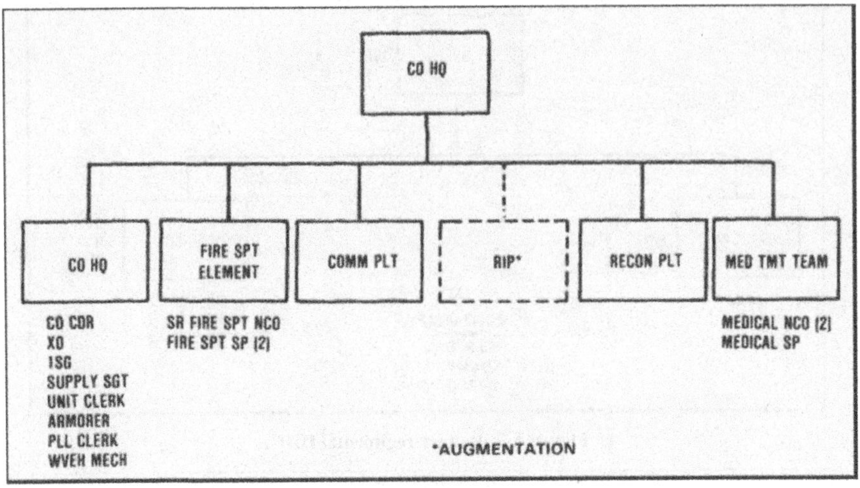

Figure A-4. Ranger regimental headquarters company.

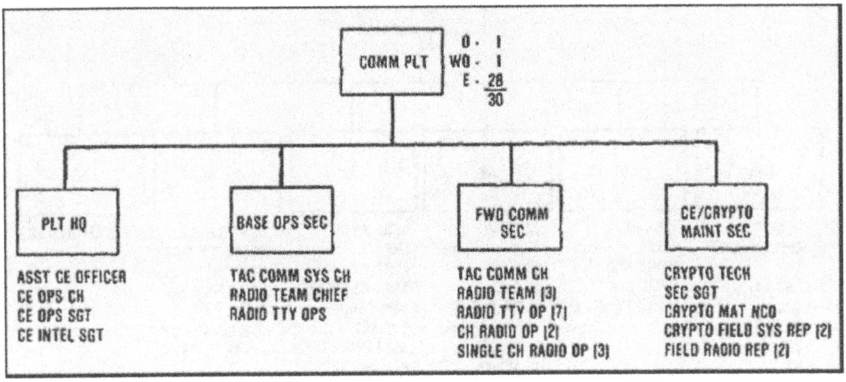

Figure A-5. Ranger regimental communications platoon.

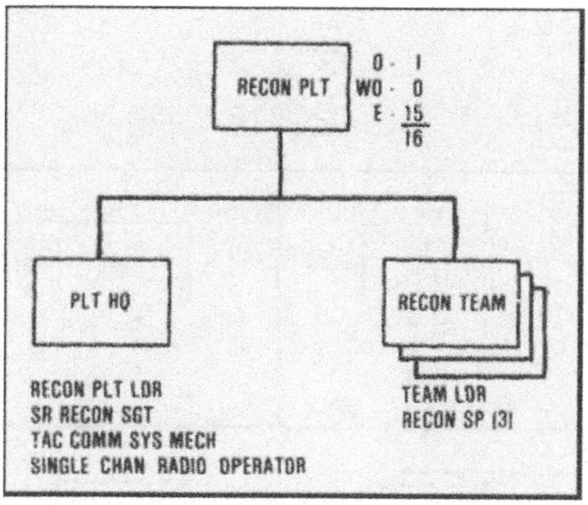

Figure A-6. Ranger regimental reconnaissance platoon.

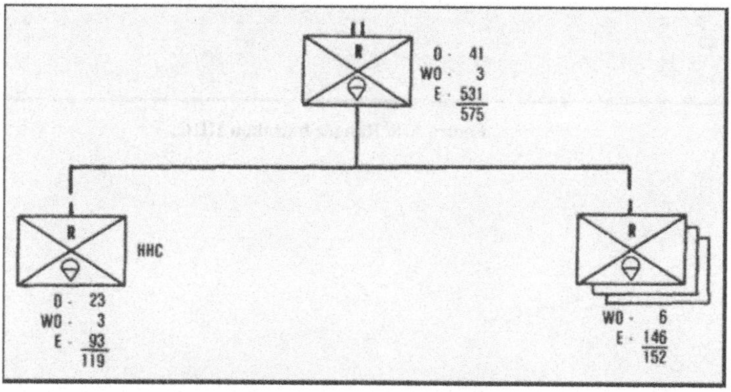

Figure A-7. The ranger battalion.

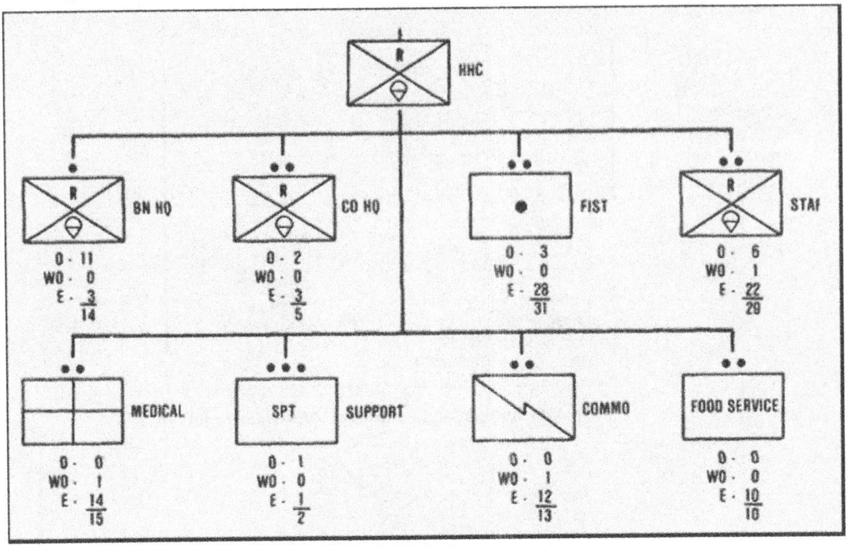

Figure A-8. Ranger battalion HHC.

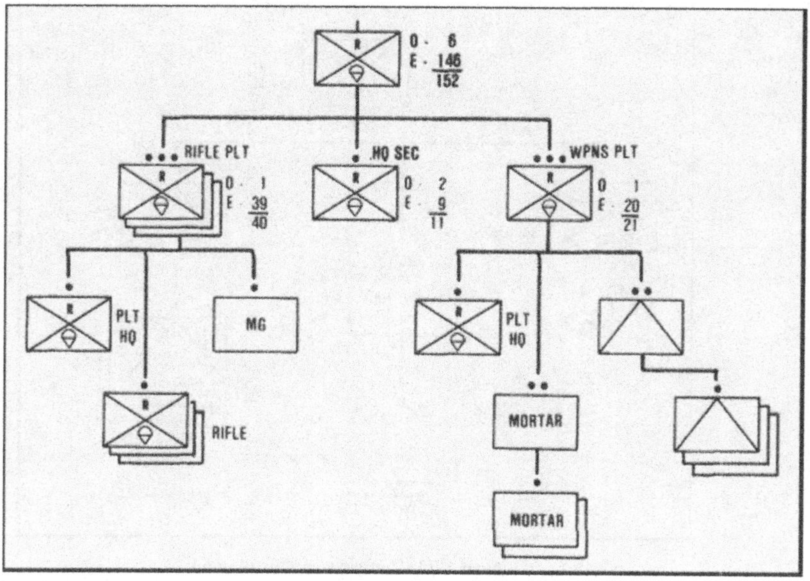

Figure A-9. Ranger rifle company.

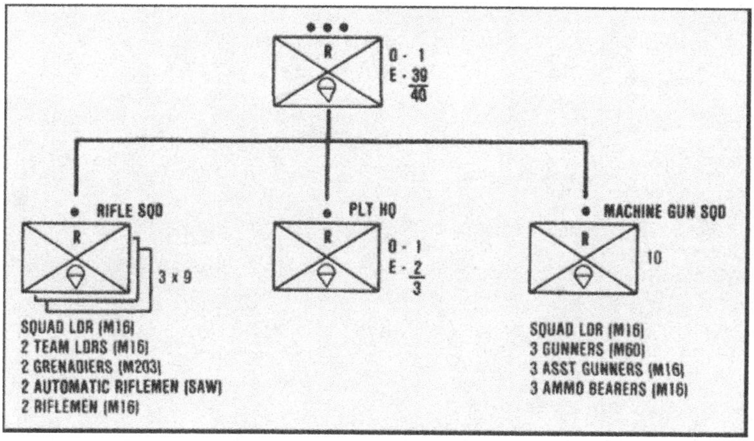

Figure A-10. Ranger rifle platoon.

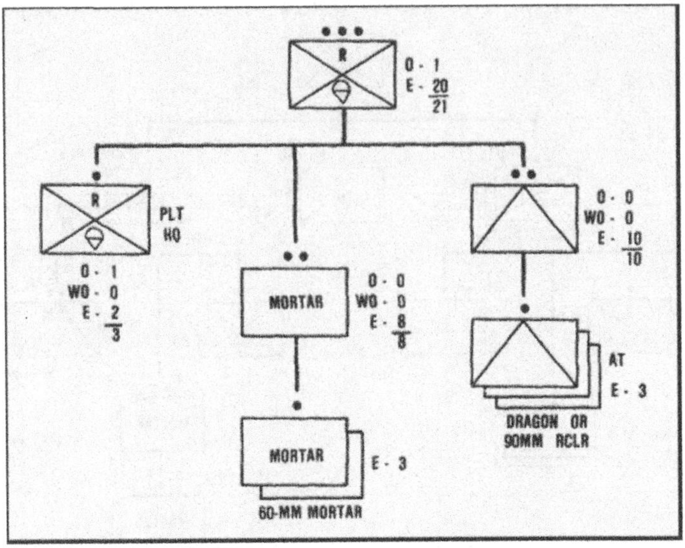

Figure A-11. Ranger weapons platoon.

APPENDIX B

Composition and Equipment of Ranger Regiment Liaison Teams

Team Red	**RANK**
Liaison Officer	MAJ
Senior Intelligence Analyst	SSG
Clerk-Typist	SP4
Tactical Communications System Operator/Mechanic	SGT
Combat Telecommunications Center Operator	SP4
Tactical Communications System Operator/Mechanic	SP4
Team White	
Liaison Officer	CPT
Senior Intelligence Analyst	SSG
Clerk-Typist	SP4
Tactical Communications System Operator/Mechanic	SGT
Combat Telecommunications Center Operator	SP4
Tactical Communications System Operator/Mechanic	SP4

Equipment (for Each Team)

HMMWV with 3/4-ton trailer	5-kw generator
Communications shelter	Tent, GP, small
AN/PCS-3 radio set (3)	AN/VSC-7 net control station (1)
MX 300 (4)	KY 57 with battery case (5)
HYX 57 with battery case (4)	HYP 57 (5)

KYK 13 (2)

KY 65 with battery case (1)

KG 84 (1)

H189 (2)

GRC 193 (1)

VGC-74 (1)

PP 1648 (1)

SB-22 switchboard (1)

KYI 18 (1)

KY 65 WLA (1)

Digital message device (1)

AN/PRC-77 radio (1)

Facsimile transmission device (1)

TA 312 with WD1 (5)

APPENDIX C

Equipment Density

C-1. Communication equipment.

	Regimental Headquarters	Battalions (x3)
AN/VSC-7 net control station	7	2
AN/PSC-3 radio set	23	13
AN/PRC-41 radio set	2	5
AN/PRC-104 radio set	9	26
AN/GRC-193A radio set	4	0
AN/PRC-70 radio set	6	0
AN/PRC-77 radio set	22	50
AN/VRC-46 radio set	2	0
AN/VRC-49 radio set	0	2
AN/PRC-68 radio set	0	66
TSEC/KY-57 speech security device	60	68
TSEC/KY-58 speech security device	6	0
TSEC/KY-65 speech security device	11	0
AO-8990/P digital message device group	15	23
AN/GXC-7A tactical FAX device	7	2
AN/UGC-74A(V)3 communications terminal	7	0
SB-22 switchboard	2	0
SB-3614(V)/TT switchboard	2	1

TSEC/KG-84	5	0
SB-993/GT switchboard	0	12

C-2. Weapons and equipment each battalion.

Infrared aiming light	96	Interrogator set, Stinger missile	3
Rifle, M16	451	Grenade launcher, M203	67
Rocket launcher, M202	13	Machine gun, M60	27
Mortar, 60-mm M224	6	Rifle, sniper, M21	24
Recoilless rifle, 90-mm	18	Viewer, infrared	9
Tracker, Dragon	9	Night vision devices (all types)	322
Truck, 1/4-ton	2	Laser target designators	11
Radio (all types)	164	Tactical satellite radios (all types)	15
Squad automatic weapon	54		
Laser range finder	14		

APPENDIX D

Standing Orders, Rogers' Rangers

1. Don't forget nothing.
2. Have your musket clean as a whistle, hatchet scoured, sixty rounds powder and ball, and be ready to march at a minute's warning.
3. When you're on the march, act the way you would if you was sneaking up on a deer. See the enemy first.
4. Tell the truth about what you see and what you do. There is an army depending on us for correct information. You can lie all you please when you tell other folks about the Rangers, but don't never lie to a Ranger or officer.
5. Don't never take a chance you don't have to.
6. When we're on the march we march single file, far enough apart so one shot can't go through two men.
7. If we strike swamps, or soft ground, we spread out abreast, so it's hard to track us.
8. When we march, we keep moving till dark, so as to give the enemy the least possible chance at us.
9. When we camp, half the party stays awake while the other half sleeps.
10. If we take prisoners, we keep 'em separate till we have had time to examine them, so they can't cook up a story between 'em.
11. Don't ever march home the same way. Take a different route so you won't be ambushed.
12. No matter whether we travel in big parties or little ones, each party has to keep a scout 20 yards ahead, 20 yards on each flank, and 20 yards in the rear so the main body can't be surprised and wiped out.
13. Every night you'll be told where to meet if surrounded by a superior force.
14. Don't sit down to eat without posting sentries.
15. Don't sleep beyond dawn. Dawn's when the French and Indians attack.
16. Don't cross a river by a regular ford.

17. If somebody's trailing you, make a circle, come back onto your own tracks, and ambush the folks that aim to ambush you.

18. Don't stand up when the enemy's coming against you. Kneel down, lie down, hide behind a tree.

19. Let the enemy come till he's almost close enough to touch. Then let him have it and jump out and finish him up with your hatchet.

MAJ Robert Rogers

APPENDIX E

The Ranger Creed

R Recognizing that I volunteered as a Ranger, fully knowing the hazards of my chosen profession, I will always endeavor to uphold the prestige, honor, and high "esprit de corps" of the Ranger Regiment.

A Acknowledging the fact that a Ranger is a more elite soldier who arrives at the cutting edge of a battle by land, sea, or air, I accept the fact that as a Ranger, my country expects me to move farther, faster, and fight harder than any other soldier.

N Never shall I fail my comrades. I will always keep myself mentally alert, physically strong, and morally straight and I will shoulder more than my share of the task, whatever it may be. One hundred percent and then some.

G Gallantly will I show the world that I am a specially selected and well-trained soldier. My courtesy to superiors, my neatness of dress, and my care of equipment shall set the example for others to follow.

E Energetically will I meet the enemies of my country. I shall defeat them on the field of battle, for I am better trained and will fight with all my might. Surrender is not a Ranger word. I will never leave a fallen comrade to fall into the hands of the enemy and under no circumstances will I ever embarrass my country.

R Readily will I display the intestinal fortitude required to fight on to the Ranger objective and complete the mission, though I be the lone survivor.

RANGERS LEAD THE WAY

APPENDIX G

Training

G-1. Application and guidance.

a. To prepare for combat, ranger units train on missions derived from their analysis of operational and contingency plans and the ranger mission training plans (MTPs). Training guidance is in doctrinal manuals, field circulars, and soldier's training publications. Training must ensure that the ranger unit is skilled in all basic light infantry skills and in ranger-unique skills. In the final analysis, good training depends upon the leadership and teamwork of ranger unit officers and NCOs.

b. Everything rangers do should prepare them to perform well in combat. Ranger units train IAW the battalion training management system (BTMS) concentrating on basic skills and on live-fire, hands-on training.

c. An effective training program breeds high morale and few discipline-related problems. Some qualities of ranger units that have good training programs are as follows:

(1) Leaders who can solve problems.

(2) A thorough and continuous analysis of the unit's mission.

(3) Validated training objectives.

(4) Attention to the basics (move, shoot, and communicate).

(5) Multiechelon sustainment training.

(6) Standardized procedures.

(7) Physical and mental strength.

(8) Live fire as an integral part of training.

(9) Safety as a basis to all operations.

(10) The use of night and adverse weather and difficult terrain as combat force multipliers,

d. When all training is planned with realism, rangers will be in training situations that require the same attention as in battle. Realism and stress in training result in faster learning and better retention of skills. Training objectives must be clear with measurable standards to determine proficiency. Rangers and ranger units must train to the standards that will be expected in combat.

e. Examples of unit training goals include:

(1) Developing each member of the chain of command to perform his duties with competence, confidence, and pride.

(2) Maintaining small unit integrity during all training and support activities. Accomplishing tasks as units.

(3) Ensuring that all tasks for training are keyed to mission needs and weaknesses. Assets--ranger's time, money, and materiel--are expended only after thorough planning.

(4) Training to standard.

(5) Developing and maintaining discipline and esprit to ensure the abilities of each ranger and all units are achieved.

(6) Emphasizing use of the ranger mission training plan (MTP) for each echelon of the organization.

(7) Emphasizing the basics.

(8) Maintaining a high level of physical fitness.

(9) Working together within the ranger regiment and other elements of the special operations forces community.

(10) Developing a unit atmosphere of quick reaction under stress. Encouraging and rewarding individual initiative. Accomplishing unit missions IAW the commander's intent.

G-2. Philosophy.

a. To be effective, training should be physically and mentally challenging. Each ranger will then have a sense of purpose, accomplishment, and satisfaction. Training is evaluated by the results rather than hours of training. Units that are well trained today need reinforcement training to sustain readiness. Training management must be continuous. Training managers must use the time available to improve training efficiency. Cyclic training will not produce the desired results. The "train-to-correct deficiencies" method is needed. The training managers must analyze weaknesses, conduct prescriptive and reinforcement training, and reevaluate their unit's training status. The process that is used to develop and maintain a continuous training and evaluation program is shown in Figure G-1.

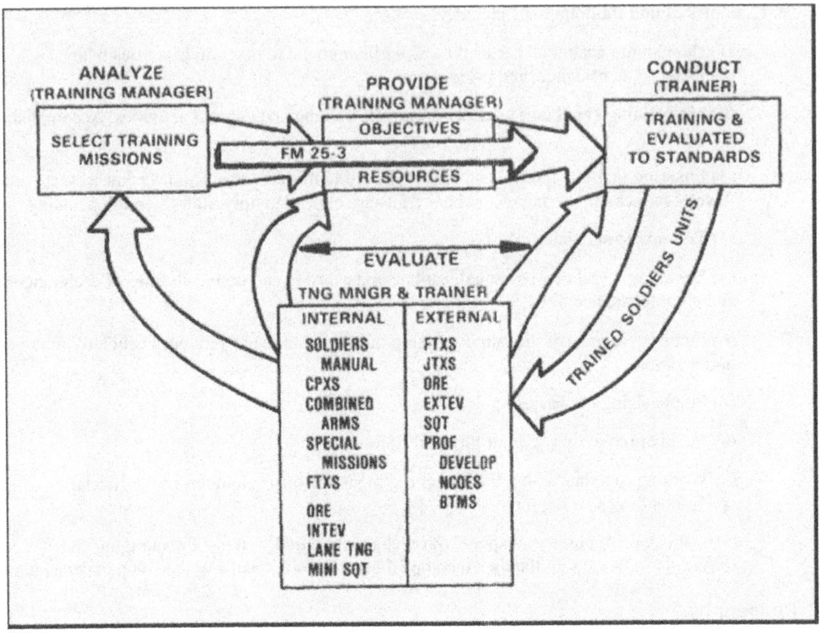

Figure G-1. The Army training and evaluation program.

b. Individual and collective training must be integrated within a unit training program. Training of the individual ranger, leaders, and units at each level must occur at the same time. The ranger commander or leader plans and conducts training for every member of his unit.

c. Ranger commanders analyze operational and contingency plans to determine unit missions and translate the associated tasks into training objectives. These objectives form the basis of the unit training program. The regimental headquarters provides training guidance in a formal regimental training plan. Commanders at lower levels program, supervise, and conduct training within their units.

d. All members of the chain of command are responsible for training. The NCO leader or first-line supervisor conducts individual and small unit collective task training. The officer plans individual training in his unit and directs and conducts collective training.

e. Centralized development of training objectives and decentralized training builds well-trained units and outstanding leaders. Authority and responsibility for the detailed planning, scheduling, conducting, evaluating, and supervising of training are delegated to the lowest element capable of managing this training. Higher headquarters is responsible for establishing objectives, monitoring and evaluating training, and issuing general guidelines.

f. Multiechelon training and evaluation means using all assets and time available to improve combat readiness. The multiechelon structure of the ranger MTP provides a basis for units to be continually trained and evaluated by internal evaluation. Each leader must know his mission tasks and train to accomplish those tasks to the set standards.

G-3. Mission analysis.

a. Mission analysis is continuous for a good training program. The selection of training missions or tasks results from analysis of the unit's overall mission. (See Figure G-1 above.) Tasks and subtasks are then placed in order of priority.

b. The trainer must identify countertasks for the opposing force (OPFOR). The countertasks should be such that an OPFOR that meets the standards for its countertask will be successful. Both sides have a chance to win.

c. The trainer should then develop combat drills and situational training exercises (STXs), if not furnished by the proponent school.

G-4. Unit training.

a. **Responsibility.** Ranger commanders at all levels are responsible for all training and management described herein and current Army training doctrine. This includes individual and collective training at all levels. Units must train to the standards specified in training objectives.

b. **Management.** The training management model for commanders is:

- Set objectives.
- Provide resources.
- Coach subordinates.
- Measure results.
- Provide feedback.

G-5. Individual training.

There are three elements basic to a well-trained individual ranger. He must be physically fit, be proficient in the use of his weapon, and have the psychological commitment to fight the enemy.

a. Individual training is the responsibility of the first-line supervisor under the supervision of the chain of command. It is conducted using performance-oriented techniques under battlefield conditions. New skills must be developed and practiced, and acquired skills must be reinforced throughout the training year. A high level of proficiency must be maintained through continuous training and practice.

b. Trainers should use peer instruction. Commanders ensure needed resources are available to conduct challenging individual training. Ranger units schedule and conduct this training and evaluate individual proficiency.

c. Flexibility in planning and scheduling is necessary. The goal of all individual training must provide for reinforcement training to offset the effects of the forgetting curve. (See Figure G-2).

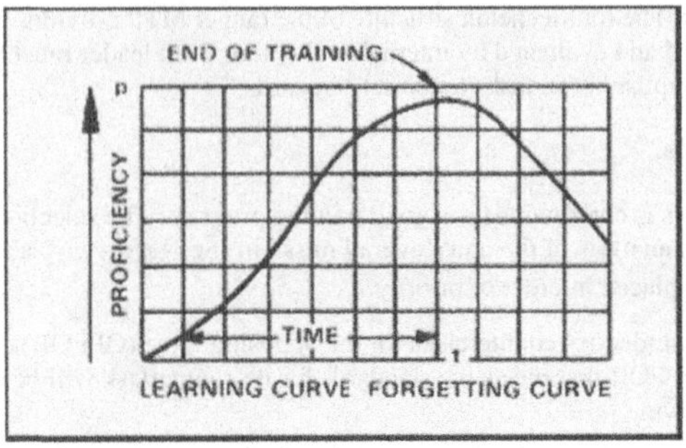

Figure G-2. The learning curve and forgetting curve.

d. The command sergeants major, unit first sergeants, and staff section sergeants are the keys to an effective training program. These senior NCOs must counsel, guide, and critique junior NCOs in training preparation and execution. Fire team leaders, section chiefs, and squad leaders coordinate and present individual training.

e. The NCO supervisors must ensure that the ranger is trained in the skills required for the specific military occupational specialty (MOS) and in those skills used daily. Not all training can be acquired in an office. For example, all rangers must maintain a high skill level with their weapons, be trained in NBC subjects, and be capable of performing contingency missions.

G-6. Collective training.

a. **Guidance.** Collective training trains crews, teams, squads, platoons, companies, and battalions to do their missions. The ranger MTP is used to plan training and evaluate the unit's weaknesses and strengths. If a ranger commander receives new training requirements or missions, they should become part of the collective training program. The ranger MTP describes the combat tasks to be performed by every echelon of the ranger battalion. It is a plan for organizing, conducting, and evaluating the year-round unit training program.

b. **Responsibility.** The ranger unit commander is responsible for the conduct of collective training and for all other training in his unit. Practical exercises emphasize the techniques of terrain use, movement, and effective employment of weapon systems. Training under stress will be conducted to move the unit to a higher plane of combat readiness.

c. **Drills.** Drills are written for small units--for example, fire team and squad, crew and section--and describe an action that is shorter than an entire STX. Drill training objectives cover both individual and collective skills (critical collective tasks) and list them together for effective group action. Drills must be mastered in order to react to rapidly changing situations in combat. The trainer can interrupt the execution of a drill to provide necessary training on individual or collective skills.

d. **Situational training exercise.**

(1) The STXs are short, mission-oriented tactical exercises in which a group of closely related collective tasks are trained. They are like drills but are more complex and flexible. They usually involve a larger unit than does a drill. They integrate leader training, leader tasks, drills, and separate individual tasks into a realistic setting. Trainers may execute STXs as written or tailor them to meet specific unit requirements. An STX is not standardized, but it must have a firm doctrinal base.

(2) Due to their design and purpose, STXs apply mainly to platoon- and company-size units. This does not preclude the development of STXs or similar exercises for unit staffs and headquarters. They combine "how-to-train" specifics for selected collective tasks with the "what-to-train" specifics found in doctrinal manuals and other documents.

G-7. Performance-oriented training.

a. Performance-oriented training emphasizes hands-on performance. Training is conducted to ensure a level of proficiency, rather than established number of hours of required training.

b. Performance-oriented training has precise training objectives and makes good use of resources. Training objectives serve as the basis for preparing, conducting, and evaluating all training.

G-8. Evaluation.

a. A training program must include an evaluation of training (planned and conducted). An evaluation plan must:

(1) Provide immediate feedback to the trainer.

(2) Recognize effective trainers.

(3) Evaluate all units in the conduct of training to provide for a comparison of training programs.

(4) Be flexible to unit needs; provide general and specific evaluations based on the level of training within the unit.

(5) Be in effect at all times.

b. Major considerations for training evaluations include:

(1) Ranger unit commanders establish and maintain an effective internal and external evaluation program.

(2) The director of each staff element is responsible for training evaluations within his area of staff responsibility with emphasis on low-density MOS groups.

(3) The operations officer of the ranger organization develops and validates training and evaluation plans.

c. Evaluation of training must be done vertically from the major task (mission) down through subtasks (collective tasks) to the leader and individual tasks. In selection of individual tasks for evaluation, priority should be given to critical individual tasks that impact on the success of the collective task or mission.

d. The critique and after-action review, after each major training event, are the most important aspects of training. The after-action review takes place immediately after completion of a training task (individual or collective). The individual(s) self-analyzes and critiques the training event. Then after a formal critique of the training, the individuals) trains to standard (not to time).

G-9. Field training.

a. **Considerations.** Ranger commanders schedule training under field conditions to permit maximum training opportunities. Field training emphasis is on mastery of a series of individual (soldier and leader) tasks and collective training tasks. These tasks are repeated until established standards are met. The goal of ranger field training is to ensure realistic, demanding, and meaningful training.

(1) When training is conducted under conditions of moderate stress, most combat critical skills are learned quicker and retained longer. (See Figure G-3 on the following page.) Stress can be incorporated in a training situation by physical deprivation (sleep, food), by live-fire exercises, by operation against an OPFOR under simulated battlefield conditions for an extended period, by the intensity of activity including demands for decision making, or by physically demanding activities. Ranger units train under stress so that they will react quickly in combat.

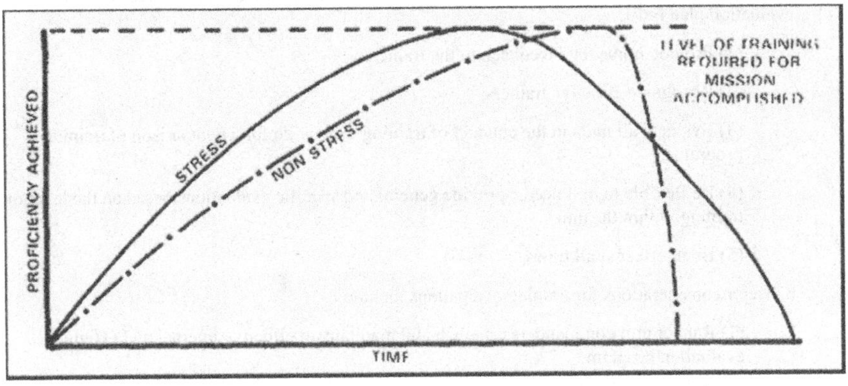

Figure G-3. The Army training and evaluation program.

(2) Antagonistic training situations are two-sided exercises conducted by opposing elements. Each element has a training objective (task and standards). The opposing force has a corresponding countertask. This training may be progressive--for example, starting with a

crawl-through, then a walk-through, and progressing to both elements operating at a fast pace. The use of multiple integrated laser equipment system (MILES) greatly enhances task and countertask training. The units should also be considered for a live-fire exercise against a realistic OPFOR target array when the basics have been mastered.

b. **Methods of training.**

(1) **Considerations.** The chance for success in combat can be increased by becoming proficient in training situations that depict realistic combat conditions. (See Figure G-4.) Meeting the standards of a realistic live-fire training exercise assures the commander that his unit could perform well in combat. Figure G-4 shows the relationship between realistic training and the challenge to the trainer.

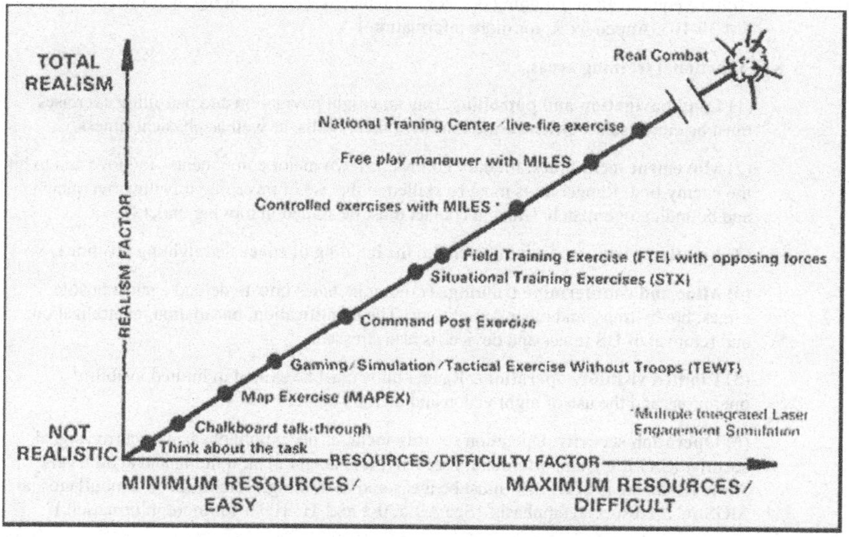

Figure G-4. Training proficiency scale.

(2) **Ranger cohesiveness training.** Ranger units promote teamwork by undergoing tough, stressful, often hazardous training. This challenges the unit, which gains in cohesiveness when it meets these challenges. This includes serving together in live-fire exercises, athletic events, precision parades, high-standard inspections, tough physical training, and combat.

(3) **Field training exercise.** An FTX is training normally conducted under simulated, realistic combat conditions. This training normally exercises command and control systems and the administrative and logistical needs of sustained combat. Stress training is part of this exercise, focusing on live-fire and MILES drills and STXS.

(4) **Combined arms training.** Combined arms training should include all means of support such as tactical air, armed helicopters, and indirect fire with emphasis on fire support coordination and airspace control.

(5) **Live-fire exercises.** Including live-fire exercises in drills and STXs offers a chance to evaluate units in their response to simulated combat missions. Ranger units train under pressure and always anticipate being selected to conduct a live-fire exercise. Live-fire exercises develop confidence and esprit, and reinforce unit discipline.

(6) **Opposing force.** Committing one unit against another adds a competitive spirit to a field training exercise (FTX). The use of MILES and squad combat operations exercises (simulation) (SCOPES) provides realism to two-sided field exercises. Trainers should use OPFOR equipment and training devices available, as well as making their own devices. (See FM 30-103, Appendix R, for more information.)

c. **Key functional training areas.**

(1) **Land navigation and patrolling.** Day and night navigation and patrolling exercises must be emphasized. Orienteering improves these skills, as well as physical fitness.

(2) **Movement techniques.** Infantry combat has two major components--to move and to hit the enemy first. Ranger units must be skilled in the use of traveling, traveling overmatch, and bounding overmatch. Units in contact must be skilled in moving under fire.

(3) **Fighting** positions. Training includes the building of concealed fighting positions.

(4) **Mine and countermine training.** Training includes how to defend against hostile mines, booby traps, and other demolitions. The identification, installation, neutralization, and removal of US mines and devices is also stressed.

(5) **Limited visibility operations.** Ranger units must be skilled in limited visibility operations and the use of night vision and thermal devices.

(6) **Operation security.** Operation security includes deception, physical security, signal security (SIGSEC), and information security. It is taught in all training and at all levels. Although all areas of OPSEC must be measured in training, techniques of camouflage and SIGSEC need special emphasis. (See AR 530-1 and TC 100-1 for more information.)

G-10. Professional development.

Ranger units at battalion and regimental levels conduct many officer and NCO professional development programs. This ensures that all officers and NCOs are skilled in their field, confident, and wellrounded. Professional readings and book reports, tactics seminars, and tactical exercises without troops are good development methods.

G-11. Physical training.

a. This program ensures that each ranger maintains the physical ability and stamina needed to do his mission. Other objectives are to promote unit cohesiveness and discipline. These are met by combining individual testing, formation, running, athletic competition, weight control, proper diet, teamwork drills, and freedom from drug and alcohol abuse. Mental fitness and toughness come from experience and success. Good physical conditioning increases a unit's chances of success.

b. Physical training (PT) programs are designed to include the elements of overload, progression, balance, variety, and regularity.

c. Ranger unit physical fitness standards are:

(1) Score 80 points for each Army physical readiness test (APRT) event and do six chinups.

(2) Pass the ranger swimming test.

(3) Complete an 8-kilometer run in 40 minutes.

(4) Complete a 12-kilometer road march in three hours (with rucksack, helmet, and weapon).

(5) Meet the Army height and weight standards.

G-12. Environmental and specialized training.

a. Rangers conduct training in environments as close to real combat missions as possible. This conditions rangers to survive and fight in a specific environment. Such training includes combat-critical tasks and technique training for that environment.

b. Each ranger battalion's environmental and specialized training includes at least the following:

(1) Jungle (once every 12 months).

(2) Desert (once every 12 months).

(3) Urban (once every 6 months).

(4) Extreme cold (once every 18 months).

(5) Mountain (once every 12 months).

(6) Amphibious (once every 18 months).

G-13. Ranger indoctrination program.

a. The RIP is a training and selection program. It is run by specially selected NCOs from the regimental headquarters. It has as its mission the training and indoctrination of newly assigned soldiers. It physically toughens and mentally sharpens the newly assigned soldier. It teaches basic skills and techniques. The RIP identifies and eliminates in three weeks any applicant who does not show dedication, motivation, physical fitness, and emotional stability.

b. An indoctrination program for newly assigned ranger-qualified officers and NCOs prepares them to assume duty as leaders within the ranger regiment. These personnel monitor and participate in RIP training, as well as undergoing individual ranger leadership training. The RIP provides them an overview of the regiment's history, its mission, and its goals. (See Appendixes C,

D, and E for further information about the history and goals of the ranger regiment.) The RIP also provides a review of the skills required in a ranger unit. Refresher training in airborne and pathfinder operations, use of indirect fires, and light infantry doctrine is conducted. During RIP all newly assigned officers and NCOs must meet the ranger physical fitness standard.

c. The RIP cadre also conducts a preranger program to prepare soldiers to attend the US Army Ranger School. It is a physically demanding course that teaches troop-leading procedures, operation orders, and small unit patrol actions.

Glossary

The following acronyms, abbreviations, and definitions are provided to aid the reader in understanding this publication.

Section I

ACRONYMS AND ABBREVIATIONS

AAT	Army assault team
ABCCC	airborne battlefield command and control center
ACL	allowable cargo load
ADA	air defense artillery
AFFOR	the USAF element of a joint task force
AGL	above ground level
AKZ	armor kill zone
ALCC	airlift control center
ALCE	airlift control element
AM	amplitude modulation
AMC	air mission commander
APERS	antipersonnel
APRT	Army physical readiness test
ARFOR	the Army element of a joint task force
ARSOC	Army special operations command
ARSOF	Army Special Operations Forces
ASG	area support group
ASPC	all-source production center

ATGM	antitank guided missile
AWADS	adverse weather aerial delivery system
BCE	battle coordination element
BTMS	battalion training management system
CA	civil affairs
CAS	close air support
CCT	combat control team
CDS	container delivery system
CE	communications-electronics
CI	counterintelligence
CINC	commander in chief
COMINT	communications intelligence
COMSEC	communications security
CONUS	continental United States
COSCOM	corps support command
CP	command post
CS	combat support; a chemical agent
CSS	combat service support
DACG	departure airfield control group
DACO	departure airfield control officer
DASC	direct air support center
DMDG	digital message device group
DOD	Department of Defense
DS	direct support
DSU	direct support unit
DZ	drop zone
EAC	echelons above corps

EACIC	Echelons Above Corps Intelligence Center
ECCM	electronic counter-countermeasure
ECM	electronic countermeasures
E&E	escape and evasion
EEFI	essential elements of friendly information
EOC	emergency operations center
EW	electronic warfare
FARP	forward area rearm and refuel point
FASCAM	family of scatterable mines
FAX	facsimile transmission
FDC	fire direction center
FEBA	forward edge of the battle area
FFL	friendly front lines
FIST	fire support team
FLOT	forward line of own troops
FM	field manual; frequency modulation
FO	forward observer
FORSCOM	US Army Forces Command
FPF	final protective fire
FPL	final protective line
FRAGO	fragmentary order
FSA	fire support area
FSCOORD	fire support coordination or coordinator
FSE	fire support element
FSO	fire support officer
FSS	fire support station
FTX	field training exercise

G3	Assistant Chief of Staff, Operations
GS	general support
GSR	ground surveillance radar
HAARS	high-altitude, airdrop resupply system
HAHO	high-altitude, high-opening
HALO	high-altitude, low-opening
HE	high explosive
HF	high frequency
HHC	headquarters and headquarters company
HSLLADS	high-speed, low-level airdrop system
HUMINT	human intelligence
IAW	in accordance with
IFF	identification-friend-or-foe
IP	initial point
IRP	initial rally point
ISB	intermediate staging base
JACC/CP	joint airborne communications center/command post
JCS	Joint Chiefs of Staff
JCSE	joint communications support element
JFC	joint force commander
JSOC	joint special operations command
JTB	joint transportation board
JTF	joint task force
kw	kilowatt
LAPES	low altitude parachute extraction system
LAW	light antitank weapon
LNO	liaison officer

LOC	logistical operations center
LRRP	long-range reconnaissance patrol
LRSU	long-range surveillance unit
LTD	laser target designator
LZ	landing zone
MAC	Military Airlift Command
MANPADS	man-portable air defense system
MCC	movement control center
METT-T	mission, enemy, terrain, troops and time available
MI	military intelligence
MILES	multiple integrated laser engagement system
mm	millimeter
MMC	Materiel Management Center
MMEE	minimum mission-essential equipment
MOS	military occupational specialty
MOUT	military operations on urbanized terrain
MRLS	multiple rocket launch system
MTP	mission training plan
NBC	nuclear, biological, chemical
NCA	National Command Authority
NCO	noncommissioned officer
NOD	night observation device
OCOKA	observation, concealment, obstacles, key terrain, and avenues of approach
OCONUS	outside continental United States
OPCOM	operational command
OPCON	operational control
OPFOR	opposing force

OPORD	operation order
OPSEC	operations security
ORP	objective rally point
PAC	personnel administration center
PDF	primary direction of fire
PIR	priority intelligence requirements
PLL	prescribed load list
POM	preparation for oversea movement
PSYOP	psychological operations
PT	physical training
PZ	pickup zone
QRE	quick-reaction element
RATELO	radiotelephone operator
REMAB	remote marshalling base
RFL	restrictive fire line
RIP	ranger indoctrination program
RP	rally point; release point
RPG	rocket-propelled grenade
RRF	ranger ready force
RRP	reentry rally point
RSE	ranger support element
RSTA	reconnaissance, surveillance, and target acquisition
S1	adjutant
S2	intelligence officer
S3	operations and training officer
S4	supply officer
S5	civil affairs officer

SAR	search and rescue
SATCOM	satellite communications)
SCIF	sensitive compartmented information facility
SCOPES	squad combat operations' exercises (simulation)
SCUBA	self-contained underwater breathing apparatus
SEAD	suppression of enemy air defense
SEMA	special electronic mission aircraft
SERE	survival, evasion, resistance, and escape
SF	special forces
SHORAD	short-range air defense
SIGSEC	signal security
SKE	station-keeping equipment
SLAR	side-looking airborne radar
SOA	special operations aviation
SOCB	special operations communications battalion
SOCOM	special operations command (usually, 1st SOCOM)
SOP	standing operating procedures
SOSB	special operations support battalion
SOTF	special operations task force
SPL	support platoon leader
STX	situational training exercise
TA	theater Army
TAACOM	Theater Army Area Command
TAC	Tactical Air Command
TACP	tactical air control party
TACSAT	tactical satellite terminal
TAMMC	Theater Army Materiel Management Center

TASG	theater Army support group
TDA	table of distribution and allowances
TOA	transfer of authority
TOC	tactical operations center
TOE	table(s) of organization and equipment
TRP	target reference point
TTY	teletype
UHF	ultrahigh frequency
US	United States
USAF	United States Air Force
USMC	United States Marine Corps
USN	United States Navy
VHF	very high frequency
WP	white phosphorus
XO	executive officer

Section II

DEFINITIONS

ADVANCE FORCE. An operation requiring the immediate deployment of US forces with the intent to negate a threat before the deployment of follow-on forces is required.

AIRLAND BATTLE DOCTRINE. An approach to military operations that realizes the full potential of US forces. Two notions (extending the battlefield and the ability to integrate conventional, nuclear, chemical, and electronics means) are combined to describe a battlefield where the enemy is attacked to the full depth of his formation.

AREA OF INFLUENCE. A geographical area wherein a commander is directly capable of influencing operations by maneuver or fire support systems normally under his command or control.

AREA OF INTEREST. That area of concern to the commander, including the area of influence, areas adjacent thereto, and extending into enemy territory to the objectives of current or planned operations. This area also includes areas occupied by enemy forces who would jeopardize the accomplishment of the mission.

AREA OF OPERATIONS. That portion of an area of conflict necessary for military operations. Areas of operations are geographical areas assigned to commanders for which they have responsibility and in which they have authority to conduct military operations.

CHALK. One specific aircraft load.

CLANDESTINE OPERATIONS. Activities to accomplish intelligence, counterintelligence, and other similar activities sponsored or conducted by governmental departments or agencies, in such a way as to ensure secrecy or concealment. (They differ from covert operations in that emphasis is placed on concealment of the operation rather than on concealment of the identity of the sponsor.)

COMMUNICATIONS INTELLIGENCE (COMINT). The interception and processing of foreign communications passed by radio, wire, or other electromagnetic means, and by the processing of foreign encrypted communications, however transmitted. Interception comprises search, intercept, operator identification, signal analysis, traffic analysis, crypto-analysis, study of plain text, the fusion of these processes, and reporting the results.

COMMUNICATIONS SECURITY. The protection resulting from all measures designed to deny to unauthorized persons information of value that might be derived from the possession and study of telecommunications, or to mislead unauthorized persons in their interpretations of the results of such study. COMSEC includes crypto-security, physical security, and transmission security.

COUNTERINTELLIGENCE. That aspect of intelligence activities, both offensive and defensive, designed to detect and neutralize or destroy the effectiveness of hostile foreign intelligence activities and to protect information against espionage, personnel against subversion, and installations and materiel against sabotage.

COVERT OPERATIONS. Operations that are so planned and executed as to conceal the identity of or permit plausible denial by the sponsor. They differ from clandestine operations in that emphasis is placed on concealment of identity of the sponsor rather than on concealment of the operation.

DECEPTION. Those measures designed to mislead the enemy by manipulation, distortion, or falsification of evidence to induce him to react in a manner prejudicial to his interests.

DEMONSTRATION. A show of force in an area where a decision is not sought. It is similar to a feint but does not make contact with the enemy.

DRILL. A drill focuses on a "chunk" or "slice" of battle and is targeted at small units. Drills are critical collective tasks that require a high degree of proficiency from a small unit. Generally, small units "drill" collective tasks that require rapid responses by unit members in the absence of detailed orders from unit leaders.

ELECTRONIC COUNTERMEASURES. That division of electronic warfare involving actions taken to prevent or reduce an enemy's effective use of the electromagnetic spectrum.

ELECTRONIC INTELLIGENCE (ELINT). The technical and intelligence information derived from foreign noncommunications electromagnetic radiations emanating from other than atomic detonations or radioactive sources.

ELECTRONIC WARFARE. Military action involving the use of electromagnetic energy to determine, exploit, reduce, or prevent hostile use of the electromagnetic spectrum and action that retains friendly use of the electromagnetic spectrum.

ELECTRONIC WARFARE SUPPORT MEASURES (ESM). That division of EW involving actions taken to search for, intercept, identify, and locate sources of radiated electromagnetic energy for the purpose of immediate threat recognition. ESM provides a source of information needed for immediate action involving electronic countermeasures, electronic counter-countermeasures, avoidance, targeting, and other tactical employment of forces.

ESPIONAGE. Actions directed toward the acquisition of information through clandestine operations.

EVACUATION/SAFEGUARD. The use of US forces to evacuate US citizens or allies from a hostile or potentially hostile area, or to safeguard US personnel or property.

EXFILTRATION. The removal of personnel or units from areas under enemy control by stealth, deception, surprise, or clandestine means. Exfiltration is a method of extracting ranger units.

EXTRACTION. The removal of personnel or units from areas under enemy control by any means, to include exfiltration, airland, airmobile, or amphibious.

FEINT. A diversionary supporting attack conducted to draw the enemy's attention from the main effort. It is normally executed by brigades and small units. Feints are usually shallow, limited objective attacks that go in before or during the main attack.

GUERRILLA WARFARE. Military and paramilitary operations conducted in enemy-held or hostile territory by irregular and predominantly indigenous forces.

HIGH-INTENSITY CONFLICT. A war between the US and any other nation or nations in which both belligerents use the most modern technology and all resources in intelligence, mobility, firepower (including NBC), command, communications and control, and service support.

HUMAN INTELLIGENCE. Intelligence derived from information provided by human resources.

INFILTRATION. The movement through or into an area of territory occupied by either friendly or enemy troops or organizations. The movement is made either by small groups or by individuals, at extended or irregular intervals. When used in conjunction with the enemy, it implies that contact is avoided. Infiltration is a subset of insertion.

INSERTION. Placement of troops and equipment into an operational area by any means.

INTERDICT. To isolate or seal off an area by military means; to prevent, hinder, or delay the use of an area or route by enemy forces.

INTERDICT LINE OF COMMUNICATIONS. An attack to seal off an area or to deny use of a route or approach.

JOINT TASK FORCE. A JTF may be constituted and designated by the Secretary of Defense or by the commander of a unified command, specified command, or an existing JTF. Normally, it performs missions having specific, limited objectives or missions of short duration. It dissolves when it has achieved its purpose. The joint task force commander is responsible to the commander, known as the establishing authority, who created the JTF. The JTF is composed of elements of two or more services operating under a single JTF commander. The JTF commander has operational control over the entire force and may have direct command of his own staff with representatives from the other services. He exercises logistics coordination or control only as necessary to meet his subordinate commanders' logistics needs.

LIAISON. That personal contact or communication maintained between elements of military forces to ensure mutual understanding and unity of purpose and effort.

LINE OF COMMUNICATIONS. All routes (land, water, and air) that connect an operating military force with one or more bases of operations and along which supplies and military forces move.

LOW-INTENSITY CONFLICT. Limited politico-military operations conducted to achieve political, social, economic, or psychological objectives. It is generally confined to a specific geographic area and is often characterized by operational constraints imposed on ranger units by political and economic considerations. It may involve intermittent combat against regular threat military forces, but it more often involves conflict with irregular, security, or paramilitary forces of a country or group involved in an operation that is counter to the best interests and national security of the United States. It is characterized by relatively low levels of combat and constraints placed on the weaponry, tactics, and movements of the ranger force. It may involve counterterrorist or antiterrorist operations and generally occurs before a pro forma declaration of war.

MID-INTENSITY CONFLICT. War, declared or undeclared, between the US and the regular forces of one or more nations and their respective allies, if any, in which the belligerents use the most modern technology and all resources in intelligence, mobility, firepower (excluding nuclear, chemical, and biological weapons), command, communications and control, and support. They are used for limited objectives under definitive policy limitations as to the extent of destructive power that can be used or the extent of geographical area that might be involved.

MILITARY STRATEGY. Art and science of using the armed forces of a nation to secure the objectives of national policy by application of force or the threat of force. Military strategy sets the fundamental conditions for operations.

NATIONAL COMMAND AUTHORITY. This includes the President and Secretary of Defense or the duly deputized alternates or successors. Commonly referred to as NCA.

OPERATIONAL COMMAND/CONTROL. Operational command (OPCOM) and operational control in joint force terminology both refer to the authority exercised by joint commanders over subordinate service components. Those terms are not interchangeable in joint operations. OPCOM applies to the authority exercised by commanders of unified commands. OPCON is the authority that subordinate joint task force commanders exercise in the conduct of specific operations. The authority that military departments exercise over their respective components is commonly referred to as command less operational command.

OPERATIONAL LEVEL OF WAR. The operational level of war uses available military resources to attain strategic goals within a theater of war. It is the theory of larger unit operations. It also involves planning and conducting campaigns. Campaigns are sustained operations designed to defeat an enemy force in a specified space and time with simultaneous and sequential battles. The disposition of forces, selection of objectives, and actions taken to weaken or to outmaneuver the enemy all set the terms of the next battle and exploit tactical gains. They are all part of the operational level of war. In AirLand Battle doctrine, this level includes the marshalling of forces and logistical support, providing direction to ground and air maneuver, applying conventional and nuclear fires in depth, and employing unconventional and psychological warfare.

OPERATIONS SECURITY. Those measures designed to protect information concerning planned, ongoing, and completed operations from unauthorized disclosure. It includes all actions a command takes to deny the enemy information about friendly units and their operations.

ORDER OF BATTLE (OB). The identification, strength, command structure, and disposition of the personnel, units, and equipment of any military force.

RAID. Usually a small-scale operation, involving a swift penetration of hostile territory to secure information, confuse the enemy, or destroy his installations. It ends with a planned withdrawal upon completion of the assigned mission.

RANGER MISSION. To conduct strike, raid, or special light infantry operations. Execution of ranger missions is normally directed by the National Command Authority. Rangers also support operational-level commanders during operations of limited duration when allocated or assigned by the theater commander.

RECONNAISSANCE. A mission undertaken to obtain, by visual observation or other detection methods, information about the activities and resources of an enemy or potential enemy; or to secure data concerning the meteorological, hydrographical, or geographical characteristics of a particular area.

SEARCH AND RESCUE. The use of aircraft, surface craft, submarines, specialized rescue teams, and equipment to search for and rescue personnel in distress on land or at sea.

SEIZURE. As an operational purpose, the capture of a voluntarily restricted portion of an enemy-controlled territory. Capture of an isolated land mass such as an island is usually categorized as a seizure.

SEIZE KEY FACILITY. An attack on an installation with the intent of holding it for a short period. Normally, rangers would make the assault to seize and secure an installation for follow-on forces.

SERVICE COMPONENT. Each service component commander is responsible for recommending the proper use of his forces and for accomplishing operational tasks assigned by the joint commander. He is also responsible for his service in matters of internal administration and discipline; training in own service doctrine, techniques, and tactics; designation of specific units to meet joint requirements; logistics functions normal to the component; tactical employment of service component forces; and service intelligence matters.

SHOW OF FORCE. A mission carried out to demonstrate US resolve, whereby US forces are deployed to defuse a situation that may be harmful to US interests or national objectives.

SIGNAL INTELLIGENCE (SIGINT). A generic term that includes COMINT and ELINT. Activities concerned with the collection and processing of foreign electromagnetic emissions for the purpose of producing intelligence information.

SITUATIONAL TRAINING EXERCISE. An STX is a short, mission-oriented exercise in which a group of closely-related collective tasks are trained, using a doctrinally preferred method of executing these tasks to established standards. An STX is drill-like in nature, but it is more complex and flexible than a drill. It usually involves a larger unit than a drill.

SPECIAL OPERATIONS. Military operations conducted by specially trained, equipped, and organized DOD forces against strategic or tactical targets in pursuit of national, military, political, economic, or psychological objectives. Special operations may support conventional military operations, or they may occur independently when the use of conventional forces is either inappropriate or infeasible. Sensitive peacetime operations, except for training, are normally authorized by the NCA and conducted under the direction of the NCA or designated commander. Special operations may include unconventional warfare, counter-terrorist operations, collective security, psychological operations, and civil affairs measures.

SPECIAL OPERATIONS FORCES. All civil affairs, psychological operations, ranger (active component), and special forces units within the US Army, US Army Reserve, and National Guard.

SPECIAL OPERATIONS COMMAND. The major subordinate command to FORSCOM exercising command and control of all active component Army special operations forces not placed under the operational command of unified or specified commands.

STRATEGIC LEVEL OF WAR. Military strategy uses the armed forces of a nation to secure the objectives of national policy by applying force or the threat of force. Military strategy sets the fundamental conditions for operations.

STRIKE. An attack that is intended to inflict damage on, seize, or destroy an objective. Sometimes referred to as direct action missions.

SURVEILLANCE. The systematic observation of aerospace, surface or subsurface areas, places, persons, or things by visual, aural, electronic, photographic, or other means.

TACTICAL. Pertaining to the employment of units in combat.

TACTICAL LEVEL OF WAR. Tactics are the specific techniques smaller units use to win battles and engagements that support operational objectives. Tactics employ all available combat, combat support, and combat service support. Tactics involve the movement and positioning of forces on the battlefield in relation to the enemy, the provisioning of fire support, and the logistical support of force before, during, and following engagements with the enemy. At corps and division, operational and tactical levels are not clearly separable, but they are guided by the same principles. An operation designed to defeat any enemy force in an extended area does so through operational maneuver and a series of tactical actions.

TACTICAL COMBAT INTELLIGENCE. That knowledge of the enemy, weather, and geographical features needed by a commander in the planning and conduct of combat operations. It is derived from the interpretation of information on the enemy and the environment.

TARGET ANALYSIS. A detailed and systematic examination of processed intelligence to identify and locate targets.

UNIFIED COMMAND. A unified command is established and designated by direction of the President. The command is organized under a single unified CINC and has a broad, continuing mission. The CINC is responsible to the Secretary of Defense for accomplishing his broad and continuing mission and operates under the strategic and operational direction of the JCS. A unified command is composed of assigned forces of two or more services. These forces, as service components, are under the operational command of the unified commander. The CINC exercises operational command authority through service component commanders. When tasked by higher authority, he may establish a subordinate JTF to conduct specific missions. Under emergency conditions, the CINC has the authority to use all facilities and supplies of assigned forces to accomplish his mission.

References

REQUIRED PUBLICATIONS

Required publications are sources that users must read in order to understand or to comply with this publication.

Field Manuals (FM)

FM 7-8	The Infantry Platoon and Squad (Infantry, Airborne, Air Assault, Ranger)
FM 7-10	The Infantry Rifle Company (Infantry, Airborne, Air Assault, Ranger)
FM 7-20	The Infantry Battalion (Infantry, Airborne, and Air Assault)

RELATED PUBLICATIONS

Related publications are sources of additional information. They are not required in order to understand this publication.

Army Regulations (AR)

AR 350-1	Army Training

Field Manuals (FM)

FM 10-52-1	Commander's Handbook for Water Usage in Desert Operations
FM 10-501	Airdrop of Supplies and Equipment: Rigging Containers
FM 10-512	Airdrop of Supplies and Equipment: Rigging Typical Supply Loads
FM 10-542	Airdrop of Supplies and Equipment: Rigging Loads for Special Operations
FM 10-547	Airdrop of Supplies and Equipment: Rigging the High Speed Aerial Delivery Container CTU-2/A
FM 10-553	Airdrop of Supplies and Equipment: Rigging Ammunition
FM 12-3-4	Echelons Above Corps (EAC) Personnel and Administrative Doctrine

FM 14-6	Comptroller/Finance Services in Theaters of Operations
FM 19-15	Civil Disturbances
FM 20-11-1	Military Diving (Volume 1)
FM 20-11-2	Military Diving (Volume 2)
FM 21-20	Physical Training Program
FM 21-26	Map Reading
FM 21-75	Combat Skills of the Soldier
FM 21-76	Survival, Evasion, and Escape
FM 21-77A(S)	Joint Worldwide Evasion and Escape Manual
FM 22-9	Soldier Performance in Continuous Operations
FM 23-67	Machinegun, 7.62-mm, M60
FM 25-1	Training
FM 25-2	Unit Training Management
FM 25-3	Training in Units
FM 25-4	How to Conduct Training Exercises
FM 25-5	Training for Mobilization and War
FM 25-7	Training Ranges
FM 26-2	Management of Stress in Army Operations
FM 29-47	Main Supply and Service Company, Forward Supply and Service Detachments, Supply and Service Battalion, Airborne Division
FM 29-51	Division Supply and Field Service Operations
FM 30-5	Combat Intelligence
FM 30-17	Counterintelligence Operations
FM 31-18	Long-Range Reconnaissance Ranger Company
FM 31-19	Special Forces Military Free-Fall Parachuting
FM 1-24	Special Forces Air Operations

FM 31-25	Special Forces Waterborne Operations
FM 31-70	Basic Cold Weather Manual
FM 34-1	Intelligence and Electronic Warfare Operations
FM 34-60A	Counterintelligence Operations (U)
FM 55-9	Unit Air Movement Planning
FM 57-38	Pathfinder Operations
FM 57-220	Basic Parachuting Techniques and Training
FM 63-3J	Combat Service Support Operations--Corps
FM 63-4	Combat Service Support Operations--Theater Army Area Command
FM 63-5	Combat Service Support Operations--Theater Army
FM 90-3	Desert Operations
FM 90-4	Airmobile Operations
FM 90-5	Jungle Operations
FM 90-6	Mountain Operations
FM 90-8	Counterguerilla Operations
FM 90-10	Military Operations on Urbanized Terrain (MOUT)
FM 90-10-1	An Infantryman's Guide to Urban Combat
FM 100-5	Operations
FM 100-20	Low Intensity Conflict
FM 100-26	The Air-Ground Operations System
FM 101-5	Staff Organization and Operations
FM 101-10-1	Staff Officers' Field Manual: Organization, Technical and Logistical Data (U)

Firing Table (FT)

FT 60-P-1	Firing Tables for Mortar, 60-mm: M224, Firing Cartridge, HE, M720, Cartridge, HE, M49A4, Cartridge, TP, M5OA3, Cartridge, WP, M302AI, Cartridge, Illuminating, M83A3

Soldier's Training Publication (STP)

STP 7-11B24-SM	Soldier's Manual MOS 11B, Infantryman, Skill Level 2/3/4

Technical Manuals (TM)

TM 10-1670-267-12&P	Operator's and Organizational Maintenance Manual Including Repair Parts and Special Tools List for High Altitude Airdrop Resupply System: 500 Pound Capacity

Training Circulars (TC)

TC 23-14	Sniper Training and Employment
TC 23-90	Mortar Training
TC 24-1	Communications-Electronics Operation Instructions, the CEOI
TC 31-20-5(C)	Special Forces Communications (U)
TC 90-6-1	Military Mountaineering

Other Related Publications

King, Michael J. Dr., *Leavenworth Papers, No. 11*, "Rangers Selected Combat Operations in World War II," Combat Studies Institute, USACGSC, Fort Leavenworth, KS

PROJECTED PUBLICATIONS

Projected publications are sources of additional information that are scheduled for printing but are not yet available. Upon print, they will be distributed automatically via pinpoint distribution. They may not be obtained from the USA AG Publications Center until indexed in DA Pamphlet 310-1.

Field Manuals (FM)

FM 7-70	Light Infantry Squad/Platoon
FM 7-71	Light Infantry Company
FM 7-72	Light Infantry Battalion
FM 7-73	Light Infantry Brigade

FM 23-36 60-mm Mortar, M224

Made in the USA
Monee, IL
05 November 2026

67810987R00203